Level 2
PLUMBING

3rd edition

NVQ/SVQ and Diploma

JTL
DELIVERING SKILLS
FOR THE FUTURE

www.pearsonschoolsandfe.co.uk

✓ Free online support
✓ Useful weblinks
✓ 24 hour online ordering

0845 630 44 44

Heinemann

Part of Pearson

Heinemann is an imprint of Pearson Education Limited,
Edinburgh Gate, Harlow, Essex, CM20 2JE.

www.pearsonschoolsandfecolleges.co.uk

Heinemann is a registered trademark of Pearson Education
Limited

Text © JTL 2011
Typeset by Tek-Art, Crawley Down, West Sussex
Original illustrations © Pearson Education Ltd 2011
Illustrated by Tek-Art, Crawley Down, West Sussex
Cover design by Wooden Ark
Cover photo/illustration © Pearson Education Ltd:
Clark Wiseman, Studio 8

The right of JTL to be identified as author of this work has
been asserted by them in accordance with the Copyright,
Designs and Patents Act 1988

First published 2011

14 13 12 11
10 9 8 7 6 5 4 3 2 1

British Library Cataloguing in Publication Data
A catalogue record for this book is available from the British
Library

ISBN 978 0 435031 31 2

Printed in Spain by Grafos

Websites
There are links to relevant websites in this book. In order to
ensure that the links are up-to-date, that the links work and
that the sites are not inadvertently linked to sites that could
be considered offensive, we have made the links available on
our website at www.pearsonhotlinks.co.uk. Search for this title
NVQ/SVQ Plumbing or ISBN 9780435031312.
The information and activities in this book have been
prepared according to the standards reasonably to be expected
of a competent trainer in the relevant subject matter. However,
you should be aware that errors and omissions can be made
and that different employers may adopt different standards
and practices over time. Before doing any practical activity,
you should always carry out your own risk assessment and
make your own enquiries and investigations into appropriate
standards and practices to be observed.

Acknowledgements

Every effort has been made to contact copyright holders
of material reproduced in this book. Any omissions will be
rectified in subsequent printings if notice is given to the
publishers. The following materials have been reproduced with
kind permission from the following organisations:
p.45 Procedures for dealing with unconscious co-workers,
©NHS Choices.
p.46 Procedures for treating shock, © the British Red Cross.

The author and publisher would like to thank the following
individuals and organisations for permission to reproduce
photographs:
The publisher would like to thank the following for their kind
permission to reproduce their photographs:

(Key: b-bottom; c-centre; l-left; r-right; t-top)

Alamy Images: AKP Photos 72, B E Eyley Construction
images 439, Beepstock 338bc, Clynt Garnham Renewable
energy 343, David J. Green 168c, DWImages 204/4, Edd
Westmacott 34/4, Forbes Stuart 349/3, FotoFacade 437, Greg
Balfour Evans 129, Insadco photography 344, Jean Schweitzer
119, Judith Collinns 59l, Lyroky 215br, Olaf Doering 105,
Realimage 137, studiomode 438, Top Notch 360bl, wiebkin
348tl; **Construction Photography:** Adrian Greeman 69c,
Buildpix 207, Jean Francois Cordella 77, Paul McMullin 68;
Corbis: 4, Fancy / Veer 162; **Getty Images:** PhotoDisc 153,
157, Photographer's Choice / Hugh Sitton 1; **Hepworth
Building Products:** 220, 378; **Lead Sheet Asscociation:** 442,
446, 451, 454, 455, 457, 459; **Masterfile UK Ltd:** 199, Mark
Peter 389; **Pearson Education Ltd:** Clark Wiseman, Studio
8 34/1, 34/2, 36/4, David Sanderson 34/3, 63, Gareth Boden
39, 45, 47, 48, 51, 52, 53, 66, 87, 92, 128, 189t, 189c, 189b,
200/1, 200/3, 201/1, 201/4, 201/5, 201/6, 201/7, 201/8, 202/1,
202/2, 202/3, 203, 204/1, 204/2, 204/3, 206, 209, 211t, 212bl,
215bl, 215bc, 216t, 218, 222, 224, 228, 234, 235, 238, 243,
248, 250, 271, 272, 274, 275, 282, 283, 286, 293, 295, 318,
320, 339/2, 339/3, 339/4, 339/5, 339/6, 342, 347, 348cl, 348bl,
349/1, 349/5, 349/6, 350, 351, 352, 353, 354tl, 355, 360tl,
362tl, 363, 366, 368, 375, 405, 407, 416tl, 416cl, 416br, 431,
Naki Photgraphy 29, 40, 59r, 60c, 100, 201/2, 202/4, 287, 319,
338cl, 339/1, 349/2, 354bl, 379/1, 380, 396, 404; **Photolibrary.
com:** 81A Productions 134; **Rex Features:** Voisin / Phanie 102;
Science Photo Library Ltd: Sheila Terry 256, Tony Craddock
109; **Shutterstock.com:** adrianmladin 146c, aldegonde
34/5, AlexussK 255, apdesign 24/5, Catalin D 23/1, cg- art
23/3, creation 23/2, djdarkflower 24/1, Dmitry Naumov 146b,
Dragana Gerasimosk 23/4, Elena Elisseeve 120br, ForoFermer
168t, Gencho Petrov 59c, humpkin 24/4, 24/6, Janet Faye
Hastings 69b, Jenson 373, Joe Gough 349/4, Kurham 12, Lack-
O'Keen 24/2, Marek R. Swadzba 338bl, mtkang 36/2, Nic Neish
60t, Norman Pogson 27, prism68 36/3, Richard Peterson 36/1,
serenethos 354cl, Serg64 201/3, tatniz 200/2, VIP DesignUSA
200/4, WDGPhoto 120tl, Workflow 143, YellowJ 305, Yobidaba
24/3; **SuperStock:** agefotostock 130, imagebroker.net 106,
Marka 329; **www.imagesource.com:** 3, 8

All other images © Pearson Education

Every effort has been made to trace the copyright holders and
we apologise in advance for any unintentional omissions. We
would be pleased to insert the appropriate acknowledgement
in any subsequent edition of this publication.

Contents

Introduction

This book is designed to support the new NVQ Level 2 Diploma in Plumbing. This new Diploma has been prepared by SummitSkills (the SSC) in consultation with employers, the main exam boards and training providers. This means that all exam boards offering this qualification will have the same unit structure and assessment strategy. Exam boards will then design their own assessment content.

This book is designed to support the following qualifications:

- 6189 City and Guilds Level 2 NVQ Diploma in Plumbing and Heating.
- 501/1752/X EAL Level 2 NVQ Diploma in Plumbing.

These qualifications are approved on the Qualifications and Credit Framework (QCF). The QCF is a new government framework which regulates all vocational qualifications to ensure they are structured and titled consistently and quality assured. SummitSkills have developed the new Diploma qualification with the awarding bodies.

Who the qualification is aimed at

The new diploma qualification is aimed at both new entrants (such as apprentices or adult career changers) as well as the existing workforce (so those looking to upskill). It is intended to train and assess candidates so that they can be recognised as occupationally competent by the plumbing industry. Learners should gain the skills to:

- work as a competent plumber
- achieve a qualification recognized by the Joint Industry Board (JIB) for professional grading to the industry
- complete an essential part of the SummitSkills Advanced Apprenticeship.

About this book

Each unit of this book relates to a particular unit of the Diploma and provides the information needed to form the required knowledge and understanding of that area.

This book has been prepared by expert JTL trainers, who have many years of experience of training learners and delivering plumbing qualifications. The content of each unit will underpin the various topics which you will be assed on by your exam board.

Each unit has knowledge tests throughout, as well as a set of multiple choice questions at its conclusion, to allow you to measure your knowledge and understanding.

This book will also be a useful reference tool for you in your professional life once you have gained your qualifications and are a practicing plumber.

Using this book

It is important to note that this book is intended to be used for training. It should not be regarded as being relevant to an actual installation. You should always make specific reference to the British Standards or manufacturer's data when designing plumbing installations.

Features of this book

This book has been fully illustrated with artworks and photographs. These will help to give you more information about a concept or a procedure, as well helping you to follow a step-by-step procedure or identify a particular tool or material.

This book also contains a number of different features to help your learning and development.

Key term

These are new or difficult words. They are picked out in **bold** in the text and then defined in the margin.

Remember

This highlights key facts or concepts, sometimes from earlier in the text, to remind you of important things you will need to think about.

Did you know?

This feature gives you interesting facts about the plumbing trade.

Safety tip

This feature gives you guidance for working safely on the tasks in this book.

Find out

These are short activities and research opportunities, designed to help you gain further information about, and understanding of, a topic area.

Working life

This feature gives you a chance to read about and debate a real-life work scenario or problem. Why has the situation occurred? What would you do?

Progress check

These are a series of short questions, usually appearing at the end of each learning outcome, which gives you the opportunity to check and revise your knowledge. Answers to the questions are supplied on the Training Resource disk.

Getting ready for assessment

This feature provides guidance for preparing for the practical assessment. It will give you advice on using the theory you have learnt about in a practical way.

Check your knowledge

This is a series of multiple choice questions at the end of each unit, in the style of the GOLA end-of-unit tests. Answers to the questions are supplied on the Training Resource disk.

Acknowledgements

JTL would like to express its appreciation to all those members of staff who contributed to the development of this book, ensuring that the professional standards expected were delivered and generally overseeing the high quality of the final product. Without their commitment this project would not have been seen through successfully.

Particular thanks to Keith Powell, who revised the content of the previous editions of this book and prepared extensive new material to cover the diploma specifications.

Thanks go to Stephen Blair for preparing the Working Life, Check Your Knowledge and Getting Ready for Assessment features and to Dave Allen for environmental information.

Pearson and JTL would also like to thank Stephen Blair for his comprehensive and painstaking review work, which has again made an invaluable contribution towards ensuring accuracy in this book.

Pearson and JTL would also wish to thank Alistair Nunn from Pearson and Martin Bullock and the rest of the staff at the JTL Training Centre in Malton for their patience, assistance, advice and support during the photo shoot.

Understand and carry out safe working practices in building services engineering

Health and safety forms a very important part of your everyday working life and is vital for all work situations, whether on a construction site or working in a client's home.

This unit will cover the following learning outcomes:

- Know the health and safety legislation that applies to the building services industry

- Know how to recognise and respond to hazardous situations while working in the building services industry

- Know the safe personal protection measures while working in the building services industry

- Be able to apply manual handling techniques

- Know how to respond to accidents that occur while working in the building services industry

- Know procedures for electrical safety when working in the building services industry

- Be able to apply basic electrical safety measures in the building services industry

- Know methods of working safely with heat-producing equipment in the building services industry

- Be able to work safely with gas heating equipment in the building services industry

- Know the methods of safely using access equipment in the building services industry

- Be able to safely use access equipment in the building services industry

- Know the methods of working safely in excavations and confined spaces in the building services industry

1. Know the health and safety legislation that applies to the building services industry

The aims of health and safety legislation

Health and safety is a very important aspect of your job. There are a number of legal health and safety requirements that have been put in place to keep you, your fellow workers and site visitors free from accidents. Having a good grasp of these requirements will help you to be aware of the potential hazards when working on site and in the plumbing workshop.

The Health and Safety Executive (HSE) is the independent watchdog responsible for enforcing health and safety laws in the workplace. Every year thousands of building-site accidents are reported to the HSE, some of which are fatal. By following health and safety guidance, you will help to minimise the risk of accidents. During this unit, you will learn more about the regulations designed to keep you safe at work. You will also learn about the various hazards you may face, safe methods of working and actions to be taken in the event of accidents.

Health and safety responsibilities

Both employers and employees are responsible for health and safety at work. The Health and Safety at Work Act 1974 (HASAWA) is the main legislation covering health and safety in the workplace. As an employee you are legally bound to cooperate with your employer to ensure they comply with the Act. You are also expected to take reasonable care of your own health and safety and to act responsibly so as not to endanger other workers or members of the public. If you break safety laws, you may be prosecuted.

Below are some of the most important aspects of the HASAWA:

- In domestic premises, the occupier is normally covered by insurance (Occupier's Liability) for visitors. However, when working on domestic premises, as the plumber you become responsible for the health, safety and welfare of the occupant, because it is your place of work.
- It is an offence to misuse or interfere with equipment provided by your employer for your health and safety or that of others. Employers have a duty to ensure that safety equipment is kept in good condition and instruct employees on its use. You should always use health and safety equipment correctly and for its intended purpose.
- Substances such as oil, grease, cutting compounds, paints and solvents are hazardous if spilled on the floor. Items such as off-cuts of pipe, cables, tools and even food are also dangerous if left underfoot. You have a responsibility both to yourself and to others to keep the workplace hazard free.

Health and safety responsibilities for many of those involved in construction projects are covered in the Construction (Design and Management) Regulations 2007 – see pages 5–8.

Employers' responsibilities

The HASAWA requires employers to ensure, so far as is reasonably practicable, the health, safety and welfare of employees at work. Employers are responsible for:

- providing and maintaining plant and systems of work that are safe and without risk to health
- safety in the use, handling, storage and transport of articles and substances
- providing information, instruction, training and supervision as necessary to ensure health and safety
- providing access to and exit from the workplace that is safe and without risk
- providing adequate welfare facilities and arrangements.

Companies employing five or more employees must also have a health and safety policy statement – see page 5.

Employers should be familiar with their responsibilities

Employees' responsibilities

Your legal responsibilities at work are listed below.

Your legal responsibilities at work

- Take care of your own health and safety and that of others who may be affected by your actions.
- Do not interfere with or misuse anything provided for your health and safety.
- Cooperate with your employer on health and safety matters.
- Tell your employer about any potentially dangerous situations.
- Alert your employer to any weakness you might spot in their health and safety arrangements.

Remember

You have a legal obligation to carry out your health and safety responsibilities throughout your apprenticeship and working career.

Personal hygiene

Good personal hygiene at work will help to keep you healthy. To maintain hygiene in the workplace:

- keep your overalls as clean as possible, and wash them regularly
- wash your hands thoroughly before contact with food
- avoid washing with solvents (e.g. white spirits), which can cause dermatitis
- use a barrier cream before starting a job, and reapply it each time you wash your hands.

Did you know?

Barrier cream fills the pores of the skin with a water-soluble antiseptic cream, so that when you wash your hands the dirt and germs are removed with the cream.

Accident prevention

The best way to reduce the risk of accidents is to remove the cause. Wherever possible, the workplace should have clearly defined passageways, good lighting and ventilation, reduced noise levels and non-slip floorings. Materials should be stored correctly, for example storing lengths of tube and pipe horizontally to minimise the risk of them falling. Dangerous substances should be locked in an approved location.

To guard against hazards that cannot be minimised:

- place safety guards and fences on or around machines
- follow safe systems of work
- wear safety equipment, such as goggles, safety helmets and safety shoes, and other protective clothing, e.g. ear defenders, respirators, eye protection and overalls.

Personal attitudes to safety training

It is essential to have a positive attitude towards safety training to safeguard your own welfare, your workmates and the general public. You should be aware of your company's health and safety policies and procedures and always act and work carefully and responsibly. You will need to know the potential dangers, the protection available and how to use it, and how to prevent accidents. Personal habits such as drinking alcohol and drug abuse can cause a worker to be a hazard not only to themselves but also to others.

Health and safety legislation

Legal status of health and safety guidance

Health and safety in the construction industry is covered by legislation, regulations, Approved Codes of Practice and guidance notes, each of which has a different legal status.

Acts of Parliament

All businesses are covered by laws that have been passed by the UK Parliament in Westminster. Acts of Parliament are the law of the land, which means it is compulsory to follow their contents.

Regulations

Regulations identify major health and safety risks and then set out the actions that must be taken to control them. Some regulations apply to every business, such as the Manual Handling Operations Regulations 1992, and others apply to specific industries, for example the Water Supply (Water Fittings) Regulations 1999. Regulations are approved by Parliament, which means that they are law and any breach may result in legal proceedings being taken.

> **Remember**
>
> Your employer must provide protective clothing, free of charge, when the work process requires its use.

Laws are passed in Westminster

Approved Codes of Practice

An Approved Code of Practice (ACOP) offers practical examples of good practice to help businesses comply with the terms of an Act and is developed after consultation with the industry concerned. An ACOP applies to anyone who has a duty of care as outlined in the code, and may include employers, employees, the self-employed, principals to contracts, owners of buildings or plant, and so on. Failure to follow an ACOP is not, of itself, an offence. However, observing a relevant code of practice may be considered as evidence of good practice in court.

HSE Guidance Notes

The HSE publishes Guidance Notes on a range of health and safety matters. The notes are designed to help businesses and individuals comply with the law and provide technical advice. Guidance is not compulsory. For further details go to www.pearsonhotlinks.co.uk, search for this title and click on this unit.

The Health and Safety at Work etc. Act (HASAWA) 1974

The Act places duties on everyone connected with health and safety at work, including employers, employees, the self-employed and manufacturers or suppliers of plant and materials. Protection is also given to members of the public affected by the activities of people at work.

In addition to setting out the basic health and safety requirements of employers and employees, the HASAWA gives more detailed guidance to employers on aspects such as health and safety policy statements, which are required for companies employing five or more members of staff. Employers must make all staff aware of the company's health and safety policy.

Under the Act employers also have a duty to:

- carry out an assessment of risks related to work activities
- identify and implement control measures
- inform employees of the risks and control measures
- regularly review the assessments
- record the assessment if the company employs more than five people.

Employers must be prepared to consult a safety representative, if one is appointed by a recognised trade union, about matters affecting their employees' health and safety. Employers must also (if requested in writing by any two safety representatives) establish a safety committee within three months of the request being made.

Construction (Design and Management) Regulations 2007

The Construction (Design and Management) (CDM) Regulations are the main set of health and safety regulations which apply to safety on construction sites. The regulations require the HSE to be notified if a 'construction project' is likely to:

- last longer than 30 days

  or

- involve more than 500 person days of construction work.

The regulations place responsibilities on the 'dutyholders':

- **Clients** – anyone having construction or building work carried out as part of their business.
- **CDM coordinators** – have to be appointed to advise the client on projects that last more than 30 days or involve more than 500 person days of construction work. The CDM coordinator's role is to advise the client on health and safety issues during the design and planning phases of construction work.
- **Designers** – those who, as part of their work, prepare design drawings, specifications, bills of quantities and the specification of articles and substances, e.g. architects, engineers and quantity surveyors.
- **Principal contractors, usually the main or managing contractor for the work** – have to be appointed for projects which last more than 30 days or involve more than 500 person days of construction work. The principal contractor's role is to plan, manage and coordinate health and safety while construction work is being undertaken.
- **Contractors** – businesses involved in construction, alteration, maintenance or demolition work.
- **Workers** – anyone who carries out work during the construction, alteration, maintenance or demolition of a building or structure, e.g. plumbers, scaffolders, and so on.

The main aim of the regulations is to ensure that all parties involved in a construction project work together in a manner that will ensure that the work is properly coordinated, so providing a safe environment for the work carried out.

The CDM Regulations were changed in 2007 to incorporate a number of previously separate regulations, including the Construction (Health, Safety and Welfare) Regulations.

Other areas covered under CDM include:

- safe places of work
- good order and site security
- stability of structures
- demolition or dismantling
- explosives
- excavations
- cofferdams and caissons
- reports of inspections
- energy distribution installations
- prevention of drowning
- traffic routes
- vehicles

Did you know?

The CDM Regulations apply to large construction projects, requiring all those involved in the project to work safely together.

- prevention of risk from fire, etc.
- emergency procedures, including emergency routes and exits
- fire detection and firefighting
- fresh air

- temperature and weather protection
- lighting
- welfare facilities – sanitary and washing facilities, drinking water, changing/rest facilities.

Summary of duties under the CDM Regulations

A summary of the duties and how they are applied is given in Table 1.1.

	All construction projects (Part 2 of the Regulations)	Additional duties for notifiable projects (Part 3 of the Regulations)
Clients (excluding domestic clients)	Check competence and resources of all appointees Ensure there are suitable management arrangements for the project holding welfare facilities Allow sufficient time and resources for all stages Provide pre-construction information to designers and contractors	Appoint CDM coordinator* Appoint principal contractor* Make sure that the construction phase does not start unless there are suitable welfare facilities and a construction phase plan is in place Provide information relating to the health and safety file to the CDM coordinator * There must be a CDM coordinator and principle contractor until the end of the construction phase
CDM coordinators		Advise and assist the client with their duties Notify the HSE Coordinate health and safety aspects of the design work and cooperate with others involved with the project Facilitate good communication between client, designers and contractors Liaise with principle contractor regarding ongoing design Identify, collect and pass on pre-construction information Prepare/update health and safety file
Designers	Check client is aware of the duties Eliminate hazards and reduce risks during design Provide information about remaining risks	Check coordinator has been appointed Provide any information needed for the health and safety file
Principle contractors		Plan, manage and monitor construction phase in liaison with the contractor Prepare, develop and implement a written plan and site rules (initial plan completed before the construction phase starts) Give contractors relevant parts of the plan Make sure suitable welfare facilities are provided from the start and maintained throughout the construction phase Check competence of all appointees Ensure all workers have site inductions, and any further information and training needed for the work Consult with the workers Liaise with CDM coordinator regarding ongoing design Secure the site

Table 1.1: Duties under the CDM regulations ▼

Contractors	Check client is aware of the duties Plan, manage and monitor own work and that of workers Check competence of all their appointees and workers Train own employees Provide information to their workers Comply with the specific requirements in Part 4 of the Regulations Ensure there are adequate welfare facilities for their workers	Check a CDM coordinator and a principle contractor has been appointed and HSE notified before starting work Cooperate with principle contractor in planning and managing work, including reasonable directions and site rules Provide details to the principle contractor of any contractor engaged in connection with carrying out the work Provide any information needed for the health and safety file Inform principle contractor of problems with the plan Inform principle contractor of reportable accidents, diseases and dangerous occurrences
Everyone	Check own competence Cooperate with others and coordinate work so as to ensure the health and safety of construction workers and others who may be affected by the work Report obvious risks Comply with requirements in Schedule 3 and Part 4 of the Regulations for any work under their control Take account of and apply the general principles of prevention when carrying out duties	

Table 1.1: Duties under the CDM regulations (cont.)

Responsibilities of the client

The client has one of the biggest influences over the way a project is run but, because they may not know much about health and safety, they are not expected to plan or manage the project. In the case of a notifiable project the client must appoint a competent CDM coordinator to advise them and manage the process. The client remains responsible for ensuring that client duties are met.

Domestic clients have no client duties under CDM 2007. This means that there is no legal requirement for the appointment of a CDM coordinator or principal contractor when such projects reach the notification threshold.

Reporting of Injuries, Diseases and Dangerous Occurrences Regulations (RIDDOR) 1995

The Reporting of Injuries, Diseases and Dangerous Occurrences Regulations (RIDDOR) set out the requirements for an employer reporting any of the following to the HSE:

- **Injuries** – fatalities (including members of the public) or injuries resulting in at least three days off work.
- **Diseases** – if a doctor advises that an employee is suffering from a work-related disease listed under RIDDOR.
- **Dangerous occurrences** – things that happened that could have resulted in a reportable injury (e.g. the collapse of an excavation).

Details of reportable accidents should be forwarded by the employer to the HSE on form F2508. If a fatality is being reported, then this must take place within ten days.

Accidents and injuries may occur in the workplace

The HSE, based on the report form, will make a decision on the level of investigation and subsequent action required, which may include legal action for breaches of health and safety regulations.

A firm should maintain an **accident book** for recording details of all types of injury, however minor, that occur at work. The accident book is used by the employer to monitor the level of accidents that occur and establish whether any additional safety controls are required to do the work.

If an accident occurs (whether it is RIDDOR reportable or not) an employee may be required to complete an accident report form. This should preferably be completed by the injured employee as soon after the accident as possible. If the form is completed by a third party, then the employee must carefully check that the details are accurate. Once signed, an accident report may be used in legal proceedings.

Regulatory Reform (Fire Safety) Order (FSO) 2005 The Fire Precautions Act 1971

The health and safety requirements for fire safety are set out in the CDM Regulations 2007. Fire safety is also covered by the Fire Safety Order (FSO), which brought together fire safety precautions from earlier legislation. The FSO applies to commercial properties including factories, shops and offices.

The employer is responsible for the safety of employees and other people on site where they may be at risk from fire, both from the work processes and activities carried out, as well as for general fire safety in the workplace. Employers must carry out a fire risk assessment, to enable them to identify – and then take steps to eliminate, reduce or control – safety risks (including risks from fire) to make sure that no one gets hurt or becomes ill. Any property at which more than five people are employed will require a written fire risk assessment.

Employers must provide the following general fire precautions:

- means of detecting and giving warning in case of fire
- escape routes
- firefighting equipment
- training of staff in fire safety.

Health and Safety (Safety Signs and Signals) Regulations 1996

These regulations standardised safety signs at work. Employers must provide specific safety signs whenever there is a risk that cannot be avoided or controlled by other means, for example by engineering controls or safe systems of work. Where a safety sign would not help to reduce the risk, or where the risk is minimal, signs are not required.

The regulations require, where necessary, the use of road traffic signs within workplaces to regulate road traffic. Employers must also:

- maintain their safety signs
- explain unfamiliar signs to employees and inform them what to do when they see a safety sign.

The regulations apply to all places and activities where people are employed, but exclude signs and labels used in connection with the supply of substances, products and equipment or the transport of dangerous goods.

Electricity at Work Regulations 1989

The Electricity at Work Regulations require all dutyholders – employers, the self-employed and employees – to take action to avoid the risk of death or injury from electricity at work. Employers and the self-employed have a duty to comply with the regulations, in respect of systems, electrical equipment and conductors, while employees have a duty to cooperate with their employers. As a plumber you will need to be especially careful when dealing with electricity. The Electricity at Work Regulations provide the full safety standards requirement; more will be said about the practical use of electricity on site in a later section. The regulations are made under the HASAWA 1974 and penalties may be imposed on anyone found guilty of malpractice or misconduct.

Definitions relevant to people who have responsibilities under these regulations

1 Employer

For the purpose of the regulations, an employer is any person or company that (a) employs one or more individuals under a contract of employment or apprenticeship, or (b) provides training under the schemes to which the HASAWA applies.

2 Self-employed

A self-employed person is an individual who works for gain or reward other than under contract of employment, whether or not he or she employs others.

3 Employee

Regulation 3(2) (b) reiterates the duty placed on employees by the HASAWA. This clause places duties on employees equivalent to those placed on employers and self-employed people where these matters are within their control. This will include apprentices – like you – who will be considered as employees under these regulations.

This arrangement recognises the level of responsibility that many employees in the plumbing trade are expected to take on as part of their job. The 'control' that they exercise over the electrical safety in any particular circumstances will determine to what extent they hold responsibilities under the regulations to ensure that the regulations are complied with.

A person may find that they are responsible for causing danger to arise elsewhere in an electrical system, at a point beyond their own area of work. This situation may arise, for example, if you are working on a circuit while somebody else is working in a different room on that same circuit. This is obviously a dangerous situation. Because such circumstances are 'within [your] control', the effect of Regulation 3 is to bring responsibilities for compliance with the rest of the regulations to you, thus making you a dutyholder.

> **Remember**
>
> You have responsibilities under the HASAWA. Can you remember them all?

4 Absolute/reasonably practicable

Duties in some of the regulations have a qualifying term: 'reasonably practicable'. Where qualifying terms are absent, the requirement in the regulation is said to be absolute. The meaning of 'reasonably practicable' has been well established in law. The interpretations below are given only as a guide.

5 Absolute

If the requirement in a regulation is 'absolute' – for example, if the requirement is not qualified by the words 'so far as is reasonably practicable' – the requirement must be met regardless of cost or any other consideration.

6 Reasonably practicable

Someone who is required to do something 'so far as is reasonably practicable' must think about the risks presented by a particular work activity or site and weigh these against the costs – in terms of the physical difficulty, time, trouble and expense – which would be involved in taking steps to reduce these risks. For example, in your own home you would expect to find a fireguard in front of a fire to prevent young children from touching the fire and being injured. This is a cheap and effective way of preventing accidents and would be a reasonably practicable situation.

If the cost or technical difficulties of taking certain steps to prevent those risks are very high, it might not be reasonably practicable to take those steps.

In the context of the Electricity at Work Regulations, where the risk is often that of death from electrocution, and where the necessary precautions are often very simple and cheap – for example, insulating surrounding cables – the level of duty to prevent danger approaches that of an absolute duty.

Provision and Use of Work Equipment Regulations 1998

The Provision and Use of Work Equipment Regulations (PUWER) require risks to a person's health and safety, from equipment that they use at work, to be prevented or controlled. Work equipment includes all tools (power or hand) and equipment such as threading machines and bending machines. Equipment must be:

- suitable for the intended use
- safe for use, maintained and, where necessary, regularly inspected
- used only by people who have received information, instruction and training
- accompanied by appropriate safety measures, e.g. protective devices, markings, warnings.

Manual Handling Operations Regulations 1992 (as amended)

These regulations lay down requirements for moving loads by hand, which could involve pushing, pulling, lowering the load, etc. They include:

- avoiding hazardous manual handling operations so far as is reasonably practicable, e.g. by using a mechanical lifting aid such as a sack trolley to move the load
- carrying out a risk assessment of any hazardous manual handling operations that cannot be avoided
- reducing the risk of injury from manual handling so far as is reasonably practicable, providing staff training in safe lifting techniques.

A risk assessment for manual handling must take into account the following five factors:

- the task
- the load
- the working environment
- individual capability (of the person carrying out the lifting)
- other factors, e.g. use of protective clothing.

Personal Protective Equipment at Work Regulations 1992 (as amended)

Personal protective equipment (PPE) is defined in the Personal Protective Equipment at Work Regulations as all equipment (including clothing affording protection against the weather) which is intended to be worn or held by people at work and which protects them against one or more risks to health or safety. PPE includes safety helmets, gloves, eye protection, high visibility clothing, safety footwear and safety harnesses. The provision of respiratory and hearing protection is covered by separate legislation.

The main requirement of the regulations is that personal protective equipment is to be supplied and used at work wherever there are risks to health and safety that cannot be adequately controlled in other ways.

Figure 1.1: A sack trolley is an example of a mechanical lifting device

PPE must be:

- suitable for use
- maintained and stored properly
- provided with instructions on how to use it safely
- used correctly by employees
- provided to employees free of charge and replaced if damaged.

Lifting Operations and Lifting Equipment Regulations 1998

The regulations aim to reduce risks to health and safety from lifting equipment (such as cranes, pulleys, ropes, slings, etc.) provided for use at work. Lifting equipment should be:

- strong and stable enough for the particular use and marked to indicate safe working loads
- positioned and installed to minimise any risks
- used safely, i.e. the work is planned, organised and performed by competent people
- subject to an ongoing thorough examination and, where appropriate, inspection by competent people.

> **Did you know?**
>
> Lifting equipment is not just cranes and machines – it is also ropes and pulleys used in lifting.

Work at Height Regulations 2005 (amended)

The regulations apply to all **work at height** where there is a risk of a fall liable to cause personal injury. They place duties on employers, the self-employed, and any person who controls the work of others. Dutyholders must ensure:

- all work at height is planned and organised
- all work at height takes account of weather conditions that could endanger health and safety
- those involved in work at height are trained and competent
- the place where work at height is done is safe
- equipment for work at height is inspected
- the risks from fragile surfaces and falling objects are controlled.

> **Key term**
>
> **Work at height** is usually described as work taking place at a height of 2 metres or more, above ground level.

Control of Substances Hazardous to Health Regulations (COSHH) 2002 (as amended)

The COSHH Regulations require employers to control exposure to hazardous substances to prevent ill health. Employers have a duty to protect employees, and others who may be exposed. To comply with COSHH an employer must:

- carry out an assessment of the risk to health arising from working with a substance
- decide what precautions are needed
- prevent exposure or control the risk

- introduce control measures to ensure equipment is maintained and the correct procedures followed
- monitor the level of exposure to the substance
- carry out health checks where required
- have procedures to deal with accidents and emergencies
- ensure employees are informed, instructed and where necessary receive training about the risks and precautions to be followed.

Control of Asbestos Regulations 2006

The regulations bring together the three previous sets of regulations covering the prohibition of asbestos, the control of asbestos at work and asbestos licensing. They prohibit the importation, supply and use of all forms of asbestos. The ban applies to new use of asbestos.

If existing asbestos-containing materials are in good condition, they may be left in place as long as their condition is monitored and managed to ensure they are not disturbed.

Duty to manage

Both employers and the self-employed have responsibilities under the regulations. In addition, regulation 4 imposes a duty on people with responsibility for non-domestic premises (owners, occupiers, managing agents, etc.) to manage asbestos. This covers anyone who has, through a contract or tenancy, an obligation to maintain or repair non-domestic premises, including any means of access to or from the premises.

Where there is no contract or tenancy, every person who has, to any extent, control of a part of non-domestic premises, or any means of access to or from them, also has a duty to manage asbestos.

Key requirements of the regulations

- Insulating materials or linings of white, blue and brown asbestos must only be removed/worked on by licensed contractors.
- On identifying high-risk white, blue or brown asbestos, work must be stopped and a risk assessment carried out in relation to working with or working close to asbestos.
- Prior to working with asbestos, a risk assessment must be carried out and suitable control measures put in place, e.g. on site washing facilities.
- Workers who may come across asbestos in the workplace (including plumbers) must attend mandatory asbestos awareness training sessions.
- Asbestos materials must be properly disposed of in approved packaging (preventing the spread of any fibres) and at approved waste sites.
- Workers removing asbestos materials or carrying out work near to asbestos must be properly informed, instructed and trained.

Asbestos cement-based materials

Work with asbestos cement-based materials is not as high a risk. This may be carried out by non-licensed contractors, provided a risk assessment is undertaken, effective control measures are used to prevent the spread of any material and workers are protected (for example by the use of effective protective clothing and respiratory protective equipment).

Control of Lead at Work Regulations 2002

The Control of Lead at Work Regulations cover the following:

- **Duties** – employers and employees are responsible for protecting themselves and other people on premises where lead work is taking place.
- **Training** – workers should be provided with information, instruction and training on the safe use of lead.
- **Assessment of risk of exposure** – if there is a risk of exposure through inhaling or ingesting (swallowing) lead, the level of risk must be assessed and recorded. Correct working practices and proper controls should be put in place and protective and safety equipment provided.
- **Control measures** – the employer and employee must ensure, as far as is reasonably practicable, that all measures are taken to restrict and control exposure.
- **Protective clothing** – employees should be provided with and wear protective clothing.
- **Respiratory equipment and ventilation** – masks or respiratory equipment must be provided and used where there is a risk of exposure to airborne lead dust or fumes.
- **Washing facilities and canteen areas** – adequate washing and changing facilities should be provided. Hands and face should be washed and contaminated clothing changed before eating, drinking or smoking and before leaving the workplace. Food and drink should not be consumed in any area where lead work is being carried out. **Spread of contamination** – both employer and employee should take such steps as are reasonably practicable to prevent lead contamination from spreading beyond the workplace or storage area.
- **Waste and scrap** – old lead sheet should be removed with care, and any scrap and dust taken to approved collectors for recycling. Vehicles transporting scrap and dust should be thoroughly washed and cleaned after use.
- **Lifting lead** – lead is heavy. Proper lifting equipment and additional staff should be used to ensure the safety of workers lifting and moving lead sheet.
- **Medical surveillance** – the blood lead level of employees working with lead should be monitored at regularly.

- **Maintenance records** – the employer must maintain records of risk assessments, information and training provided, precautionary measures taken, medical surveillance and ventilation and respiratory equipment provided.

Enforcing authorities and inspectors

Enforcing authorities

HASAWA is enforced by the Health and Safety Executive (HSE). It is an independent regulator responsible for encouraging, regulating and enforcing health and safety in the workplace in the UK. It also has responsibility for researching any occupational risks in England, Wales and Scotland.

The HSE's duties include:

- helping and encouraging anyone who is working with the objectives in HASAWA
- carrying out research, publication and training for health and safety at work
- ensuring that employers, employees, suppliers and all other relevant people are kept informed and advised on health and safety matters.

The HSE's enforcement role is usually delegated to the local government bodies, such as county or district councils.

Local authorities can be **enforcing authorities** for several types of workplaces including offices, shops, retail and wholesale distribution, hotels, petrol filling stations, residential care homes and the leisure industry.

Powers of inspectors

An enforcing authority may appoint **inspectors**, who have the power to:

- enter any premises in order to enforce the Act, or in a dangerous situation
- be accompanied by the police if it is likely that they may be stopped from carrying out the enforcement
- bring any equipment or materials required and any other person authorised by the enforcing authority
- examine and investigate any circumstance in order to enforce the Act
- give orders that the premises, or anything inside, should be left undisturbed while the investigation takes place
- take measurements, photographs and make recordings
- take samples of any items or substances and of the atmosphere in or close to the premises
- have an item or substance which appears to be a danger to health or safety, dismantled, tested or destroyed if necessary
- ensure items or substances are not tampered with and kept for use as evidence in any **prosecution**

- interview or question any person believed to have information, and ensure all statements are signed as being true
- inspect and take copies of any book or document required for the investigation
- invoke any other power that is necessary to enforce the Act.

Inspectors will also advise on improvements and changes that could be made to a workplace in order for it to meet statutory requirements. Inspectors will meet and speak with both employers and employees during a visit and provide them with information related to health, safety and welfare. This can be done more informally, through oral communication (especially when a breach is very small) or more formally through written communication.

Inspectors have three clear procedures they can follow when they encounter a serious breach of health and safety law. These are outlined below.

Improvement notice

When a breach of the law is serious, the inspector may issue an improvement notice to tell the dutyholder to comply with the law. The note will explain what needs to be done, why and by when, usually within 21 days. After this, the inspector can take further action.

Prohibition notice

Where an activity involves a risk of serious personal injury, the inspector may serve a prohibition notice. This will order that the action is stopped (prohibited) immediately or after a specified time period. The action will not be allowed to resume until remedial action has been taken. As with the improvement notice, the prohibition notice will explain why the action is necessary, and the dutyholder will be told in writing about the right of appeal to an industrial tribunal.

Prosecution

For some very serious breaches of the law, the inspector may decide to prosecute the dutyholder as set down in the HSE's *Enforcement Policy Statement*. Health and safety legislation gives the courts power to punish offenders, as this will help to deter other offenders.

> **Did you know?**
>
> The dutyholder can appeal to an industrial tribunal about the improvement notice if they believe it is unfair or incorrect.

> **Remember**
>
> A failure to comply with improvement and prohibition notices can be punished by a fine of up to £20,000 or six months' imprisonment.

Progress check

1. Which regulations cover who is responsible for the health and safety of those involved on construction projects?
2. What is an employer's legal responsibility with regard to the HASAWA?
3. What should an accident book contain?
4. Under the Fire Safety Order 2005, what are the main general fire precautions that an employer must provide?
5. Who do the Electricity at Work Regulations 1989 apply to?
6. Whose responsibility is it to supply equipment under the Personal Protective Equipment at Work Regulations 1992?
7. What does the phrase 'working at height' mean?
8. When welding lead indoors, what precautions must be taken?
9. What is an 'enforcing authority'?
10. When would a prohibition notice normally be served?

2. Know how to recognise and respond to hazardous situations while working in the building services industry

General site hazards

Most accidents are caused by human error. Failure to keep the workplace tidy and in good condition can create a number of hazards. Below are several general types of hazard that you may encounter.

Trips and fire hazards

Tools, equipment and materials left lying about, trailing cables and welding hoses, spilled oil and so on may cause people to trip, slip or fall. Clutter and debris, oily rags and paper should be removed to prevent fire hazards. As an individual always make sure your own work area is kept clear and tidy.

Using equipment and tools

When carrying out practical work check that you are using the correct tools and safety equipment. Using the wrong, damaged or inadequate personal protective equipment (PPE) may lead to injuries and endanger your health. PPE should be regularly maintained and CE marked. PPE is covered in depth on pages 33–37.

Plumbing tools and equipment are potentially dangerous if misused or neglected. Practical instruction in the proper use of tools and equipment will form part of your training on site and in your training centre. (Tools and equipment are covered in greater detail on pages 200–207.) You should always use the right tool for the job. Below are some guidelines to follow.

Safety tip

Never be tempted to just 'make do' with whatever tool you have to hand.

Safe use of hand tools and manually operated equipment

- Never use a hammer on a tool with a wooden handle (e.g. wood chisel) as you may damage the wooden handle and release dangerous splinters. If the handle splits as you hit it, the hammer could slip off and damage your hand.
- Cutting tools, saws, drills and similar tools must be kept sharp and in good condition. As a plumbing apprentice, you will frequently be asked to use cutting tools such as hacksaws and wood saws: you should ensure that the blades are always fitted properly and are sharp. Hacksaw teeth should be pointing in the forward direction of cut. After use, guards should be fitted where possible.
- Handles should be properly fitted to tools such as hammers and files, and should be free from splinters. Hammerheads should be secured correctly using metal or wooden wedges.

- Mushroom heading' of chisels is a dangerous condition, which can lead to serious eye injury. Unprotected file tangs present a serious danger of cuts and puncture wounds.
- The plugs and cables of hand-held electrically operated power tools must be kept in good condition. Frayed cables and broken plugs should be replaced.
- Electric power tools of 110 or 230 volts must be PAT tested in accordance with your employer's procedures. It is good practice to check all electrical equipment and test labels to ensure they are in safe working order.
- Other common items of equipment, e.g. barrows, trucks, buckets, ropes and tackle, are all likely to deteriorate with use. If they are damaged or broken they will sooner or later fail in use and may cause an accident. Unserviceable tools and equipment should not be used. They should be repaired or replaced and the unsafe equipment removed from the site.
- You may have to use cartridge-operated tools during your career. If you do, you will be given the necessary instructions on the safe and correct methods for using them. These can be dangerous, especially if they are operated by accident or used as toys; this could cause richochets, which may lead to serious injury.

Key term

PAT (Portable Appliance Testing) tests – Checking for safety and keeping maintenance records of all portable electrical equipment to ensure it is in safe working order.

Personal conduct

To be safe when working as a plumber, you will need to follow health and safety advice and guidelines, in particular when manual handling (see pages 37–40) and working at height (see pages 64–70).

Potential dangers to workforce and members of the public

On construction sites

Construction sites are one of the most dangerous areas you will work in. You will need to watch out for trip hazards, scaffolding and moving traffic to name a few.

In industrial commercial premises (occupied and unoccupied refurbishment)

Empty premises that are being refurbished may be similar to a construction site. Where premises have been empty for some time, they may have been vandalised and there may be other dangers such as broken windows or contamination from vermin.

Manufacturing, processing or production activities may be taking place in occupied premises, so you will need to be aware of additional hazards such as machinery operating, chemicals and forklifts.

Find out

You have been asked to go to an occupied house to fit a new bathroom suite. There are two children aged 16 and 12 and there is a pet dog. Make a list of potential dangers you might come across.

In dwellings (occupied and unoccupied refurbishment)

A dwelling can have similar hazards to that of a construction site or commercial/industrial premises. While these are likely to be on a smaller scale, potential dangers are no less serious. In occupied properties you may need to take into account other hazards such as pets or children.

The list below shows some of the most common risks faced by plumbers.

Possible hazardous operations

- Working with electrically powered plant, e.g. power transformers; extension cables; plugs and sockets; portable tools; electric arc welders; threading machines; large hammer drills; specialist equipment; and fixed equipment.
- Working with non-electrically-powered plant, e.g. hand tools, specialist tools, pneumatic tools, hydraulic tools.
- Working with non-powered tools, e.g. specialist equipment, manual handling of loads, specialist tools, general lifting.
- Working with hazardous substances.
- Working in excavations, e.g. think about physical conditions and interruption of gas, water or electrical services.
- Working with powered industrial trucks, e.g. forklift trucks; dump trucks; JCB/tractors; tail-lift vehicles; and road vehicles.

- Working with highly flammable liquids and liquefied petroleum gases.
- Working with lead, e.g. think about manual handling and lead hygiene.
- Working at heights, e.g. ladders, scaffolds, ropes/harnesses.
- Working with demolitions.
- Working with electrical installations.
- Controlling work with fumes, noise or dust.
- Working with asbestos.
- Working within vessels.
- Working in confined spaces.
- Clearing hazardous waste.
- Working with cartridge fixing devices.
- Working on suspended timber floors.

Methods to prevent accidents or dangerous situations

Method statements

Method statements are used for high-risk activities such as working at height to line a flue system. The method statement draws together the findings and control measures of all the risk assessments associated with the activity (for instance, working at heights, working with flue liner materials, working with mortar mixes, etc.) into one document, which is then given to employees, co-contractors, etc. to provide guidance on how the activity is to be carried out.

Permit to work systems

A permit to work system is used as a safety control for very high-risk or potentially fatal activities such as working with gas heating equipment in potential fire risk areas, working on electrical equipment or in confined spaces such as sewers. The permit to work is a checklist of tasks that must be completed before an activity is carried out and is usually countersigned by a company supervisor.

Risk assessments

The main aim of a risk assessment is to keep you safe at work and ensure no one falls ill or is hurt by your actions. It involves looking at the possible hazards you may face while performing a specific working task, and then taking precautions to minimise the risks. At this stage of your apprenticeship you will not be required to write risk assessments, but you must have a working knowledge of them and understand their purpose.

Role of the employer

Employers have a legal duty to assess risks in the workplace. They need to decide whether a hazard is significant, and if it is covered by satisfactory precautions so that the risk is minimised. The Management of Health and Safety at Work Regulations 1999 set out the requirements of risk assessments.

Working life

Risk assessment

Jane, a new apprentice plumber, was carrying out a risk assessment with the senior plumber. She spotted several hazards, including redundant building materials and litter, and, carcassing timbers with nails sticking through them. The senior plumber told Jane not to worry as no one would be silly enough to tread on an unprotected nail and the rubbish would be cleared away the next day by the casual labourers employed to start the clean-up process.

Discuss the following points:

- What should Jane do as she is well aware of the dangers involved?

- What are the dangers involved, and what injuries could be sustained in this type of environment?

- Has the senior plumber contravened any health and safety legislation by not completing the risk assessment correctly? If so, what legislation has been contravened?

- What should have been done to the nails in the carcassing timbers to prevent this type of accident?

- Is there any type of footwear that could help reduce the risk of puncture wounds to the feet?

Hazard and risk identification using a generic risk assessment

To assess the potential of a risk to cause harm, hazardous operations are usually 'ranked'. The example that follows is from a model developed by the British Plumbing Employers' Council (BPEC) specifically for the plumbing industry. It is based on a number of factors, which are presented on a generic risk assessment form. These are used to assess any risk associated with the plumbing industry.

Figure 1.2 shows an example of a generic risk assessment form for working with specialist tools.

Most of the form is self-explanatory, but you should focus on the following items from the form:

- Likelihood of accident occurring
- Maximum consequences (severity) of an accident
- Risk factor (likelihood of accident occurring × maximum consequences (severity) of an accident).

TASK		
Manual handling of loads		
Specialist tools		

APPLICATION OF EQUIPMENT	APPLICATION OF SUBSTANCE
Pipe-bending machines, stilsons, ropes, lead dressers, bending springs, block and tackle, spanners etc.	N/A

ASSOCIATED HAZARDS

Risk of muscle strains

Risk of sprains

Risk of musculo-skeletal injury

LIKELIHOOD	CONSEQUENCE	RISK FACTOR
3	3	9

RISK EXPOSURE	SAFEGUARDS HARDWARE
Employees	Nil

CONTROL MEASURES

1 Specific training and instruction to employees - kinetic lifting
2 Individual assessment to be performed for all tasks
3 Workplace inspections conducted at 3 monthly intervals
4 Random safety inspections
5 Suitable and sufficient personal protective equipment
6 Medical screening for staff at risk

Figure 1.2: Risk assessment form

Likelihood	Scale value
No likelihood	0
Very unlikely	1
Unlikely	2
Likely	3
Very likely	4
Certainty	5

Table 1.2: Likelihood of an accident occurring

Injury or loss	Scale value
No injury or loss	0
Treated by first aid	1
Up to 3 days off work	2
Over 3 days off work	3
Specified major injury	4
Fatality	5

Table 1.3: Maximum consequences of an accident

Likelihood of accident occurring

The likelihood of an accident occurring is assessed using Table 1.2.

Maximum consequences of an accident

The maximum consequences of an accident can be assessed using Table 1.3.

Risk factor

Risk factors are calculated using a simple formula:

$$\text{Likelihood} \times \text{Consequence} = \text{Risk}$$

This calculation will reveal a risk factor of between 1 and 25, which can then be placed in one of the following categories:

- **Minor** risk (risk factor between 1 and 7) – can be disregarded but closely monitored.
- **Significant** risk (risk factor between 8 and 15) – requires immediate control measures.
- **Critical** risk (risk factor between 16 and 25) – activity must cease until risk is reduced.

Other factors contained on the assessment form are discussed below.

Risk exposure – the individuals or groups of people that may be affected by the work activity or process. Control measures must take account of all those people.

Safeguards hardware – the in-built safety features of work equipment. For example, on powered machines this would include machine guards or trip switches. In this particular example safeguards hardware is nil.

Control measures – the additional safeguards that underpin your arrangements. Where these are identified they must be followed through, and a record kept of any outcomes.

Safety signs

Safety signs are designed to keep workers and visitors to the site safe, to pass on information (regarding exits, fire extinguishers, etc.) and to warn personnel about the possible dangers they may be exposed to during their time on site.

There are four categories of safety sign as shown in Table 1.4:

- prohibition signs
- mandatory signs
- warning signs
- information signs or safe condition signs.

> **Find out**
>
> Find out what the common warning, information and prohibition signs found in construction environments are. You will need to remember these. When looking at the signs, try to make a mental note of which group they fit into.

	Prohibition signs	Mandatory signs	Warning signs	Information or safe condition signs
Shape	Circular	Circular	Triangular	Square or rectangular
Colour	Red borders and cross bar. Black symbols on white background	White symbol on blue background	Yellow background with black border and symbol	White symbols on green background
Meaning	Shows what must not be done	Shows what must be done	Warns of hazard or danger	Indicates or gives information on safety provision
Example	No smoking	Wear eye protection	Danger electric shock risk	First-aid kit

Table 1.4: Safety signs

Remember

COSHH stands for Control of Substances Hazardous to Health.

a) Toxic

b) Harmful

c) Corrosive

IRRITANT

d) Irritant

e) Oxidising

EXTREMELY FLAMMABLE

f) Extremely flammable

Figure 1.3: Categories of hazardous substance

In addition to the four categories of safety sign, BS 5499 recommends avoiding excessive use of signs. Several messages may be merged into one clear, concise sign – known as a combination sign – as long as the information remains clear and is not confusing.

Firefighting signs

Under the Fire Precautions (Workplace) Regulations 1997, any non-automatic firefighting equipment should be indicated with a firefighting sign in line with the BS 5499 requirements. These signs include fire warden details and extinguisher location.

Hazardous substances and precautions to be taken when working with them

COSHH

Before working through this section, you may find it helpful to re-read the information on the Control of Substances Hazardous to Health Regulations (COSHH) 2002 on pages 13–14. The regulations provide a legal framework for the control of hazardous substances in all types of business, including factories, farms, offices, shops – and plumbing.

Employers are required to make an assessment of all work where employees may come into contact with hazardous solids, liquids, dusts, fumes, vapours, gases or micro-organisms. Any risks to health must be evaluated and a decision taken on the action to remove or reduce the risks. For example, if you were replacing waste pipework in a college laboratory, your employer should provide you with protective clothing to prevent injury to yourself while carrying out the work.

What is a substance hazardous to health?

Substances that are 'hazardous to health' are labelled as dangerous (e.g. toxic, harmful, corrosive or irritant) (see Figure 1.3). They also include micro-organisms, substantial quantities of dust, or any material, mixture or compound used at work, or arising from work activities, that can harm a person's health. A wide variety of substances used in plumbing fall under the COSHH Regulations, including irritants such as loft insulation, and human waste such as urine. Any work involving contact with these substances will require specific actions to enable plumbers to carry out their job safely.

What precautions are required?

In order to comply with COSHH employers have a duty to take precautions – see pages 13–14. Working with hazardous substances usually involves the use of protective clothing and equipment such as respiratory equipment masks, dust masks, goggles, gloves and so on.

Chemical safety

Chemicals are substances that form the basis of many common items. However, they can be harmful to anyone working with them if they come into contact with the skin by accident. When not contained or handled properly, chemicals can be:

- inhaled as a dust or gas
- swallowed in small doses over a long period
- absorbed through skin or clothing
- touched by or spilled on unprotected skin.

Some chemicals can cause:

- injury to eyes, skin, organs – from fires and burns, etc.
- silent illnesses – effects are not noticeable immediately after exposure, but can cause medical problems many months or years after exposure (e.g. asbestosis)
- allergies – these may cause skin irritation or rashes, or coughing and breathing problems
- death – some poisonous chemicals can kill outright.

When working with chemicals you must understand the dangers involved. This information will usually be communicated to you using a risk assessment similar to the one shown in Figure 1.4. The packaging of hazardous substances usually provides essential safety information – you must follow all recommended safety rules and procedures and know what to do in an emergency.

The example in Figure 1.4 shows an assessment sheet for using sulphuric acid. You should follow all the guidance to use the substance safely, paying particular attention to the control measures.

SULPHURIC ACID (H_2SO_4) Located: drainage waste soil-pipe discharges main sewer systems	CORROSIVE	HAZARDS Fumes Contact
HAZARDOUS OPERATIONS Fumes – inhalation of fumes where insufficient ventilation Contact – ingestion via exposed skin or into the eye		
HEALTH RISKS Sulphuric acid is a strongly corrosive liquid which will cause severe burns to skin and could on entry to the eye cause permanent damage. Respiratory irritant causing shortness of breath and, as a result of long-term contact, delayed pulmonary oedema.		

SPILLAGES	FIRST AID
Cover with soda ash and lift, rinse with copious amounts of water	Seek medical advice if ingested from skin contact or entry into the eye. Wash with plenty of water.

FIRE	STORAGE
N/A	Never add water to sulphuric acid. Store in a secure location.

RISK FACTORS Duration	PROCESS	OUTDOORS
0–1 minute	Unblocking drains	Critical
1–10 minutes	Unblocking drains	Critical
10+ minutes	Unblocking drains	Critical

CONTROL MEASURES
Eye protectors, impervious gloves, impervious overalls and cartridge-type respirator capable of removing fumes.

Figure 1.4: Risk assessment sheet for working with sulphuric acid

Why do chemical accidents happen?

- Hurrying, over confidence, fooling around or not adhering to instructions can lead to accidents.
- Spills and leaks can be dangerous if not wiped up.
- Vapours may build up where there is no proper ventilation.
- The exposure of some chemicals to heat or sunlight can cause explosion, fire and poisonous reactions.
- Contact between a chemical and the wrong material can cause harmful reactions.
- Neglect or failure to dispose safely of certain old chemicals is dangerous, as chemical changes can happen over time.

Types of chemical

There are four main types of chemical: toxic agents, corrosives, flammables and reactives. Table 1.5 describes the chemicals, their health hazards and how to protect yourself.

Safety tip

More haste, less speed! Try not to rush, that's when accidents happen.

Types of chemical	Description/health hazards	To protect yourself
Toxic agents	Poisons, e.g. hydrogen sulphide and cyanides – can cause injury, disease and death	• Close containers tightly when not in use. • Ensure work area is well ventilated. • Wear personal protective equipment. • Wash hands often. • Carry cigarettes in a protective packet. • Safely dispose of contaminated clothing. • Keep any antidotes handy.
Irritants	Acids and alkalis – dangerous to the skin, eyes and respiratory tract	• Wear personal protective equipment – goggles, breathing devices, protective gloves. • Ensure good ventilation. • Run for the nearest water if you come into contact with corrosives; use a safety shower if one is available. • If your eyes are affected, flush with water for 20 minutes and seek medical aid. Many larger first-aid kits contain eye-irrigation kits.
Flammables	Liquids and gases that burn readily, e.g. methanol, ethanol, ether, liquefied petroleum gas (LPG) and petrol	• Keep flames, sparks, or cigarette lighters away from flammables. • Keep only a small quantity of flammables in the work area. • Store and dispose of flammables safely. In an emergency: • evacuate the area, if possible • turn off all flames and sparking equipment • clean up flammables and ventilate the area thoroughly • call the emergency services.
Reactives	Substances that can explode, e.g. nitrogen compounds	• Know your chemicals before working with them: read about them and test them for stability. • Handle reactives with great care, e.g. certain cleaning products mixed with bleach can give off hydrogen peroxide. • At the first sign of a problem, close the doors and evacuate the room through doors that do not lead to the danger area.

Table 1.5: Chemical health hazards and how to protect yourself

General precautions needed for working with common substances

There are several substances that you will frequently encounter in your work as a plumber. For each of these, there will be some general precautions you will need to take to be sure you are working safely.

Lead (solid and fume)

Most pipework and tube that you come into contact with will be copper, steel or plastic, but you may also come across lead. Lead pipework has largely been removed from dwellings, but lead is still used on roof weatherings, and plumbers sometimes carry out this work although it is now usually carried out by lead specialists. Therefore, you need to be aware of the hazards involved when working with lead.

The Control of Lead at Work Regulations 2002 set out measures similar to those concerning safe work with asbestos – see page 31.

Working with lead takes two forms:

- handling tasks – controlled by simple protective methods
- heating-process tasks – which require more stringent control methods.

Table 1.6 summarises the effects of working with lead and the protective measures you will need to take. Your employer should also provide areas to eat, drink and smoke well away from the worksite.

Lead safety is revisited in Unit 11, pages 447–48.

> **Did you know?**
>
> Extended unprotected exposure to lead working can lead to chronic illness.

> **Safety tip**
>
> You should never smoke while working with sheet lead, as the lead can be ingested via the cigarette.

Process	Effects	Protection
Handling lead materials	Harmful lead particles can be inhaled, ingested (swallowed) or absorbed through the skin	• Use PPE such as barrier cream, gloves, eye protection and overalls to prevent lead entering the body through the skin • Wash hands, arms and face after a lead-working session
Heating of lead during lead welding	Poisonous fumes given off	• Use respiratory protective equipment or local extraction ventilation systems to avoid breathing in fumes • Wash hands, arms and face after a lead-working session

Table 1.6: Effects of working with lead and protective measures to be taken

Solvents and lubricants, fluxes, jointing compounds and sealants

Table 1.7 looks at a range of common substances used in plumbing, their purpose and the precautions you will need to take to work safely.

Figure 1.5: Typical sealant cartridge and gun

Substance	Purpose	Precautions
Solvent	Used to make a joint on some plastic materials	• Use in open space • Keep away from naked flame • Avoid contact with skin • Read manufacturer's information and/or COSHH sheet
Lubricant	Used to make a joint push together	• Usually harmless, few restrictions on use • Read instructions on container before use
Fluxes	Used for soldering copper pipe and fittings	• Avoid contact with skin – apply with small brush • When heated, gives off fumes – use in well-ventilated area • Remove excess flux with damp cloth • Read instructions on container before use
Jointing compounds	Commonly used for making joints when installing low carbon steel pipes and fittings	• For wholesome water or gas joints, ensure compound is of an approved type • Always check compound is suitable for the purpose • Usually harmless, few restrictions on use – usually applied with a finger • Read manufacturer's guidance before use
Sealants	Widely used, e.g. to seal round a bath	• Supplied in cartridge and applied using a 'gun' (see Figure 1.5) • Use in well-ventilated area • Avoid contact with skin

Table 1.7: Working with common substances and the precautions to be taken

Gases

LPG

Did you know?

LPG is heavier than air. If it leaks, it will not disperse into the air, but will sink to the lowest point and form an explosive mixture, which could be ignited by a spark. Leakages are especially dangerous in trenches and excavations because the gas cannot flow out of these areas.

Liquefied petroleum gas (LPG) is widely used in construction and building work as a fuel for burners, heaters and blowlamps. It may be any concentration of commercial propane or butane used in heating processes. Even small quantities of LPG, when mixed with air, create an explosive mixture. If 1 litre of LPG is boiled or evaporated, it becomes 250 litres of gas (LPG is a gas when its temperature rises above –42°C). This is enough to make an explosive mixture in a large shed, room, store or office.

LPG is supplied in cylinders and containers, and is highly flammable. It needs careful handling and storage.

Storage and handling of LPG

• Keep all cylinders upright in open-air, secure compounds.

• When not in use close cylinder valves and ensure protective dust caps are in place.

• When handling cylinders do not drop them or allow them to come into heavy contact with other cylinders.

• When using a cylinder with an appliance, including a blowlamp, ensure it is connected properly, in accordance with the instructions you have been given. The amount of draw-off from LPG cylinders MUST be limited because, if this rate is too high, the control valve may freeze; this could lead to leakage.

• Never allow gas to leak. When assembling blowlamp equipment or if you suspect a leak from gas heating equipment, check the equipment with a leak detection solution. If there is a leak, the leak detection solution will bubble and identify the leak source for you to repair.

Transportation of LPG

The transportation of LPG is covered by the Carriage of Dangerous Goods and Use of Transportable Pressure Equipment Regulations 2009. When LPG is being transported in an enclosed van, only small quantities should be carried, and two dry-powder fire extinguishers must be available at all times. (It is also preferable that the vehicle is ventilated and there must be an LPG hazard warning sign displayed on the vehicle.) If significant quantities are being transported, an open vehicle should be used, and the driver will need to undertake special training.

Oxy-acetylene

Oxy-acetylene is a fuel and oxygen mixture for welding or cutting metal. Pure oxygen is used instead of air within a mixture of oxygen to acetylene to increase flame temperature to around 3500 °C. Acetylene is a mixture of carbon and hydrogen. When welding, the flame of a welding torch heats the metal of the two pieces to be joined until a pool of molten metal forms between them, joining them together as it cools. When cutting, a similar process is followed, heating the metal until red and then introducing a stream of oxygen from a trigger. This oxygen reacts with the metal to produce heat and burns in the oxygen as a liquid iron oxide, which then flows out of the 'cut'.

An oxy-acetylene rig has two cylinders (oxygen and acetylene) attached by hoses and regulators to a torch. The torch has two valve knobs at the bottom of the handle that connect to the cylinders, allowing the user to adjust the mixture flowing to the tip (where the flame forms).

When using an oxy-acetylene rig, always take the following precautions:

- Use the correct eye protection.
- Do not overuse the acetylene cylinder contents in any one hour. Overuse can cause acetone inside the acetylene cylinder to leak out of the cylinder and contaminate the hose and possibly the torch. Acetylene is unstable and explosive above 15 psi pressure. Correct ventilation is important to avoid chemical exposure.
- Ensure fuel and oxygen cylinders are fastened securely upright within a suitable trolley as shown in Figure 1.6.
- The purpose of the regulators on the cylinders is to turn a high-pressure gas to a low-pressure stream. Never attempt to use high-pressure gas directly.
- Make sure the flashback arrestor is in place, as this device (think one-way valve) prevents the flame from burning backwards into the hose and cylinder, potentially causing an explosion.

Carbon dioxide

Carbon dioxide (CO_2) is a colourless, odourless gas or an extremely cold liquid under pressure or refrigeration. It is a compressed gas and as it is slightly heavier than air it can build up in low areas, posing a

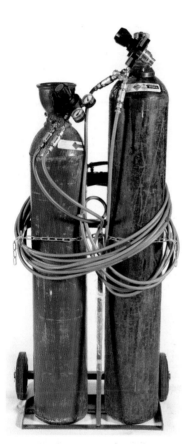

Figure 1.6: Fuel and oxygen cylinders should be fastened securely upright within a suitable trolley

Did you know?

An oxygen cylinder is especially dangerous as, if the valve is knocked off, the cylinder will effectively become an airborne missile. Therefore, never move an oxygen cylinder without its valve cap being in place.

hazard in confined spaces. High concentrations of CO_2 gas may cause headaches, nausea, dizziness, sweating, restlessness and disorientation. Liquid CO_2 may cause frostbite.

Liquid CO_2 is used as a refrigerant, for low temperature testing of electronic components, a fire-extinguishing agent and for controlling temperature in chemical reactions. People working with CO_2 should be properly trained regarding its hazards and its safe use. Always take the following precautions:

- Use the smallest, practical cylinder size in a well-ventilated area separate from the storage area.
- Never use with incompatible materials such as powdered metals.
- Always leave the cylinder cap on until the cylinder is secured and ready for use.
- Always secure cylinders in an upright position.

Cleaning agents

There are many different cleaning products available in the workplace. They contain a wide range of chemicals, and under COSHH, employers have a duty to:

- carry out a risk assessment
- prepare a permit to work for cleaning in confined spaces
- provide good washing facilities
- provide employees with protective gloves, e.g. 'low-protein, powder-free'
- ensure workers are fully trained in using the product.

Below are some general rules to follow when using cleaning products.

Did you know?

Contact with many cleaning products can cause dermatitis. Some can damage the eyes; others may cause asthma.

General rules for the safe use of cleaning products

- Where possible remove heavy dirt first.
- Never transfer concentrated products into an unlabelled container.
- Immediately after use, put the cap back on the container.
- Never re-use a product container – dispose of it safely or return it to the supplier.
- Follow the manufacturer's instructions on the product label.
- Use skin creams to help protect your skin and to replace skin oils.
- Store products securely in a cool, dry, dark place. Apply COSHH regulations. Never store more than 50 litres of flammable liquid indoors – use a specially designed storage facility.
- If products come into contact with the skin, wash off. Avoid contact with surfaces until they are dry.

Asbestos in the workplace

Before working through this section, you may find it helpful to re-read the information on the Control of Asbestos Regulations 2006 on page 14.

What is asbestos?

Asbestos is a mineral found naturally in certain rocks. When separated from rock it becomes a fluffy, fibrous material. In the past it was used for many purposes in the construction industry, for example in cement products, roofing, plastics, insulation, floor and ceiling tiling, and fire-resistant board for doors and partitions.

Although the use of asbestos is now banned, a large number of buildings still contain asbestos materials, including asbestos cement wall and roof panels, roof soffits and window seals, ceiling tiles, textured coatings and insulation lagging around boilers. This applies to buildings built or refurbished before 2000, and especially before 1980.

Asbestos becomes a health hazard if the dust is inhaled; some of the fine rod-like fibres may work their way into the lung tissue and remain embedded for life. This can lead to lung diseases (including asbestosis, lung cancer and mesothelioma), particularly if a person is repeatedly exposed over a number of years. There is no cure for asbestos-related diseases.

Asbestos is perfectly safe in its solid form and only begins to pose a risk when it breaks down or is disturbed. Where possible, it should be left undisturbed or, if necessary, removed only by a licensed contractor.

> **Remember**
>
> It is estimated that 20 tradespeople die from asbestos-related diseases every week. As a plumber you are likely to come across asbestos during your career. Following the correct procedures will help to keep you safe.

Types of asbestos

There are three main types:

- chrysotile – white (accounts for about 90% of asbestos found)
- amosite – brown or grey
- crocidolite – blue.

However, they cannot be identified by colour alone; laboratory analysis is required. It is a legal requirement to know the hazards of your job; any asbestos on site should have been identified before you start work.

Procedures for working with asbestos cement-based materials

It is possible to work with the following asbestos cement-based materials:

- flue pipes
- gutter systems
- soil pipes
- Artex ceilings containing asbestos.

Find out

The HSE provides guidance notes on working with asbestos. Find out the procedures you will need to follow by going to www.pearsonhotlinks.co.uk, searching for this title and clicking on this unit.

To protect yourself you should:

- avoid working with asbestos if possible – if you are unsure whether asbestos is present don't start work until your employer has confirmed it is safe to do so
- never work if the asbestos materials are in sprayed coatings, boards or lagging on pipes and boilers – only licensed contractors should work on these
- know the hazards, avoid exposure and always follow recommended controls
- wear a properly fitted, suitable mask (e.g. disposable FFP3 type), suitable disposable overalls (type 5) and boots without laces or disposable boot covers
- minimise dust
- use hand tools instead of power tools
- keep materials damp but not wet
- clean up as you go using a class H vacuum cleaner
- double bag asbestos waste and label the bag properly
- report any hazardous conditions, e.g. unusually high dust levels, to your supervisor.

When work is finished:

- wipe down your overalls with a damp rag and remove them before removing your mask
- dispose of overalls, etc, properly as with asbestos waste
- don't take overalls home to wash
- don't smoke, eat or drink on site.

Remember

The removal of asbestos, or work in an area of asbestos other than asbestos cement-based materials, must only be carried out by specialist contractors. Never put your health or that of others at risk.

Your employer should provide all the necessary information, instruction and training if you are going to work on an asbestos cement-based product.

Actions to be taken when asbestos is encountered at work

If you come across a substance that you think might be asbestos, you should not proceed any further with your work until you have taken specialist advice.

Protection of the workforce and members of the public

Ensure that the area is closed off to everyone until it is confirmed that it is not a danger, or it has been removed. The type of asbestos will determine how it will be removed and disposed of, and by whom.

Licensing requirements for asbestos-removal organisations

The HSE issues specialist companies with a licence to allow them to operate and specialise in the safe removal of asbestos. They will need to work to high standards, have the correct facilities, such as a dirty work and clean work area, and dispose of asbestos correctly.

Safe disposal requirements

Carrying asbestos material and disposing of it is classified as special waste and many restrictions apply. A specialist contractor will be able to undertake this. In some cases, such as Artex, you may have to place the substance in a bag and place the bag in a container, which will then be collected by a registered waste disposal company.

Working life

Caroline and John are working in a former nursing home on an old-fashioned large boiler, a calorifier (a large hot-water storage cylinder) and associated pipework. The whole installation has been lagged with a very generous layer of something thick and bulky, and is painted in bright colours.

The calorifier has a leak but it is not obvious where the water is coming from. They decide to investigate further by removing the lagging.

Discuss the following points:

- What is the brightly painted lagging (insulation) made from?
- What should Caroline and John do when they realise what they are dealing with?
- There is a lot of powder all around the boiler house, which has fallen off the installation over the years. Are Caroline and John in any immediate danger?
- Who should the installation be reported to, as it will have to have work carried out on it?

Progress check

1. What is the commonest cause of accidents?
2. What does PAT testing involve?
3. What is the purpose of a method statement?
4. What is the formula used to calculate a risk factor?
5. Which category safety sign is circular in shape, has a red border and cross bar with a black symbol on a white background?
6. What are the six categories of hazardous substances?
7. When using sealants, what should you do to ensure your safety?
8. What should be done if any gas storage cylinder is damaged or if there is a likelihood of it having been involved in a fire?
9. What would a plumber use carbon dioxide (CO_2) gas for?
10. What are the three main types of asbestos?

---

3. Know the safe personal protection measures while working in the building services industry

Using personal protective equipment (PPE)

Before working through this section, you may find it helpful to re-read the information on the Personal Protective Equipment at Work Regulations 1992 on pages 12–13. PPE includes most types of protective clothing and equipment such as eye, hand, foot and head protection.

Figure 1.7: Safety glasses

Figure 1.8: Safety goggles

Figure 1.9: Gloves used for general work

Figure 1.10: Specialist gloves used for working with chemicals

Did you know?

Over 700,000 work days per year are lost due to dermatitis.

Figure 1.11: Safety helmet/hard hat

Clothing protection

When working on site your whole body needs to be safe and protected. You will usually wear:

- a high-visibility jacket, which will ensure that you are clearly visible at all times; or
- overalls, which will protect you against dirt and minor cuts.

In wet or cold conditions, you may also require waterproof or thermal clothing. There are also chemical resistant overalls, which you may need to wear in certain situations.

Eye protection

Every year, thousands of workers suffer eye injuries, which result in pain, discomfort, lost income and even blindness. Your eyes are the parts of your body that are most vulnerable to injury at work. Following safety procedures correctly and wearing eye protection can prevent these injuries. There are many types of eye protection equipment available such as safety glasses, safety goggles and welding goggles. Eye protection that is used to guard against the threat created by flying objects should include high-impact resistant lenses.

Hand protection

Almost one in four work-related injuries happens to hands and fingers. One of the most common health problems – other than cutting, crushing or puncture wounds – is dermatitis, an inflammation of the skin normally caused by the hands coming into contact with irritating substances. Symptoms include sores, blisters, redness or dry, cracked skin which can easily become infected.

To protect your hands from irritating substances, keep them clean by regular washing using approved cleaners. Use barrier creams where provided, and wear appropriate personal protection (usually a strong pair of gloves) when required. The type of glove used depends on the work being carried out:

- general purpose gloves for the prevention of cuts
- rubber-type gloves when working with hazardous substances (e.g. the contents of a toilet bowl).

Head protection

Head protection is important because it guards your most vital organ: your brain. A head injury can disable a person for life or even be fatal. One study revealed that over 80% of industrial head injuries occurred as a result of people not wearing 'hard hat' protection.

On building sites there is a real danger of falling objects, as well as other serious hazards such as objects at head height, for example scaffolding poles. You need to wear a safety helmet whenever it is required, and you should ensure it is worn correctly.

Head protection

Fazal is a first-year apprentice who is starting his first day on a construction site. The plumber in charge gives Fazal a hard hat to wear. Fazal notices that the hat has a crack down one side and reports this to his supervisor. The supervisor tells Fazal that he must wear this hat because there are no more in the store and he needs to get to work.

- What should Fazal do in this situation?
- Was the supervisor correct in telling Fazal that he would have to put up with the hard hat until a new one was purchased?
- What could the supervisor have done to rectify the problem?

Fazal carried on working, but not long after he was accidentally hit on the head by a falling hammer. Due to the fracture in the hard hat, he was knocked unconscious and taken to hospital. Fazal had six stitches and a week off work. On his return he was given a new safety hat.

Discuss the following points:

- Who would be at fault for the injury Fazal received?
- How would this incident need reporting?
- Would Fazal be entitled to any compensation?
- What would have been the consequences to the company should the accident have resulted in a fatality?

Checklist

Safety helmets

- Adjust the fit of your safety helmet so that it is comfortable.
- All straps should be snug but not too tight.
- Do not wear your helmet tilted or back to front.
- Never carry anything inside the clearance space of a hard hat, e.g. playing cards.
- Never wear an ordinary hat under a safety helmet.
- Do not paint your safety helmet as this could interfere with electrical protection or soften the shell.
- Handle it with care: do not throw it or drop it, etc.
- Regularly inspect and check the helmet for cracks, dents or signs of wear.
- Check the strap for looseness or worn stitching and also check your safety helmet is within its specified 'use by' date.

Safety tip

Always wear your safety helmet the correct way round – despite what your workmates might be doing with theirs.

Remember

- Notify your supervisor of unsafe conditions and equipment.
- Get medical help promptly in the case of head injury.

Foot protection

Did you know?

Toe and foot injuries account for 15% of reported accidents in the workplace. More than 30,000 injuries to the feet are reported each year, and many more accidents go unrecorded.

Although foot protection refers to items designed to protect your toes, ankles and feet from injury, wearing correct footwear can also protect your whole body, for example from electric shock. It can also lessen the severity of injuries should they occur. Accidents to the feet have serious consequences and can result in pain and suffering, disability and loss of work and income.

The basic universal form of foot protection is the safety shoe, which will usually include some kind of metal toe protection, rubber soles and sturdy leather uppers.

Figure 1.12: Ear defenders

Figure 1.13: Earplugs

Hearing protection

You may have to wear ear protection when carrying out a noisy job, for example using a large hammer drill to bore a hole through an external wall for an overflow pipe. Ear protection ranges from simple earplugs to earmuffs.

Respiratory protection

There are many types of respiratory protection, each used to guard against a particular circumstance. In plumbing, respiratory protective devices are used primarily to guard against the following:

● Breathing in dusts, mists or fibre particles – e.g. in a loft space where loft insulation particles can cause irritation, or during the removal of a solid-fuel fireback boiler where the dust and soot created can require the use of a respirator.

● Breathing in dangerous gases – e.g. in situations where you have to undertake lead welding indoors or where there is no mechanical extraction system. In these situations, a respirator would typically be used.

● Breathing in harmful sewer gases – occasionally a plumber may have to enter a sewer system: this is a dangerous situation where a plumber may be required to wear full breathing apparatus.

The consequence of not protecting yourself is the increased risk of lung problems in later life, including potentially fatal conditions such as lung cancer and emphysema.

Safety tip

Remember that the filters on respirators require regular replacement. Full breathing apparatus requires specialist training in its use.

Figure 1.14: Respirator

Figure 1.15: Dust mask

Dave has recently started at college studying for the Level 2 Plumbing & Heating Diploma. He noticed that most people in his group were wearing corporate overalls or work trousers and jackets/tops and that their safety footwear for the workshop was of high quality.

He was told at his interview that he would have to provide his own protective clothing including overalls and suitable footwear and anything he may need.

From his studies and talking to his friends in the group, he now realises that his employer is treating him differently from other employees in the group. He feels a little hurt; Dave struggled to put the money together to buy his work clothing. He was saving to buy a hard hat and some eye protection next.

Discuss the following points:

- Whose responsibility is it to provide personal protective equipment (PPE)?
- What should Dave do about the situation? Is he still entitled to PPE paid for by his employer, or has that changed now he has bought his own?
- Should he approach his employer or should he leave things as they are?
- Would Dave be entitled to any reimbursement from his employer?
- Some of the work he has been involved in required him to wear ear defenders and eye protection when core drilling. What would have been the consequences if he had an accident while carrying out such a task, without PPE?
- Who would have been at fault if an accident had occurred? Dave, his employer or both?

Progress check

1. Almost one in four work-related injuries involve which parts of the body?
2. What is dermatitis and how is it usually caused?
3. Approximately what percentage of industrial head injuries are the result of not wearing a safety helmet?
4. Why should safety helmets not be painted?
5. Other than the obvious protection to toes, ankles and feet, what else does safety footwear give the wearer?
6. What are the consequences of not wearing the correct respiratory protection when required?

4. Be able to apply manual handling techniques

Before working through this section, you may find it helpful to re-read the information on the Manual Handling Operations Regulations 1992 on page 12.

Introduction to manual handling

The manual handling or lifting of objects is the cause of more injuries on worksites than any other factor. Back strains and associated injuries are the main source of lost hours in the building services industries. Manual handling can involve: pushing, pulling, and the lifting and lowering of loads (tools, cylinders, boilers, radiators, etc.).

The movement of loads requires careful planning in order to identify potential hazards before they cause injuries. You should follow safety

1 Does the **task** involve: • stooping? • twisting? • excessive lifting or lowering distances? • excessive carrying distances? • excessive pushing or pulling distances? • frequent or prolonged physical effort? • the sudden risk of the load moving?	2 Is the **load**: • heavy? • bulky or unwieldy? • difficult to grasp? • unstable or with contents that are likely to shift? • sharp, hot or otherwise potentially damaging?
3 Does the **working environment** have: • space constraints? • slippery or unstable floors? • variation in levels? • poor lighting? • hot/cold/humid conditions?	4 Does the **individual** have: • any restriction on their physical capability? • the knowledge and training for manual handling?

Figure 1.16: Basic risk assessment for moving a heavy load

precautions and codes of practice at all times. Always ask yourself whether the load could be moved another way with less risk of personal injury, for example using a sack trolley or a cart.

Before moving a heavy load, carry out the basic risk assessment shown in Figure 1.16. If, after doing a risk assessment, you decide it is not a safe load for a person to lift, you could use a mechanical lifting device such as a sack trolley.

Factors to consider when manual handling

Take a few moments to consider the information in the basic risk assessment above and the impact each factor could have on the safe movement of tools and materials.

Extreme care must be taken when lifting or moving heavy or awkward objects manually – see the step-by-step guide opposite. The load generally accepted as maximum for a fit person to lift is 20 kg for males and 15 kg for females.

Team lifting

In cases where team lifting (involving two or more people) is required:

- team members should ideally be of similar height and build
- all team members must know the lifting sequence
- one member must be nominated to act as coordinator
- good communication when lifting should reduce the risk of accidents happening.

Movement of loads and methods of transport

Using various methods of transport can make the movement of loads much simpler and safer. Flat trailers, sack trolleys and forklift trucks are some of the moving aids you will come across on site.

Untrained people must **never** use equipment such as dumper trucks and similar mechanical devices to move equipment on construction sites; these should only be used by people who are properly trained. It is also vital to ensure that the item of mechanical equipment is safe to use for the task to be carried out.

Correct lifting procedure

Step 1: Ensure that the path along which you want to move the load is clear from obstructions, that any doors you have to pass through are opened and that you have a clear area for placing the load. Test the load by gently applying force with your foot – if it feels heavy or difficult, you may need to seek help for a double lift.

Step 2: If it feels comfortable to move, start from a good base and stand with the feet hip-width apart. Grip the load firmly. (Use gloves to avoid injuries if the load has sharp or rough edges.) Balance the load, using both hands if possible.

Step 3: Take account of the position of the centre of gravity of the load when lifting. Maintain a straight back; bend your knees and let the strong muscles of your legs and thighs do the work. Keep your arms straight and close to the body.

Step 4: Avoid sudden movements and twisting of the spine. Move slowly and evenly. Never obstruct your vision with the load that you are carrying.

Step 5: Place the load carefully at the end of the move. Do not drop or slam the load down.

> **Remember**
>
> A safe lifting technique requires thinking before acting, total concentration, correct use of body power and a smooth, rhythmical sequence of activity. Safe handling is a skill.

> **Remember**
>
> Forklift trucks and certain other methods of transport can only be operated by a qualified person.

5. Know how to respond to accidents that occur while working in the building services industry

First-aid provision

First-aid kits

The Health and Safety (First Aid) Regulations 1981 require employers to provide 'adequate and appropriate' equipment (including first-aid kits), facilities and personnel to ensure their employees receive immediate attention if they are injured or taken ill at work. The regulations apply to all workplaces including those with less than five employees and to the self-employed.

What is 'adequate and appropriate' will depend on the circumstances in the workplace. This includes whether trained first-aiders are needed, what should be included in a first-aid box and if a first-aid room is required. Employers should carry out an assessment of first-aid needs to determine what to provide.

The kit should only contain basic first-aid items. Additional items such as medication for pain relief should never be stored in first-aid kits.

Where plumbers work in domestic properties, it is usual practice for a first-aid kit to be provided with every company vehicle.

Figure 1.17: First-aid kit

Actions when an accident or emergency is discovered

It is recommended that your plumbing apprenticeship training course includes a certificated first-aid course, but in case it does not, here are a few pointers about dealing with accidents.

First actions

Consider a situation where you are working with another plumber who has just cut her hand. There is a lot of blood and your workmate is in distress. What are you going to do *first*? There are a number of things that need to be done immediately following an accident:

- Seek or administer immediate first aid.
- Get help if necessary, e.g. phone for an ambulance.

- Report the accident to the site supervisor.

The order in which you carry out these actions will depend on several things:

- Are you a qualified first-aider?
- How severe is the injury?
- Is your supervisor or a qualified first-aider immediately available?

Later you will need to:

- write down the details in an accident report book
- complete a company accident report form.

There are several important things to remember. In this scenario, we are dealing with a cut hand; the patient is bleeding quite heavily so it would be necessary to administer first aid as soon as possible.

However, if the scenario were, for example, an electric-shock casualty, you must first make sure that the area is safe – to ensure that whatever caused the accident is not going to injure you. In this case, you should isolate the supply first and call for an ambulance if necessary; then someone can attend to the casualty and inform the supervisor.

Remember

People with back injuries should only be moved by trained personnel.

Remember

After an accident – seek first aid, inform your supervisor, complete the accident report form.

Working life

Accident report form

Alan, a second-year apprentice is working with Ishmail, who is the senior tradesperson in charge of the site. Alan was balancing on the top of a stepladder trying to secure some guttering when he overbalanced and fell to the ground. The ground was hard concrete and as a result of this fall Alan has sustained a broken wrist and bleeding to his hand and forehead. Ishmail accompanies Alan in the ambulance. An accident form must now be completed.

- How could the accident have been avoided?
- What information do you need to fill in the form as well as you can?
- Would the accident require any further paperwork to be completed, and if so who should it be sent to and in what period of time after the accident?
- Who, if anyone, is to blame in this scenario?
- Do you think it likely that a risk assessment was carried out on this occasion?

If someone has fallen from a height, do not move them: if they have a serious back injury, your actions could make things worse. If they are conscious, make them as comfortable as possible until help arrives; this may include treating minor wounds. If the person is unconscious, they may be put in the recovery position, but only if you are confident that there is no back injury.

If someone is injured in an accident, the details have to be entered in the accident book. You may have to enter details on a casualty's behalf if they are unable to do so themselves. If the accident involved a piece of faulty equipment, do not tamper with it as it may be subject to an investigation by the Health and Safety Inspectorate.

Accidents that result in three or more days' absence from work must be reported to the Health and Safety Executive. An example of an accident report form is shown on page 42.

Full name of injured person:		
Home address:	Sex: Male/Female	
	Age:	
Status: Employee Contractor Visitor		
Date of accident:		
Time of accident:		
Precise location:		
What was the accident and its cause? (You may have to give a detailed written description.)		
Name and addres of witness if any:		
Details of apparent injuries:		
Summary:		

Figure 1.18: Typical accident report form

Did you know?

Any act of aggression by a colleague at college or at work that causes a situation of conflict is also considered a dangerous occurrence.

Other emergencies

Emergencies other than a fire or an accident could include a bomb threat or chemical spillage. These events are rare, but you still need to be aware of the procedure for dealing with them.

- Find a telephone in a safe environment well away from the emergency. Make sure you are not going to be trapped while making the call.

- Dial the emergency services number, which is normally 999. However, this may not always be the case, especially if you are on an internal exchange – make sure you know any special procedures that may apply.

- Keep calm and listen to what the emergency-service operator has to say.

- When asked, give your name and the name of the service you require.

- You will then be connected to the emergency service you require; again, stay calm, listen to what the operator has to say and answer the questions they ask.

- You will be required to explain the nature of the emergency and where it occurred. Try to give the operator the exact location of the incident, such as the name of the company, its address, etc.

- When you have completed the call, arrange for someone to meet the emergency services to show them where the incident has occurred.

Evacuation procedures

If you are working in an environment where there is an increased risk of an emergency situation occurring, your employer will usually provide an evacuation procedure which gives details of where to exit the building and which assembly point to go to once you are outside. Here a specified warden will check that everyone who should be at the muster point is there. If someone is missing, you should report it to the emergency services. Do not attempt to go back into the building to find them.

Every organisation has unique evacuation procedures – you should make sure you are familiar with them.

Procedures for dealing with minor injuries on other people

Treatment of cuts

Assess the injury. If the cut is minor it should be treated as follows:

- Clean your own hands and put on disposable gloves.
- Gently clean the cut in running water if it is dirty.
- Pat the area dry using a sterile dressing or clean cloth.
- Apply a plaster or sterile dressing.

Severe bleeding

For more severe bleeding:

- Assess the injury. Is loss of blood the main issue or could there be other medical problems?
- Controlling the bleeding should be the first priority.
- Apply pressure to the injury using a clean cloth or sterile dressing. If the wound is large, you may have to bring the edges of the wound together.
- Keep applying pressure to the wound and raise the injured area. This reduces blood flow to the wound.
- Place the injured person at rest. This brings down the heart rate, reduces blood flow and the risk of shock.
- Cover the wound with a dressing (preferably sterile) and bandage this firmly in place. Avoid tightening the bandage so much as to stop circulation to fingers or toes.
- After bandaging, monitor the dressing to check that blood is not still seeping through. If it is, apply another dressing on top (do not remove the first dressing as this can disturb any blood clot forming).
- Treat for shock (see page 46).
- In all cases of severe bleeding, dial 999 for an ambulance or seek urgent medical attention.

Treatment of burns

The treatment of burns is based on the severity of the burn. All burns should be treated as soon as possible, to reduce pain and scarring.

Severe burns and scalds

Severe burns carry a large risk of shock. Only attempt to resuscitate an unconscious person once you have started to cool the injury.

- Cool the burn immediately by applying running water for 10-20 minutes. This can be done using a tap, or a shower or hose if the area is large. If the burnt area is large, ensure that the person keeps warm during this process by using a blanket or layers of clothing (do not disturb the injured area). This will avoid the risk of hypothermia.
- Dial 999 for an ambulance.

- Lay the injured person down and reassure them.
- Carry on cooling the injured area using cold water and while you are doing this, remove any jewellery in the affected area, unless it is stuck to the injury.
- Wearing disposable gloves if possible, apply a sterile dressing or clean non-fluffy cloth to the burn. Cling film or a clean plastic bag can be used as alternatives.
- Treat for shock (see page 46).

Do not:

- Overcool the injured person – this may lead to hypothermia.
- Apply butter, ointment, lotions or creams to the burn.
- Interfere with the injury or remove anything sticking to it – this may cause infection.

Minor burns and scalds

- Cool the burn with cool or lukewarm running water for 10–20 minutes, ideally within 20 minutes of the injury occurring. While you are cooling the burn, remove any jewellery in the affected area.
- Once the temperature of the burn has been lowered, cover the area with a sterile, non-adhesive, dressing. If this is not available, use a clean cloth, clean plastic bag or cling film

Do not:

- Use ointments, butter, lotions or creams on the burn.
- Apply adhesive dressings to the area.
- Break any blisters that have formed.

Electrical burns

Electrical burns may not always look serious, but they can be very harmful.

- If the burn is caused by a low-voltage supply, such as in a house, switch off the power supply or remove the person from the electrical source using something that does not conduct electricity, e.g. a wooden chair or a stick.
- If a person is connected to a high-voltage source (1,000 volts or more), do not approach them. Dial 999.
- In all cases, seek medical attention by dialling 999 or going to accident and emergency (A&E).

Chemical burns

Chemical burns can be very harmful.

If someone else has been injured:

- Put on appropriate protective clothing for yourself.
- Remove any affected clothing from the person who has been burnt.
- Brush away any dry chemical from the skin.

- Wash off any traces of the chemical from the affected area using running water.
- Try and identify the chemical in order to tell the medical professionals.
- Dial 999 for an ambulance or go to accident and emergency (A&E).

Treatment for eye injuries

All eye injuries are potentially serious. The casualty will be experiencing intense pain in the affected eye with spasms of the eyelids.

Before attempting to treat, wash your hands. If there is something in the eye, irrigate the eye with clean, cool water or sterile fluid from a sealed container to remove loose material. Do not attempt to remove anything that is embedded. If chemicals are involved, flush the open eye with water or sterile fluid for at least 10–15 minutes. Apply an eye pad and send the casualty to the hospital.

Exposure to fumes

If you find someone suffering from the effects of fume inhalation or asphyxiation, then **provided that it is safe to do so,** get them outside into the fresh air as soon as possible. Loosen any clothes around their neck or chest that may impair their breathing.

Call the emergency services. Remember that some fumes, such as welding fumes, involve exposure to carcinogens (cancer-producing substances) and therefore there is a need for urgent first aid and medical management following accidental exposure.

Procedures for dealing with major injuries

Treatment for broken bones

If a broken bone is suspected, get expert help. Do not move the casualty unless they are in a position which exposes them to immediate danger.

Unconscious co-workers

Any unconscious casualty should be placed in the recovery position, so long as they have no obvious back injury. This position prevents the tongue from blocking the throat, and, because the head is slightly lower than the rest of the body, it allows liquids to drain from the mouth, reducing the risk of the casualty choking on their own vomit.

The head, neck and back are kept in a straight line, while the bent limbs keep the body propped in a secure and comfortable position. If you must leave an unconscious casualty unattended, they can be left safely in the recovery position while you get help.

Figure 1.19: The recovery position

To place someone in the recovery position:

- Place the person on their side so they are supported by one leg and one arm.

- Open their airway by tilting the head back and lifting the chin.
- Monitor their breathing and pulse continuously.
- If injuries allow, turn the person onto their other side after 30 minutes.

National Health Service (www.nhs.uk)

A casualty who has received an electric shock, or who has been unconscious for any reason, should be sent to hospital in an ambulance.

Treatment for shock

Shock can develop when there is a serious reduction in blood flow caused by the heart failing to work properly. Symptoms of shock include 'shallow, fast breathing, a rapid pulse becoming weaker and pale, cold sweaty skin' (British Red Cross, www.redcross.org.uk). The person may also feel weak or nauseous. The most common cause of shock is a heart attack, but shock can also be caused by severe external or internal bleeding, severe diarrhoea or vomiting, or severe burns. Key actions are as follows:

- Treat any obvious injuries.
- Lay the person down on a blanket or rug to keep them warm.
- Reassure them.
- Raise and support their legs above the level of their heart.
- Loosen any tight clothing.
- Call an ambulance.
- Do not give them anything to eat or drink because they may later need a general anaesthetic in hospital.

British Red Cross (www.redcross.org.uk)

Electric shock and its treatment

Electric shock occurs when a person becomes part of the electrical circuit. The severity of the shock will depend on the level of current and the length of time it is in contact with the body.

The lethal level is approximately 50 miliamperes (mA), above which muscles contract, the heart fibrillates and breathing stops (the victim dies). A shock current above 50 mA could be fatal unless the person is quickly separated from the supply. Below 50 mA, only an unpleasant tingling sensation may be experienced. However, this may cause someone to fall from a roof or ladder and the resulting fall may lead to a serious injury.

Action in the event of an electric shock

- First of all, check for your own personal safety to ensure that you will not be putting yourself at risk by touching the casualty – they are part of the electrical circuit and they are 'live' too.
- Break the electrical contact to the casualty by switching off the supply, removing the plug, or wrenching the cable free. If this is not possible, break the contact by pushing or pulling the casualty free using a piece of non-conductive material, e.g. a piece of wood.

Remember

It is not the volts that give you the jolts; it's the amps that put out your lamps!

Did you know?

To assess someone's consciousness level, ask a question or give a command – for example, 'What happened?' or 'Open your eyes'. Speak loudly and clearly, close to the casualty's ear. Carefully shake the casualty's shoulders. A 'slightly conscious' casualty may mumble, groan or make slight movements. A fully unconscious casualty will not respond at all.

Guide the casualty safely to the ground – do not allow them to bang their head on the way down.

- Talk to the casualty and gently shake them to assess their consciousness and level of response.

- If the casualty appears unharmed, they may only be shaken and should be advised to rest.

- If there is no movement or any sign of breathing, summon help immediately. If there is someone with you, tell them to get help, i.e. ring 999. If you are on your own with the casualty, you will have to leave them while you get help.

- As soon as you return to the casualty you need to start cardio-pulmonary resuscitation (CPR).

Cardio-pulmonary resuscitation

- *Open the airway.* An unconscious casualty's airway may be narrowed or blocked, making breathing difficult and noisy or impossible. The main reason for this is that muscular control in the throat is lost, which allows the tongue to sag back and block the throat. Lifting the chin and tilting the head back lifts the tongue away from the entrance to the air passage.

- *Remove any obvious obstruction from the mouth.* Place two fingers under the point of the casualty's chin and lift the jaw. At the same time, place your other hand on the casualty's forehead and tilt the head well back.

- *To check for breathing*, put your face close to the casualty's mouth, listen for the sound of breathing, and feel for breath on your cheek. At the same time, look for chest movements. Look, listen and feel for 10 seconds before deciding that breathing is absent.

- *Check for a pulse.* If the heart is beating adequately, it will generate a pulse in the neck (the carotid pulse) where the main carotid arteries pass up to the head. These arteries lie on either side of the larynx (throat), between the Adam's apple and the strap muscle that runs from behind the ear across the neck to the top of the breastbone.

- To check for the pulse ensure the head is tilted back and feel for the Adam's apple with two fingers. Slide your fingers back into the gap between the Adam's apple and the strap muscle, and feel for the carotid pulse. Feel for 10 seconds before deciding that the pulse is absent.

Mouth-to-mouth ventilation is given with the casualty lying flat on their back.

- Open the airway by tilting the head and lifting the chin.

- Remove any obvious obstruction, including broken or displaced dentures, from the mouth.

- Close the casualty's nose by pinching it with your finger and thumb.

- Take a full breath, and place your lips around the casualty's mouth, making a good seal.

> **Remember**
>
> CPR is based on the ABC of resuscitation: Airway, Breathing, Circulation. These are the three crucial things your casualty needs to stay alive.

Figure 1.20: Remove obvious obstructions from the mouth

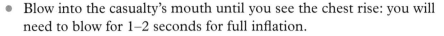

- Blow into the casualty's mouth until you see the chest rise: you will need to blow for 1–2 seconds for full inflation.
- Remove your lips and allow the chest to fall fully.

Deliver two complete breaths in this manner. Thirty chest compressions are now given as follows:

- With the casualty still lying flat on their back on a firm surface, kneel beside them, and find the middle of the chest. To do this, you will need to locate one of the lowest ribs using your index and middle fingers, and then slide your fingers upward to the point where the rib margins meet the breastbone.
- Place your middle finger over this point and your index finger on the breastbone above.
- Place the heel of your hand on the breastbone, and slide it down until it reaches your index finger. This is the point at which you will apply pressure.
- Place the heel of your first hand on top of the other hand, and interlock fingers.
- Leaning well over the casualty, with your arms straight, press down vertically on the chest to depress it approximately 4–5 cm, then release the pressure without removing your hands.
- Give 30 chest compressions, aiming for a rate of approximately 100 compressions per minute.

Figure 1.21: Chest compressions

Return to the head and give two more ventilations, followed by a further 30 compressions. Continue to give two ventilations to every 30 compressions until help arrives; you should be aiming to complete roughly 3 cycles per minute. Do not interrupt CPR to make pulse checks unless there is any sign of returning circulation. If a pulse is confirmed, check breathing, and if it is still absent continue with ventilation. Check the pulse after every 10 breaths, and be prepared to re-start chest compressions if it disappears. If the casualty starts to breathe unaided place them in the recovery position. Re-check breathing and pulse every two minutes.

Recording accidents and near misses

Before working through this section, you may find it helpful to re-read the information on the Reporting of Injuries, Diseases and Dangerous Occurrences Regulations (RIDDOR) 1995 on page 8.

Reporting accidents

Accidents and emergencies must be reported to the relevant authorised people. These can be:

- First-aiders – all accidents need to be reported to a first-aider. If you are unsure who the first-aiders are or have no direct way of contacting them, you must report it to your supervisor.

- Supervisors – you must inform your supervisor of any accident as it is vital that they can act immediately to inform the relevant first-aider or their manager, and stop the work if necessary to prevent any further accidents.

- Safety officers – your supervisor or the site manager will alert the safety officer who will assess the area to check if it is safe, investigate what may have caused the accident and prepare reports for the Health and Safety Executive (HSE) (if needed).

- HSE – if death or major injury occurs to a member of staff or a member of the public is killed or taken to hospital the accident must be reported to the HSE immediately, and followed up by a written report within ten days. The written report is made on form F2508. If an employee suffers an 'over-three-day' injury it must be reported on the F2508 form within ten days.

- Managers – managers should be informed by either the supervisor or safety officer as they may need to report to head office. They may also be the one tasked with contacting the HSE.

- Emergency services – the emergency services should be called as soon as possible. Usually the first-aiders will call the ambulance and the supervisors will call the fire service, but if in doubt you should also call.

The nature and seriousness of the accident will determine who it needs to be reported to. There are several types of documentation used to record accidents and emergencies.

The accident book

The accident book is completed by the person who had the accident or, if this is not possible, someone who is representing the injured person. The accident book will ask for some basic details about the accident, including:

- who was involved
- what happened
- where it happened
- the day and time of the accident
- any witnesses to the accident
- the address of the injured person
- what PPE was being worn
- what first-aid treatment was given.

Near misses

As well as reporting accidents, 'near misses' must also be reported. A 'near miss' is when an accident nearly happened but did not actually occur. Reporting near misses might identify a problem and can prevent accidents from happening in the future.

> **Remember**
>
> An accident that falls under RIDDOR should be reported by the safety officer or site manager. It can be reported to the HSE directly on the phone or via the RIDDOR website. Go to www.pearsonhotlinks.co.uk, search for this title and click on this unit.

> **Safety tip**
>
> The emergency services would rather be called twice than not at all.

> **Safety tip**
>
> Near misses must be recorded because they are often the accidents of the future.

Progress check

1. Who should supply and maintain first-aid kits in the workplace. What is the name of the regulations that indicate this?

2. There are a number of things to be done if an accident has occurred. What are they?

3. What should be done as soon as possible after the casualty has been treated or dealt with?

4. What must be a priority with regard to burns and scalds?

5. If a casualty is suspected of having a broken bone, what is the only reason to move them before help arrives?

6. With regard to electrical shock, the lethal level is approximately 50 mA. What effect would this have on a casualty, if they were not removed from the electrical supply immediately?

7. When administering mouth-to-mouth resuscitation, what is the number of breaths followed by the number of chest compressions, before repeating the process as required?

8. Under RIDDOR an employer must report to the HSE any accident that results in what?

9. When reporting an accident in an accident book, what sort of basic detail would be required?

10. What is meant by the term 'near miss'?

6. & 7. Know and be able to apply the procedures for electrical safety when working in the building services industry

Electricity on site

The safe use of electricity on site is covered by the Electricity at Work Regulations 1989, which impose specific duties on employers to protect their employees against death or injury from the use of electricity at work. Before working through this section, you may find it helpful to re-read the information on the regulations on page 10.

Employers are required to have specific codes of practice for their employees including keeping maintenance records for all portable equipment (PAT tests). The HSE suggests that portable electrical equipment should be tested every three months for construction-site applications. Records must show that the equipment is tested regularly by a competent person using suitable test equipment. You must, however, make a visual inspection of a power tool to establish whether it is safe to use **every time you use it**!

The electricity supply to a work site may be from a generator or from the local public supply. Care should be taken to site a generator so that noise and fumes are reduced to a minimum. Whatever the source of supply, it must be routed to where it is needed on the site. This may involve electric cables being buried underground or, more often, suspended overhead on poles.

Both methods of distribution present hazards. Sites will often have **residual current devices (RCDs)** in place for added protection.

Did you know?

The scope of what is classed as portable equipment is very wide and includes everything from kettles to 110 volt industrial portable equipment.

Key term

Residual current device (RCDs) – a highly sensitive unit that measures changes in the electrical current between different electrical conductors in a system and automatically disconnects the circuit if a small change occurs.

These are intended to 'trip out' the electrical supply if there is any current leakage.

You may work on a site where a site distribution system is needed for the temporary supply of electrical power to the various locations around the site. A number of different units are used for this purpose, all of which should comply with the relevant British Standard.

All the units will be clearly marked with details of the output voltage and the 'Danger–electricity' symbol. The installation of this system must conform to the requirements of the current edition of BS 7671. You should never interfere with or alter any installation: only an authorised person may carry out any alterations.

Checklist

Electricity on site safety

- Do not use lighting circuits for power tools.
- Power tools should be double insulated.
- Never carry a portable electric tool by its cable.
- Ensure that equipment is not damaged before you plug it in.
- Always have enough light for the job.
- Keep lights clean.
- Check that no cables are buried or hidden before beginning work.
- Check that all cables are correctly insulated and not damaged or frayed.
- Check that plugs and sockets are clean and in sound condition.
- Check location of pipework and their proximity to cables.
- Check for current PAT labels.
- Check that RCD protection is provided where necessary.

Methods of safely using electrical tools and equipment on site

Battery-powered supplies

Battery-powered tools are quite commonly used on site. The danger of electrical harm from this equipment is low compared to electrical power tools. However, battery-powered tools have to be charged and this is normally done by the means of a battery charger supplied from a mains electrical supply. It is important that the charging unit is checked to ensure it is not damaged and there is no damage to the electrical lead. The charging unit will need to be PAT tested like other electrical equipment. Care needs to be taken when using battery-powered tools to ensure your own safety and that of others.

110 volt supplies

110 volt power supplies are common on larger construction sites and are identified by yellow cables, plugs and sockets. See Figure 1.22.

Figure 1.22: Typical 110 volt extension lead and plugs

A supply is taken from the mains supply (230 volts) and is then transformed down to 110 volts. This small voltage offers greater protection to the user in the event of a fault and electric shock.

230 volt supplies

The electrical current supplied to domestic properties in the United Kingdom is 230 volts. In the event of receiving an electrical shock it has the potential to cause serious harm or fatality. Great care is needed when using 230 volt power tools and, where possible, an **RCD** should be used. However, it is best to use a 110 volt or battery-powered supply.

What is an RCD?

A residual-current device (RCD), similar to a residual current circuit breaker (RCCB), is an electrical wiring device that disconnects a circuit whenever it detects that the electric current is not balanced between the energized conductor and the return neutral conductor. Such an imbalance is sometimes caused by current leakage through the body of a person who is grounded and accidentally touching the energized part of the circuit. A lethal shock can result from these conditions. RCDs are designed to disconnect quickly enough to reduce the harm caused by such shocks although they are not intended to provide protection against overload or short-circuit conditions.

Visual inspections of a power tool for safe condition

Before using electrical power tools you should always carry out a visual check to make sure that the casing, cable, or the plug are not damaged. Where applicable there should be a PAT label attached to the tool.

According to the Health and Safety Executive 25% of reportable electrical accidents involve portable appliances. The Electricity at Work Regulations place a legal responsibility on employers, employees and self-employed people to take reasonably practicable steps to ensure that no danger results from the use of such equipment. This in effect requires the implementation of a systematic and regular programme of maintenance, inspection and testing. The Health and Safety at Work Act 1974 places such an obligation in the following circumstances:

- where appliances are used by employees
- where the public may use appliances in places such as hospitals, schools, hotels, shops, etc.
- where appliances are supplied or hired
- where appliances are repaired or serviced.

The level of inspection and testing depends upon the risk of the appliance becoming faulty, which is in turn dependent upon the type of appliance, the nature of its use and the environment in which it is used.

Figure 1.23: 110 volt transformer

Remember

PAT testing or portable appliance testing is an important part of any health and safety policy.

Procedures that should be applied for tools and equipment that fail safety checks

If any electrical equipment is not fit for purpose then it should not be used. If you carry out an inspection and find that the equipment is unsafe for any reason you should ensure that it is removed from use straight away. If possible label it 'unsafe' and ensure that it cannot be used by others. Do not try to repair it but report it to your supervisor who should get it repaired by a competent person or replace it.

Electrical industry safe isolation procedure before carrying out work on building services equipment

Before working on any electrical supply you must make sure that it is completely dead and cannot be switched on accidentally without you knowing. Not only is this a requirement of the Electricity at Work Regulations 1989, but it is *essential* for your personal safety and that of your client or co-workers.

The proper way to test if a circuit is live is to use an approved voltage-indicating device, similar to the one shown in Figure 1.24.

Voltage-indicating devices may use either an illuminated lamp to indicate the presence of a voltage or a meter scale. Test lamps are normally fitted with a 15 watt lamp and must be constructed so that they are not dangerous if the lamp is broken. They must also be fitted with protection against excess current, either by a fuse not exceeding 500 mA or by a current-limiting resistor and a fuse. The test leads should be held captive and sealed into the body of the voltage detector. Test lamps and voltage indicators should be clearly marked with the maximum voltage that may be tested by the device and the maximum voltage that the device will withstand.

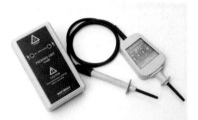

Figure 1.24: An approved voltage-indicating device

Testing

Most fatal accidents involving electricity occur at the isolation stage. This is when you must be most careful and fully aware of what you are doing, as you may have no idea of the type of supply you are confronted with. Do not take any risks, and if you are not sure, seek assistance. Below you will look at how to do the following:

- identify sources of supply
- isolate
- secure isolation
- test that the equipment or system is dead
- begin work.

Standardised procedures have been drawn up by the Electrical Contracting Industry and these are also used as the standard for safe working in the plumbing industry.

Identify sources of supply

It is important that you identify both the *type* of supply and the *source* of supply. In domestic dwellings, this will be a single-phase 230 volt supply to the circuits.

To identify the source of supply:

1. Locate the plug socket nearest to the point where you intend to work on the component or appliance.
2. Make sure the socket is live by testing it – this can be done by plugging in a power tool to check that it works.
3. If it is sound, a check at the consumer unit should clearly indicate which circuit breaker (or fuse on an older property) will isolate the supply.

Now that the type and source of supply have been identified, the supply needs to be isolated. Regulations require that a means of isolation must be provided to enable skilled people to carry out work on or near parts that would otherwise be energized (live).

Isolating devices (fuses, miniature circuit breakers, RCDs) must comply with British Standards, and the isolating distance between the contacts must comply with the requirements of BS EN 60947–3 for an isolator. The position of the contacts must be externally visible or clearly, positively and reliably indicated.

Secure isolation

To prevent the supply being turned on accidentally by your client or co-workers, remove the fuse or circuit breaker and keep it in your pocket, or lock off the isolator with a locking-off device. As an extra precaution, a sign saying 'work in progress and system switched off' must be left at the consumer unit or the area in which you are working.

Test equipment: checking the system is dead

Any circuit you work on *must* be tested to ensure that it is dead. Test equipment must be checked regularly to make sure it is in good and safe working order. Your test equipment must have a current calibration certificate, indicating that the instrument is working properly and providing accurate readings. If it is not calibrated, test results could be inaccurate. Before starting work:

- check the equipment for any damage – look to see if the case is cracked or broken, indicating a recent impact, which could result in false readings
- check that the insulation on the leads and the probes is not damaged, and that it is complete and secure.

If you have any doubts about an instrument or its accuracy ask for assistance. These instruments are very expensive and any unnecessary damage caused by ignorance should be avoided.

Test the voltage indicator on a proven supply before you start; this will confirm that the kit is working. The best piece of equipment for doing this is a **proving unit**.

Only now can you use the voltage-indicating device to establish that the circuit you are intending to work on is dead. You should check phase (live)-to-neutral conductors, phase-to-earth conductors and neutral-to-earth conductors to make sure all connections are dead.

You are not quite ready to begin work – you should again check the test equipment on a known supply to make sure it is working correctly and has not become damaged during the testing procedure.

Only if all the above procedures have been followed correctly should you consider the circuit to be dead and safe to work on.

Begin work

Make use of warning notices: 'Plumber at work'. It may be helpful to put your name and contact number on the notice, so that if you have to leave the job while the client is out, they can contact you to find out why the power has been turned off.

The flow chart in Figure 1.25 shows the procedure for isolating an individual circuit or item of fixed equipment.

Use of temporary continuity bonding when working on pipework components

To prevent the build-up of potentially dangerous voltages in the event of a fault on the electrical system, all the metalwork is connected together using conductors and 'earth clamps' (or clips) in a process known as **bonding**. It is sometimes called **'equipotential bonding'** because all the metalwork is kept at equal potential so that dangerous potentials cannot exist.

It is a requirement of the electrical wiring regulations BS 7671 that all exposed metalwork in a building is bonded together and connected to the earthing block within the consumer unit.

The following list covers the steps that must be taken to ensure the safe earthing of all metallic materials within domestic properties.

- Gas, oil and water pipes can provide a path for stray electrical current. This could lead to corrosion of the pipework and the potential of electric shocks for anyone touching or removing a section of the pipework.

Key term

Proving unit – low-voltage, inverted DC testing device.

Find out

What is the difference between circuit protection and safe isolation?

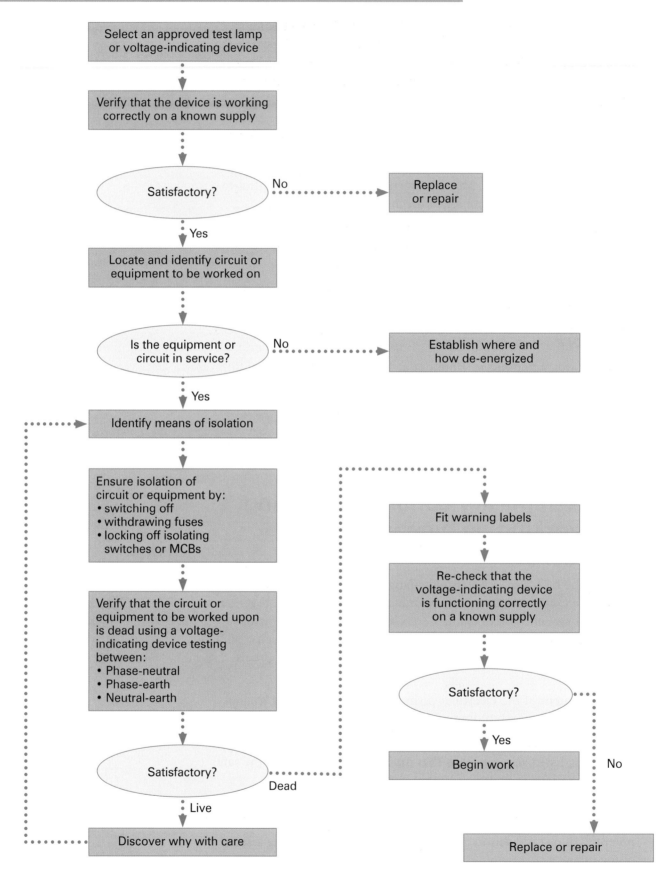

Figure 1.25: Isolating a circuit

- The bonding of all exposed metal components in a dwelling that are not part of the electrical installation is known as equipotential bonding. The equipotential bonding conductor should be close to the consumer unit.

- In certain areas of domestic property, supplementary bonding may be required to link sections of central heating or cold or hot water pipework where the metal pipework has been separated by a plastic fitting or length of pipe. This will ensure earth continuity throughout the property.

- When maintenance processes necessitate the removal of a length of metal pipework, it is essential that the earth continuity be maintained. This is achieved by 'bridging' the gap exposed by the removed section of pipe with a temporary bonding wire. It is vital that the temporary bond is securely fixed in place before the length of pipe is removed.

- Earth clips should be used when connecting bonding wire to pipework. These are designed to indicate clearly the importance of the connection and to show that it ensures a safe electrical connection.

Figure 1.26 shows a typical bonding arrangement for domestic premises. All of the incoming services to the property, such as gas, water and electricity, are bonded together close to the point of entry to the building and are connected to the main earthing terminal.

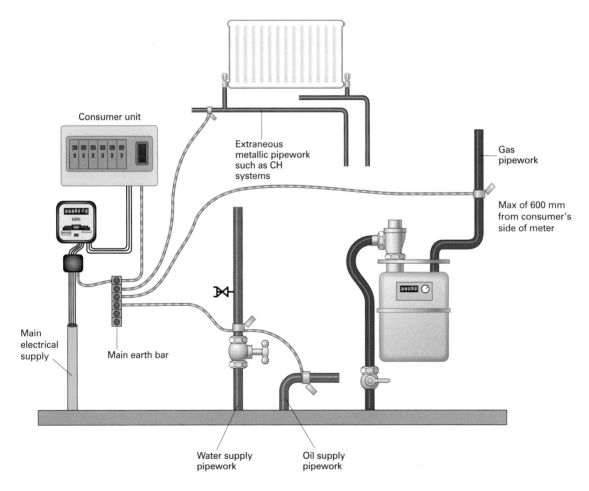

Figure 1.26: Example of the main bonding system in a domestic property

Working life

Brendon and Luke have been sent to investigate a central heating system, which was not working at the start of the heating season. They quickly realised that the circulation pump located in the airing cupboard was the problem. After several attempts to free the impellor they decided that a replacement pump was required.

They then found the service valves either side of the pump also needed replacing as they were badly corroded and would not operate sufficiently. While Luke cleared a space to work, Brendon turned off the boiler and removed the fuse from a fused spur outlet near to the boiler and programmer in the kitchen and started to drain the heating system.

They both vented radiators on the first floor, Luke asked if the power had been isolated and Brendon confirmed that the fuse was safely in his pocket. Luke started to remove the cables from the old pump but received a mild electric shock.

Brendan had removed the fuse from an intruder alarm – the correct isolation switch was located on the other side of the central heating programmer.

Discuss the following points:

* Who was to blame for this potentially dangerous accident?
* What should Luke have done first, before starting to remove the cables from the circulation pump?
* What equipment should Luke have used to carry out the required procedure?
* Should Brendon and Luke have asked a qualified electrician to carry out the wiring on the circulation pump replacement?
* Should the incident be reported, and if so how and to who?

Progress check

1. How often does the HSE recommend that portable appliance testing (PAT) is carried out?
2. Regardless of PAT testing, what should you always do before using a portable appliance?
3. What is the purpose of a residual current device (RCD)?
4. How are 110 volt extension leads, sockets and plugs identified?
5. What is the correct way of testing if an electrical circuit is live?
6. What should you do with the circuit fuse or circuit breaker while you are carrying out work connected to an electrical circuit?
7. What is a proving unit?
8. Which parts of the circuit are checked using a voltage-indicating device?
9. What is meant by the term 'equipotential bonding'?
10. How is earth continuity maintained when it is necessary to remove a portion of metal pipework from a plumbing/heating system?

8. & 9. Know and be able to apply the methods of working safely with heat-producing equipment

Types of gases used in pipe and sheet jointing processes

Propane

As a plumber the most common heat-producing item in your toolkit will be a blowlamp, which is mainly used for soldering of copper pipe and fittings. Blowlamps are available in different sizes and will have either a disposable canister or a refillable cylinder filled with propane gas. Propane cylinders are usually orange coloured.

When using propane on construction sites, always:

● test blowlamps when assembling them.
● keep blowlamps upright during use to prevent liquid coming out of the nozzle.

Further guidance on the use, transportation and storage of LPG is provided on pages 28–29.

Oxygen and acetylene

The other gases you may come across are acetylene and oxygen which are used to carry out lead welding (see Unit 11, page 448). These are available in different sizes of cylinders. Oxygen cylinders are coloured black and acetylene maroon.

Types of heat-producing equipment and safety checks

Hoses

Propane hoses should be visually checked for damage and to ensure that they are within date. Hoses are dated at regular intervals but this may not be visible due to wear. If in doubt, they should be replaced with hoses that arc suitable for high-pressure propane.

> **Did you know?**
>
> Oxygen is a colourless, odourless gas, which is found in air.
>
> Acetylene is an extremely flammable, colourless gas, made up of carbon and hydrogen.

> **Remember**
>
> Whatever equipment you are using remember to have some form of firefighting equipment such as an extinguisher close by. In some cases you may need a permit to carry out hot works.

Figure 1.27: Blowtorch with disposable cylinder and large refillable cylinder

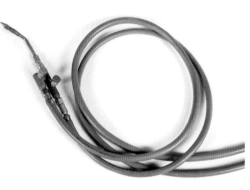

Figure 1.28: Typical set of oxygen and acetylene hoses

You will need to carry out a similar inspection on acetylene and oxygen hoses. These types of hoses are normally much longer than blowlamp hoses and are colour coded red for acetylene and black, green, or blue for the oxygen.

Also remember to check cylinders for signs of damage to the threads.

Figure 1.29: A single-stage regulator

Control valves and gauges

The regulator shown in Figure 1.29 is the type normally used on oxygen and acetylene cylinders. The large threaded part will be different depending on the gas. The two gauges show the pressure setting and the contents. The large handle in the middle is screwed clockwise to increase the outlet pressure or anticlockwise to reduce it.

When using a blowlamp for soldering, the pressure of the propane gas in the cylinder will need to be reduced using either a pre-set regulator, which is labelled to show the pressure it is set at, or an adjustable type, which has a device in the middle that you will be able to turn to the pressure required. The large left-handed thread (denoted by the notching around the nut) is screwed into the cylinder and the seal is a face type joint so care is required to make sure it is not damaged. This should be screwed hand tight into the cylinder and then finally small pressure applied using a spanner.

Before use, regulators and gauges should have the threads inspected and where they are adjustable checked that they operate.

Blowlamp and welding torch

The welding torch has separate controls for oxygen and acetylene so the gases can be controlled to ensure you get the correct flame. When turned on, the gases pass through the inside and are mixed in the mixing chamber and pass out at the tip. Different sizes of tips are available depending on the flame required. For lead welding a neutral flame is required; both gases running equally together.

The blowlamp has a handle for the operator to hold and a control device, which will turn the gas on or off. Depending on how much you turn, the flame will vary in size.

Figure 1.30: Blowlamp and hose

Blowlamps and welding torches should be checked before use to ensure the control valves operate and handles are not damaged.

Safe assembly and use of gas heating equipment

There is a wide range of equipment assembly sequences for gas heating equipment. The sequence will vary depending on the type of tool being used. Before beginning work with gas heating equipment, make sure you have made full reference to the manufacturer's instructions and guidance to make sure that you are following the correct procedure.

Remember

If you find that you are not happy with any equipment, you should raise this with your supervisor and it should not be used.

Bottle location and position

Bottles and cylinders should always be stood upright and be on a firm base, well away from any heat source. Oxygen and acetylene cylinders should be chained back so they will not fall over. When not in use they should be turned off at the control valve.

Leak-detection procedures

Joints on hoses and torches should be checked with leak detection solutions prior to using the equipment. If a leak is suspected with any equipment but cannot be located, then use an approved leak detection fluid. When checking for leaks, do not ignite – simply allow the gas to pass through at a slow rate.

Safe purging procedure

Before lighting the blowpipe, purge the hose by opening the gas for a few seconds. This will flush out (purge) any flammable mixtures of gases in the hose. If there is more than one gas involved, purge one hose at a time and close the blowpipe valve after purging.

Be careful to purge oxygen and fuel lines one at a time – open the gas valve slowly and then close the valve of the purged system before opening the next valve. Never carry out the procedure close to people, clothing or combustible materials or near a flame, lit cigarette or other source of ignition. After purging and before lighting the torch, check all connections, valves and regulators for leaks.

> **Safety tip**
>
> Always carry out leak-detection and safe purging procedures in a well-ventilated area.

Safe lighting and extinguishing procedure

When lighting the welding torch the acetylene should be turned on first and ignited, followed by the oxygen. Adjustment is then carried out to establish the correct flame pattern. To extinguish the flame, the lighting procedure is carried out in reverse.

When lighting a blowlamp, turn on the control valve slowly and place a spark igniter a few centimetres away from the nozzle. Once ignited the control valve can be operated to get the correct flame required. To extinguish, close the control valve.

Fire safety

Fire is a constant risk in the plumbing industry where you will regularly come into contact with electricity, gas and heating equipment.

The fire triangle and combustion

Fire or burning is the rapid combination of a fuel with oxygen (air) at a high temperature. A fire can reach temperatures of up to 1000 °C within minutes. For a fire to start there are three requirements: a combustible substance (called the fuel), oxygen (usually as air) and a source of

Figure 1.31: The fire triangle

heat (spark, friction, match). When these come together in the correct combination, a fire occurs.

Fires can spread rapidly. Fire prevention is largely a matter of 'good housekeeping', and the workplace should be kept clean and tidy.

Dangers of working with heat-producing equipment and how to prevent fires occurring

In plumbing, the use of blowlamps and welding equipment – often used near combustible materials (e.g. timber skirting, joists, etc.), and sometimes in tight or difficult-to-access areas – means you are particularly vulnerable. Because of this, employers are required to lay down strict working methods. These are largely based on common sense, but are also a requirement of insurance companies. They include:

- providing a fire extinguisher in the working area
- completing work with a blowlamp or welding torch a minimum of an hour before leaving a site.

Other precautions you can take include the following.

- If working close to combustible material, protect the area around the fitting you are soldering with a heat-resistant mat.
- If the pipe is insulated, remove the insulation for about 300 mm on each side of the area where you are using the torch.
- If you think you may have caught any material with the flame, wet the area and check again after a few minutes to make sure it is not alight.
- When working in a client's home, pull back carpets or remove curtains from the working area.
- Make sure the area where you are working is clean before you start, e.g. no timber shavings, as well as after you finish.

Another major cause of fires is electrical faults. All alterations and repairs in electrical installations must be carried out by a qualified person, and must meet the standards laid down in BS 7671.

You may sometimes work in an occupied building, such as an office block. You must be aware of the building's fire safety procedures, escape routes, and where your assembly point is located and who you should report to.

If you discover a fire:

- raise the alarm immediately
- leave by the nearest exit
- call the fire service.

In addition, windows should be closed to help starve the fire of oxygen, but they should not be locked.

Safety tip

If in doubt, stay out.

Classes of fire

Fires are commonly classified into four groups, according to fuel type and how the fires are extinguished:

- **Class A** – fires involving solid materials, extinguished by water
- **Class B** – fires involving flammable liquids, extinguished by foam or carbon dioxide
- **Class C** – fires involving flammable gases, extinguished by dry powder
- **Class D** – fires involving flammable metals, extinguished by dry powder.

Firefighting equipment

If a fire is small, it may be possible to put it out quickly and safely using a range of firefighting equipment, including extinguishers, buckets of sand or water and fire-resistant blankets, which should be readily available in buildings. In larger premises, you will find automatic sprinklers, hose reels and hydrant systems.

Fire extinguishers are red with a colour-coded panel identifying the types of fire it should be used for – see Table 1.8. Only trained personnel should use fire extinguishers. Care should be taken when choosing which extinguisher to use as the wrong one can make the fire worse – always check operating instructions before use.

> **Remember**
>
> Personal safety must always come before your efforts to contain a fire. Smoke and fumes can be as lethal as the fire itself.

Colour code of panel	Content of extinguisher	Main use of extinguisher	Do not use:
Red	Water	Wood, paper or fabrics	On electrical fires – risk of electrical shock or explosion On oils and fats – risk of explosion
Cream	Foam	Petrol, oil, fats and paints	
Black	Carbon dioxide	Electrical equipment	In a confined space – may cause suffocation
Blue	Dry powder	Liquids, gases, electrical equipment	

Table 1.8: Types of fire extinguishers

Figure 1.32: Fire extinguishers

1. Why are odours added to gases?

2. Before using gas equipment, what should you do regarding the hoses?

3. What is a 'neutral flame'?

4. How should oxygen and acetylene cylinders be positioned when in use?

5. How should potential leaks be identified before using gas-heating equipment and/or if a leak on the equipment is suspected?

6. What does the term 'purging' mean?

7. Which gas would be ignited first on an oxy-acetylene torch?

8. What can you do to protect work areas, particularly where they are flammable (timber, skirting, joists etc.), when carrying out 'hot working'?

9. What are the three requirements for a fire to start?

10. Why should water-based fire extinguishers not be used in the vicinity of, or on, electrical fires?

10. & 11. Know and be able to safely use access equipment in the building services industry

A safe system of work

Under the Health and Safety at Work Act 1974, employers are legally required to provide systems of work that are, as far as is reasonably practicable, safe and without risk to health. (See also the Work at Height Regulations 2005 on page 13.) The process required to develop a safe system of work for specific work activities is outlined below.

Developing a safe system of work

- Carry out a risk assessment of the work activity.
- Work out how hazards can be removed and then remove them.
- Where hazards cannot be removed, develop a safe system of work to control them.
- Produce a method statement showing how the activity is to be carried out (see page 20) – the working procedure.
- Where necessary, the working procedure should be accompanied by a permit to work system and lock-off systems for electrical supplies (see pages 21 and 54).
- Monitor the procedure.
- Feed back any information on weaknesses or failures in the system to your supervisor.
- Modify the system of work as required. Continue to monitor/modify where necessary.

Using access equipment to work at height

As a plumber it may be necessary to work at height for a variety of reasons and in various situations. Access equipment forms an important part of site work and refers to items such as ladders, stepladders, roof ladders, trestle scaffolds, independent and putlog scaffolds, and tower scaffolds. It is vital that access equipment is kept in good order and is regularly checked and the inspection recorded.

Ladders

Plumbers need to use ladders frequently, either working from the ladder directly or using it to gain access to the place of work or scaffold.

A ladder should be used only for:

- gaining access to a work platform, e.g. a scaffold
- relatively short-term working (duration of 30 minutes or less) such as repairs to an external waste pipe.

A ladder is not suitable for long-term working; for this type of work another form of access equipment should be used.

Ladders can be manufactured from wood, aluminium or fibreglass. Because ladders are used frequently, their condition tends to be neglected, which can lead to defects. It is advisable to inspect any ladder before use.

You should check the following to ensure the ladder is safe for you to use:

- the **stiles**/strings are not cracked or warped
- the **rungs** are not split or dirty
- tie-rods are not missing and are not damaged
- there is no wood rot
- there are no temporary repairs
- the ladder is not painted as the paint may be hiding defects.

Short ladders can be carried by one person, on the shoulder in either the horizontal or vertical position. Longer ladders should be carried horizontally on the shoulders by two people, with one person at either end holding the upper stile. When carrying ladders you should take care in rounding corners or passing between/under obstacles.

There are certain rules for erecting ladders, which must be followed to ensure safe working:

- The ladder should be placed on firm, level ground. Bricks or blocks should not be used to 'pack up' under the stiles to compensate for uneven ground.

Key term

Stiles – the upright bars of a ladder, into which the **rungs** are set

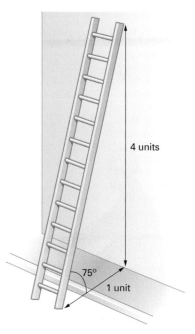

4 units

75°

1 unit

Figure 1.33: Angle of ladder to the building

Did you know?

There are three classes of ladder: 1, 2 and 3. A Class 1 ladder is intended for industrial use.

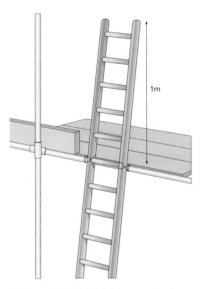

1m

Figure 1.34: Ladder attached to platform

- If using extension ladders, these should be erected in the closed position and extended one section at a time. When extended, there must be at least three rungs overlap on each extension.
- If the ladder is placed in an exposed position it should be guarded by barriers.
- The angle of the ladder to the building should be in the proportion of 4 up to 1 out, or 75° (see Figure 1.33).
- The ladder should be secured at the top and, as necessary, at the bottom to prevent unwanted movement. Alternatively, the ladder may be 'steadied' by someone holding the stiles and placing one foot on the bottom rung; this is commonly known as 'footing' the ladder. This person must not, under any circumstances, move away while someone is on the ladder.
- When the ladder provides access to a roof or working platform, the ladder must extend at least 1 metre or five rungs above the access point.
- Ensure that the ladder is not resting against any fragile surface (e.g. a glass window) or against fittings such as gutters or drainpipes – these could easily give way, resulting in an accident.
- When climbing up ladders you must use both hands to grip the rungs. This will give you better protection if you slip.
- All ladders, stepladders and mobile tower scaffolds should be tested and examined on an annual basis by a competent person. The results of the tests should be recorded, and ideally the item tested should be marked to show it has been tested.

Figure 1.34 shows a correctly positioned ladder, securely lashed to the scaffold with sufficient extension past the access point (five rungs).

Stepladders

Plumbers use stepladders extensively. The first essential check before using a stepladder is to make sure the ground is level and firm. If it is not, you should not use the stepladder on it. All four legs of a stepladder should rest firmly and squarely on the ground (see Figure 1.35). They will do this provided that the floor or ground is level and that the steps themselves are not worn or damaged.

When using the steps, ensure that your knees remain below the top of the steps. The top of the steps should not be used unless the ladder is constructed as a platform.

On wooden stepladders, check that the hinge is in good condition and that the ropes are of equal length and not frayed. A stepladder should only be used face on to the work: if it is positioned sideways on, this increases the risk of the ladder overturning.

Low level work platforms

Low level platforms (commonly referred to as 'hop-ups' or 'odd-job' platforms) are now being used extensively throughout the industry.

Figure 1.35: Stepladder

They are commonly made from aluminum and come in a variety of types and sizes. They offer excellent portability and stability and allow for a better working posture as well as enabling the user to carry out a wider range of movements. They are also easy to manoeuvre. Work from low level work platforms should also follow the general guidance of working from stepladders and ladders. As in the case of other forms of access equipment they should only be used when accompanied by the specific risk assessment and a method statement covering the scope of operations.

Figure 1.36: Low level work platform/hop up

When using low level work platforms, the following should be kept in mind:

- Whoever assembles and uses the equipment must be trained and competent to do so.
- Only low risk, very light duty, short duration tasks should be carried out when using them.
- Only carry light materials and tools when using them.
- They should not be used for overhead work or to elevate the user above a handrail or to gain extra height on a scaffold.
- Each platform should be of the correct type – check classification to make sure that they are designed for their use i.e. industrial applications.

Scaffolds

Independent and putlog scaffolds

As a plumber you will not be expected to erect either an independent or a putlog scaffold. However, you will have to work from one at some stage, maybe to install guttering, or to access roof weatherings. It is important, therefore, that you are happy that the scaffold has been erected correctly and is safe for you to work from. As a rule of thumb, stand back and look at the scaffold and ask yourself the following questions:

- Does it look safe?
- Are the scaffold tubes plumb and level?
- Are there sufficient braces, guard rails and scaffold boards?
- Is it free from excessive loads such as bricks?
- Is there proper access from a ladder?
- Is there protection to prevent objects falling off the edge?

Make sure there are ledger-to-ledger braces on each lift for independent scaffolds, and that putlog scaffolds are tied to the building. Look for any gaps between the boards of the platform from which you will be working, and make sure a toe board and guard rail are fitted.

Mobile scaffold towers

Mobile scaffold towers may be constructed of basic scaffold components or may be specially designed 'proprietary' towers made from light alloy tube. The tower is built by slotting the sections together until the required height is reached. Mobile towers are fitted with wheels, static towers are fitted with base plates.

When working with mobile scaffold towers the following points must be followed:

- The working platform must be provided with suitable edge protection and toe boards. Guard rails should be at least 950 mm high and an intermediate guard rail should be provided so the unprotected gap does not exceed 470 mm.
- When the platform is being used, all four wheels must be locked.
- The platform must never be moved unless it is clear of tools, equipment and people. It should be pushed at the bottom of the base and not at the top.
- The stability of a tower depends upon the ratio of the base width to height. A base to height ratio of 1:3 gives good stability.
- Outriggers can increase stability by effectively increasing the area of the base but, if used, must be fitted diagonally across all four corners of the tower and not on one side only. When outriggers are used they should be clearly marked (e.g. with hazard marking tape) to indicate that a trip hazard is present.
- Access to the working platform of the tower should be by a ladder securely fastened inside the tower, or forming part of the tower structure.
- Towers should only be erected by trained and competent people. There are a number of organisations that provide training for the safe erection and use of tower scaffolds against a PASMA competence standard.
- There are specific requirements laid down for the inspection of a mobile tower scaffold. This inspection must be carried out by a person who has been adequately trained.

Roof ladders

Roof ladders are used by plumbers to gain access to chimneys in order to carry out work such as small repairs to sheet weathering. For a full sheet weathering installation, a scaffold would be needed. Roof ladders are also used to gain access to a roof when installing chimney flue liners for gas fire or gas boiler installations in existing flues. They are made of aluminium, which is extremely light, so that they can easily be manoeuvred up the roof. A roof ladder is positioned by turning it on its wheels, pushing it up the roof, and then turning it through 180 degrees to hook it over the ridge tiles.

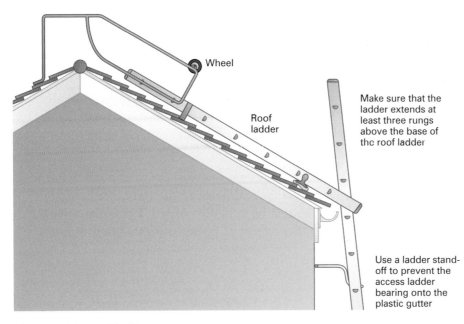

Wheel

Roof
ladder

Make sure that the
ladder extends at
least three rungs
above the base of
the roof ladder

Use a ladder stand-
off to prevent the
access ladder
bearing onto the
plastic gutter

Figure 1.37: Roof ladder

The key safety points are as follows:

- The supports under the ladder must all fully rest on the roof surface.
- The ladder must be long enough for the roof on which it is being used.
- The ladder used to gain access to the roof from the ground must be securely fastened at the top and must be next to the roof ladder.
- Gaining access to a roof and completing the work using a roof ladder usually requires two people: one to access the roof, the other to assist with providing equipment.

Safety checklist for the roof ladder:

- Stiles must be straight and in sound condition.
- Rungs must be sound.
- The ridge hook must be firmly fixed.
- Wheels must be firmly fixed and running freely.
- Pressure plates must be sound (these are the parts that rest on the roof surface).

Mobile elevated work platforms

This access equipment is often referred to as mobile elevating platforms (MEWPs) or 'powered platforms'. Those mounted on trucks are also often referred to as 'cherry pickers'.

From a safety and cost perspective, scissor lifts can offer quick and efficient access solutions for a wide range of installation and maintenance tasks. Compact dimensions and tight turning circles give these machines great versatility. There is a wide range of machines

Figure 1.38: Scissor lift and boom lift platforms

available, including rough-terrain applications. Access equipment can be used for various installations from large factories to repairs to external lighting.

However, scissor lifts can only extend upwards. Sometimes it is necessary to 'reach' over objects to be able to carry out the work, for example repairing a street-lighting column. In these circumstances a telescopic boom platform (the cherry picker) is likely to be more appropriate.

In both types of lift, it is essential that the workers wear a safety harness. This must be attached to the lift and never to the structure being worked on. All operatives must be suitably trained. Particular attention must also be paid to whether overhead power supplies are present.

Working life

A lead chimney flashing on a detached property is to be replaced, and Jordan has been instructed to carry out the work. The chimney is mid-roof.

His employer has told him he is to use a roof ladder and a double extending ladder to gain access. The work is estimated to take two days including the formation and installation of the new flashing.

He erects the double extension ladder and finds it is only long enough when fully extended to pass the soffit of the building by one rung. Jordan contacts his boss, who tells him that it is the longest ladder they have and to continue working.

Jordan is also concerned with regard to how he will get the old lead down and, of course, the new lead up, after he has carried out the formation work at ground level.

Discuss the following points:

- What should Jordan do regarding the length of the ladder? Is the employer within his rights to insist Jordan carries on?
- Are a roof ladder and an extension ladder suitable equipment for the work to be carried out?
- What alternative means of access to the chimney stack ought to be made available?
- How should the old lead be moved to the ground level safely and how should the new lead be taken up to the stack?
- What legislation/regulations would be infringed if the work were to carry on?
- If an accident occurs who could be prosecuted?

Progress check

1. What are the only purposes for which a ladder should be used?

2. Before using a ladder, what checks should be carried out?

3. When a ladder is extended, how many rungs should overlap each extension?

4. If the ladder is providing access to a roof or working platform, how far should the ladder extend above the access point?

5. What two types of scaffold are used to provide access to buildings?

6. What is the difference between a mobile and a static scaffold tower?

7. A ladder will be used to provide access to the working platform of a scaffold tower. Where should this ladder be located?

8. When using a roof ladder, how close should this be located to the access ladder?

12. Know the methods of working safely in excavations and confined spaces in the building services industry

Situations where work in excavations or confined spaces may be required

As a plumber you may need to work in a trench but in most cases it will not be more than 1.5 m in depth as the type of work you may be doing is laying water supply pipes usually between 750 mm and 1350 mm. You are unlikely to work on underground pipework, although you may have to help with installing underground pipe for ground-source heating – this does not usually involve working in the trench as it will be laid in from above and the depth will not normally be more than 2 m.

However, you are likely to be working in confined spaces, of which excavations are but one. The HSE defines a confined space as 'a place which is substantially enclosed (though not always entirely), and where serious injury can occur from hazardous substances or conditions within the space, or nearby' (www.hse.gov.uk). Examples include grain storage silos, chambers in plant rooms, duct-rooms, storage tanks, cylinders, boilers or cisterns, under suspended timber floors, roof spaces, drains and sewers or cramped areas under work benches.

A confined space may become dangerous because of:

- a lack of oxygen
- poisonous gas, fumes or vapour
- liquids or solids suddenly filling the space
- fire and explosions, e.g. from flammable gases
- high concentrations of dust, e.g. in flour silos
- hot conditions, which may lead to a dangerous rise in body temperature
- lack of light and space that lead to physical problems
- presence of machinery or nearby work operations.

Remember

Ultimately, try to avoid entering confined spaces and check whether the work can be done differently. If this is not possible and you cannot avoid entry into a confined space make sure you have a safe system for working inside the space.

Work in excavations

Work in excavations can be particularly hazardous and therefore a risk assessment must be conducted before the work is commenced to ensure that proper control measures are applied to safely do the work. The following are typical measures:

- An assessment of the ground material type must be undertaken to determine the need for shoring.
- Only properly trained plumbers must work in excavations.
- Any work carried out should be in pairs, one plumber carrying out the work in the trench, the other passing tools and materials.

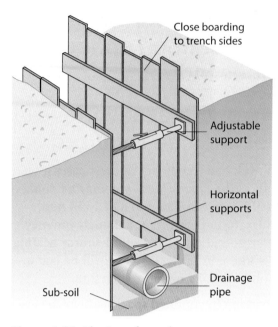

Figure 1.39: Shoring of trenches

- Fencing must be erected around trenches over 2 m in depth to prevent people falling into the trench. Safety signs should also be in place in advance of any excavation.
- Material removed from the trench should be safely stored so that it may not fall back into the trench.
- Barriers such as vehicle stop blocks may be needed if dump trucks are being used in the vicinity of the trench. These should prevent vehicles falling into the trench.

Measures needed to prevent people or equipment falling into excavations

Use of warning signs

Warning signs should be placed in a prominent position to warn people of the danger and risk. However, the use of a sign alone would not be acceptable – it must be used alongside other measures. An example of a sign is shown in Figure 1.40.

Use of barriers for pedestrians

A barrier must be erected around the trench to keep any unauthorised people away from it. The type of barrier can vary from a ridge type to a plastic net type.

Vehicle proximity to excavation edges

If vehicles are operating near to the trench, care needs to be taken that this does not cause the trench walls to collapse. Vehicles should be kept as far away as possible and good practice would be to have an observer directing the operator.

Figure 1.40: An example of a warning sign

Potential dangers when working in confined spaces

Many of the same rules for working at height also apply to working in excavations. Every year people are killed or seriously injured when working in excavations.

They are at risk from:

- excavations collapsing and burying or injuring people working in them
- material falling from the sides into any excavation
- people or plant falling into excavations.

Excavation work has to be properly planned, managed, supervised and carried out to prevent accidents. Before digging any excavations, it is important to plan against the following:

- collapse of the sides
- hazardous entry and exit points
- materials, people and vehicles, falling onto people working in the excavation
- people being struck by plant
- undermining nearby structures
- fumes
- contact with underground services
- inflow of water
- accidents to members of the public.

Make sure the necessary equipment is available on site before work starts. Once work starts, adhere to the following.

- Wear a hard hat when working in excavations.
- Prevent the sides and the ends from collapsing by battering them to a safe angle or supporting them with timber, sheeting or proprietary support systems.
- Never go into an unsupported excavation.
- Never work ahead of the supports.
- Never store spoil or other materials close to the sides of excavations. The spoil may fall in and the extra loading will make the sides more prone to collapse.
- Always make sure the edges of the excavation are protected against falling materials. Provide toe boards where necessary.
- Take steps to prevent people falling into excavations. If the excavation is 2 m or more deep, provide substantial barriers, e.g. guard rails and toe boards.
- Keep vehicles away from excavations wherever possible. Use barriers if necessary.

Remember

Liquefied petroleum gas (LPG) is heavier than air, and therefore poses a particular danger when working in excavations. If LPG is being used in an excavation, any leakage could result in the formation of pockets of LPG. This is a potentially dangerous situation, as LPG is a highly explosive gas. (For more information on LPG, see pages 28–29 and 58–59.)

- Where vehicles have to tip materials into excavations, use stop blocks to prevent them over-running. Remember that the sides of the excavation may need extra support.

- Keep workers separate from moving plant such as excavators. Where this is not possible, use safe systems of work to prevent people being struck.

- Make sure that any plant operators are competent.

- Make sure excavations do not affect the footings of scaffolds or the foundations of nearby structures. Walls may have very shallow foundations and these can be undermined by even small trenches.

- Look around for obvious signs of underground services, e.g. valve covers or patching of a road surface.

- Use locators to trace any services. Mark the ground accordingly.

- Make sure that the person supervising the excavation work has service plans and knows how to use them.

- Everyone carrying out the work should know about safe digging practices and emergency procedures.

- Provide good ladder access or other safe ways of getting into and out of the excavation.

- Fence off all excavations in public places to prevent people and vehicles falling in.

- Take precautions (e.g. securely covering excavations) where children might get onto a site out of hours, to reduce the chance of them being injured.

- Make sure that a competent person supervises the installation at all times.

Progress check

1. When must fencing be provided around an open excavation?
2. What is the main danger of using LPG equipment in and around trenches?
3. What are the main risks to people working in excavations?

Check your knowledge

1. Who is responsible for providing safety helmets for plumbers to use on a construction site?
   a The individual plumber
   b The plumbing employer
   c The construction company
   d The customer

2. Which is the correct statement under the Health and Safety at Work Act 1974?
   a All safety on a site is the responsibility of the owner
   b All operatives and employers have a responsibility for their own and others' safety in the workplace
   c Plumbing employers have sole responsibility for the safety of all those involved
   d Operatives do not have any responsibility for health and safety – it is down to management

3. Which document provides information regarding the general safety requirements for scaffolds on building sites?
   a The specification for the construction work
   b Construction (Health, Safety and Welfare) Regulations 1996
   c Construction (Lifting Operations) Regulations 1961
   d Health and Safety at Work Act 1996

4. Which is the safest method of providing temporary lighting when working in a building under construction?
   a 110-volt temporary electrical supply
   b Paraffin lamp
   c 3-phase electrical supply
   d 230-volt temporary electrical supply

5. If a ladder is placed against a wall, with a vertical height of 4 m above ground level, how far from the wall does the base need to be?
   a 100 mm
   b 700 mm
   c 900 mm
   d 1000 mm

6. What is the correct term to describe a hazardous substance that causes inflammation through contact with the skin?
   a Toxic
   b Corrosive
   c Harmful
   d Irritant

7. If blue asbestos were identified while carrying out work, what should be done?
   a Report it to your supervisor so they could get a licensed contractor to remove it
   b Ask someone over 18 to remove it
   c Let your supervisor remove it with a mask on
   d Remove it yourself with a mask on

8. When carrying out excavations, which of the following is correct?
   a Operatives must wear a safety harness
   b Fencing must be erected around trenches under 2 m deep
   c The sides of the excavations deeper than 1200 mm should be sloped or shored
   d A plumber can remove shoring to install pipes

9. When should waterproof industrial gloves be used?
   a When welding
   b When installing pipework
   c When solvent-welding plastic pipework
   d When working on drainage systems

10. Which is the correct statement in relation to LPG heating gases?
    a Butane is lighter than air and any leaking gas will accumulate at ceiling level
    b Propane is lighter than air and any leaking gases will accumulate at ceiling level
    c Propane is heavier than air and any leaking gas will accumulate at ground level
    d Butane is the same weight as air and any leaking gas will mix with the air in the room

11. If there is a fire in a paint store, which fire extinguisher should **not** be used?
    a Water
    b Foam
    c Powder
    d Carbon dioxide

12. Which of the following must only be changed by a person with an appropriate qualification?
    a The fuse in a 13 A plug
    b A drill bit
    c An abrasive wheel on a bench grinder
    d The dies on a threading machine

Getting ready for assessment

The information contained in this unit of the book, as well as the continued practical assignments that you will carry out in your college or training centre, will help you with preparing for both your end-of-unit test and the diploma multiple-choice test. It will also support you in preparing for the practical assignments you will need to complete to demonstrate your understanding of carrying out safe working practices in building services engineering.

There are opportunities throughout the unit to test your progress in and understanding of the required underpinning knowledge; this will enhance your preparation for the forthcoming assessments, so make good use of them.

This unit will be assessed in a combination of assessment methods:

- simulated practical assessments
- externally set knowledge assignment.

With regard to the building services industry, you will need to be familiar with and have knowledge of:

- the health and safety legislation that applies to the industry
- how to recognise and respond to hazardous situations
- safe personal protection measures including manual handling techniques

- how to respond to accidents
- the procedures for electrical safety and the application of basic electrical safety measures
- methods of working safely with heat producing equipment
- methods of using access equipment safely
- methods of working safely in excavations and confined spaces.

Always make sure that you are working safely throughout the tasks and assessments. Make sure you are working to all the safety requirements given throughout the assessments and wear all appropriate personal protective equipment. When using tools, make sure you are using them correctly and safely.

Check it out – using your knowledge from this unit, practise and test yourself on:

kinetic lifting (one person and two person)	mechanical lifting
continuity bonding	safe isolation
safe use of power tools	using LPG equipment
fire extinguishers (recognition, selection and use)	access equipment

Good luck!

Understand how to communicate with others within building services engineering

As well as having excellent technical skills – and you should always aim to be a top-class plumber – you need different skills when dealing with people. These are known as 'personal skills'.

You will deal with a wide range of professionals in the building services industry, in a variety of different job roles. You need to know what each of these professionals does, to enable you to give them the information they need.

Communication is an important part of all working life. When dealing with anyone professionally, you need to make sure that any information you give is delivered clearly and in an appropriate format.

This unit will cover the following learning outcomes:

- Know the members of the construction team and their role within the building services industry
- Know how to apply information sources in the building services industry
- Know how to communicate with others in the building services industry

1. Know the members of the construction team and their role within the building services industry

Key roles in the site management team

When working on site it is important to know about the roles and responsibilities of the site management team as this will enable you to understand what everyone is doing and who to report to if there is a problem.

Architect

Architects plan and design buildings. The range of their work varies widely and can include the design and procurement (buying) of new buildings, alteration and refurbishment of existing buildings and conservation work.

An architect's work includes:

- meeting and negotiating with clients
- creating design solutions
- preparing detailed drawings and specifications
- obtaining planning permission and preparing legal documents
- choosing building materials
- planning and sometimes managing the building process
- liaising with the construction team
- inspecting work on site
- advising the client on their choice of a contractor.

Project manager/clerk of works

Remember

A project manager may also be known as a clerk of works.

The project manager or clerk of works takes overall responsibility for the planning, management, coordination and financial control of a construction project. They work for architects, for clients such as local authorities, or as consultants.

A project manager will ensure the client's requirements are met, and that the project is completed on time and within budget. Depending on the project, their responsibilities can start at the design stage and continue through to completion, and handover to the client.

A project manager's work includes:

- representing the client's interests
- providing independent advice on the management of projects
- organising the various professional people working on the project
- making sure that all the aims of the project are met
- ensuring quality standards are met

- keeping track of progress
- accounting, costing and billing.

Structural engineer

Structural engineers are involved in the structural design of buildings and structures such as bridges and viaducts. Their primary role is to ensure that these structures function safely. They can also be involved in the assessment of existing structures, perhaps for insurance claims, to advise on repair work or to analyse the viability of alterations and adaptations.

Building surveyor

Building surveyors are involved in the maintenance, alteration, repair, refurbishment and restoration of existing buildings.

A building surveyor's work includes:

- organising and carrying out structural surveys
- legal work including negotiating with local authorities
- preparing plans and specifications
- advising people on building matters such as conservation and insulation.

Building services engineer

Building services include: water, heating, lighting, electrical, gas, communications and mechanical services such as lifts and escalators. Building services engineering involves designing, installing and maintaining these services in domestic, public, commercial and industrial buildings.

A building services engineer's work includes:

- designing the services, mostly using computer-aided design packages
- planning, installing, maintaining and repairing services
- making detailed calculations and drawings.

Most building services engineers work for manufacturers, large construction companies, engineering consultants, large mechanical and electrical contracting companies, architects' practices or local authorities. Their role often involves working on the design of buildings as part of a team of professionals, such as architects, structural engineers and contractors.

When working for a consultant, the building services engineer is mainly office based at the design stage. Once construction starts, they will make site visits to liaise with the contractors installing the services. When working for a contractor, the building services engineer is likely to be site based, overseeing the job and even managing the workforce. When working for a services supplier, the role will require involvement in design, manufacture and installation and may involve a lot of travelling between the office and various sites.

Quantity surveyor

A quantity surveyor advises on and monitors the cost of a project. It is their responsibility to make sure that any costs are allocated effectively to any specialist subcontractors and also arrange for payments to be made. The quantity surveyor then continues to manage costs throughout the project, negotiating with the client's quantity surveyor about payments and the final account.

Buyer

The buyer identifies supplies for any materials needed on site. They then obtain quotes from the suppliers before deciding where to purchase the construction materials required. As part of this, the buyer negotiates on prices and delivery costs, and also resolves any quality problems.

The information the buyer collects is used as the basis of the tender. The contractor submits this to a client prior to getting a contract.

Estimator

The estimator's role is to calculate how much a project will cost. This includes all the plant, materials and labour that may be used. The estimator will identify the most cost-effective construction methods that can be used, calculating all cash flows and final margins. The estimator also seeks clarification on any contract issues affecting costs.

Contracts manager

The contracts manager will be responsible for running several contracts at any one time and works closely with the rest of the site management team. The contracts manager is often the link between the other sections of the business and the managing director or CEO of the company. They make sure that the project is running to cost and to schedule.

Construction manager

Also known as the site manager or site agent, the construction manager is responsible for running the site (or a section of a large project). The construction manager develops a strategy for the project, planning ahead to solve any problems before they happen. They also motivate the workforce and make sure that any site and construction processes are carried out safely. The construction manager communicates with the client to report progress and seek further information.

Working life

You are working on a house under construction on a small housing development, installing a bathroom suite. A couple arrives, claiming to be the purchasers of the property. They ask you to position the appliances in a different way from the building plan.

- How should you handle the situation?
- You will need to think about who should be informed about any changes.
- Should the client's request be carried out?

Key roles of individuals reporting to the site management team

There are several key individuals on site who report into the site management team. The key positions are explained below.

Subcontractors

These are workers hired by the main contractor (the company in charge of the construction project) to carry out specific work on site. This is usually specialist work. For example, a general builder may hire a plumber as a subcontractor, because none of their staff is able to carry out the plumbing work.

Plumbing businesses often work for construction companies on a subcontracting basis. This means that the main contractor is a construction company, and that they contract directly with the customer or client. The main contractor then sublets a part of the contract, for example the plumbing work, to a plumbing contractor (subcontractor).

The plumbing subcontractor deals directly with the construction company and not the client. The tender price, contract details and payment of work are between the plumber and the construction company.

Plumbing is not the only part of the contract that is subcontracted. Other areas include:

- electrical
- heating and ventilating – domestic, industrial and commercial
- refrigeration and air-conditioning
- service and maintenance
- ductwork
- gas – domestic, industrial and commercial.

Many companies in the plumbing industry will seek specialist support and assistance through involvement in, or membership of, the various industry trade bodies.

Find out

Use the Internet to find out about the role of industry trade bodies: JTL, the Association of Plumbing and Heating Contractors (APHC) and the Joint Industry Board for Plumbing and Mechanical Engineering Services (JIB-PMES).

Site supervisor

The site supervisor oversees the day-to-day running of the job on site. Reporting to the construction manager, the site supervisor is responsible for the work of the trade supervisor and the trade operatives.

Trade supervisor

On a particularly large job, there may be trade supervisors overseeing the work of operatives from a specific trade on site (e.g. overseeing bricklayers). For smaller jobs, this position is often not needed.

Trades

Any construction job is likely to involve several craft trades. Table 2.1 outlines the major trades and the tasks they are generally responsible for.

Craft or trade	Main aspects of job	Typical job details
Carpenter and joiner	Positions and fixes timber materials and components from roofs and floors to doors, kitchens and stairs.	Work on site is usually divided into two phases or 'fixes'. The first fix is usually completed before the building has been made watertight and includes fixing floor joists, boards and sheets, stud partitions, and roof trusses and timbers. The second fix takes place when the building is watertight. Staircases, doors, kitchen units, architraves and skirting boards are common second-fix items.
Roof slater and tiler	Creates a waterproof covering for a building by applying individual slates and tiles to a basic framework.	Roofing felt is laid over the roof timbers and tacked down. Timber battens are then fixed horizontally at centres to suit the roof dimensions and the type of slate or tile. The roof is loaded with slates or tiles and then they are laid. Tiles have to be cut to fit at valleys, hips and gable ends. Ridge tiles are bedded on mortar. On re-roofing or maintenance work, tiles and slates are removed, and timber checked and replaced where necessary. Reclaimed tiles and slates must be checked and sorted before reuse.
Built-up roofer	Built-up felt roofing is used mainly on flat roofs and sometimes on sloping roofs and vertical surfaces.	It is called 'built-up' because it involves putting layers of felt on top of each other using bitumen to form a waterproof surface.
Bricklayer	Uses bricks and blocks to construct interior and exterior building walls as well as tunnel linings, archways and ornamental brickwork.	Works on new buildings, extensions, maintenance and restoration of older buildings. Builds foundations, bringing brickwork and blockwork up to damp-proof course level. Works at height from trestles, hop-ups and scaffolds. Involves construction of drainage and concrete work.
Construction general operative (labourer)	Provides varied support on site in a number of areas.	Skills include concreting, form working, steel fixing, kerb laying and drainage work. Some operatives specialise in areas such as spraying or repairing concrete.
Plasterer	Applies plaster to interior walls so that decoration can be applied.	Renders external walls and sometimes lays floor screeds. Some plasterers specialise in fibrous plasterwork. Others work in the maintenance, conservation and restoration of existing buildings.
Painter and decorator	Applies paint, wall coverings and other materials to the inside and outside of buildings.	Work can include maintenance, conservation and restoration of existing buildings as well as on new-build, commercial and industrial sites. Wall coverings need to be accurately measured, cut and hung using the correct adhesive. Some decorators help clients with interior design.
Wall and floor tiler	Fixes tiles on exterior and interior walls, as well as on other surfaces such as floors and swimming pools.	On maintenance and restoration work, most surfaces need to be repaired. Tiles are normally laid on adhesive, cut along the edges and around obstacles, and finally grouted up. Setting out the tiles is critical when intricate and complex patterns are being produced. Tiles come in a variety of shapes, sizes, textures and colours and materials (e.g. ceramic, stone, terracotta).

Table 2.1: Main tasks of construction trades ▼

Craft or trade	Main aspects of job	Typical job details
Floor layer	Prepares and levels floors for new and old buildings. The most common materials are carpet, cork, plastic and timber.	Prepares floor measurements and plans to avoid waste. Different techniques are used for different materials. These include gripper fixings, glues, adhesives and secret nailing.
Plant operator	Uses cranes and other plant for moving and transporting construction materials.	Cranes can be mobile, track mounted or tower cranes. Transporting plant includes excavators, specialised earth-moving equipment, forklift trucks and power-access equipment.
Scaffolder	Erects scaffolding and working platforms for construction workers, and fixes edge protection. Scaffolders need to be properly certificated to do their work.	The scaffold needs a firm foundation and careful planning. It can be made from traditional tubes and fittings or may be a purpose-made system. Working platforms are made using scaffold boards. Access is usually provided with ladders. Hoists, special loading platforms, safety nets and guard rails are all features of scaffolding erection. Most scaffolds need to be lifted and modified as construction work progresses.
Electrician	Installs wiring and fixtures such as electrical sockets and lights. Can carry out maintenance to electrical installations or repairs.	In a new domestic property, work involves installing the cables for the power sockets, lights and associated fixtures, which would be done at the first fix. Once other trades have completed their work, electricians come back to fit the items like sockets and light fittings at the second fix.
H&V fitter	Installs industrial/commercial heating/plumbing, ventilation, refrigeration and air conditioning	Most H&V fitters do industrial or commercial work and can specialise in the different areas. Those working on the plumbing and heating will carry out similar work to the plumbing and heating engineer but this is generally with larger pipework.
Gas fitter	Installs, maintains or repairs gas pipework and appliances. A qualified person who has undertaken training and assessment.	A gas fitter or gas engineer is often a plumber who has taken further qualifications enabling them to work on gas. The work can vary depending on what qualifications have been obtained. Remember: before working on gas, you **must** be Gas Safe Registered.

Table 2.1: Main tasks of construction trades (cont.)

Key roles of site visitors

As well as the people working on site, you may also encounter some important visitors. These people are often sent by local authorities to inspect the working practices used on site.

Building control inspector

Working for local authorities, building control inspectors ensure that buildings conform to regulations on public health, safety, conservation and access for the disabled. The job involves inspecting plans and work-in-progress at various stages relating to Building Regulations.

A building control inspector's work includes:

- checking plans and keeping records of how each project is progressing
- carrying out inspections of foundations, drainage and other major building elements
- issuing a completion certificate when projects are finished

Did you know?

There are approximately 8000 floor layers in the UK. They usually work in pairs for specialist firms or are self-employed.

- carrying out surveys of potentially dangerous buildings
- meeting with architects and engineers at the design stage
- using technical knowledge on site.

Water inspector

The water inspector is employed by the **water undertaker** and may have a number of roles as part of their job. One of these roles is to carry out inspections of any work that may affect the quality, waste, or undue consumption of water. These inspections are covered in the Water Regulations. If, after work has been carried out, it is found and proven that the regulations have been breached, the inspector will collect the required evidence, on behalf of the water undertaker, which may lead to a prosecution.

Electrical services inspector

The Electricity Safety, Quality and Continuity Regulations 2009 require the relevant distribution network operator (the electricity supplier) to ensure adequacy of their generation, distribution network and equipment. This includes the metering and consumer's terminals, which are provided at the main intake position of an installation.

The role of the electrical inspector involves inspecting and testing an electrical installation within the premises. It is covered by BS 7671. This activity will normally be carried out by the original installer upon completion but periodic inspections are also required, and the client may ask any relevantly qualified company to undertake these.

HSE inspector

The role of the HSE inspector is covered in Unit 1 pages 16–17. The HSE inspector has the right to prosecute any site they find working unsafely, or infringing the duties laid down in the Health and Safety at Work Act.

> **Key term**
>
> **Water undertaker** – the company responsible for the supply of water to properties.

Working life

A building control officer visits a property you are working on and informs you the soil stack you are erecting does not conform to Building Regulations. He asks you to remove the work you have installed and gives you a sketch of how it should have been installed. You have completed several other above-ground drainage systems using the same layout. These are in several other buildings on the same site.

- How should this situation be handled?
- Should you carry on after the inspector has left?
- Should you tell the inspector that you have fitted other systems similarly?
- Who should you inform (if anyone) about the other systems?

Progress check

1. List five members of the construction site management team.
2. Choose one of the five people you have listed above (Question 1). Explain what this role typically involves.
3. List five trades associated with a construction site.
4. Choose one of the five trades you have listed above (Question 3). Explain what this role typically involves.
5. List three official site visitors.

2. Know how to apply information sources in the building services industry

Types of statutory legislation and guidance information

There are many important pieces of legislation that govern working in the building services industry. Some of the key areas are covered in the section below.

Data protection legislation

The Data Protection Act 1998 provides a common-sense set of rules that prevent the misuse of your personal information without preventing it from being used for legitimate purposes.

The details of the Data Protection Act are quite complex. The following eight rules outline the key points of the Act. These require personal information to be:

- fairly and lawfully processed
- processed for limited purposes
- adequate, relevant and not excessive
- accurate
- kept no longer than necessary
- processed in accordance with your rights
- kept secure
- not transferred abroad without adequate protection.

Organisations using personal information must comply with these requirements and, moreover, are required to register under the Act. The Act provides stronger protection for sensitive information about an individual's ethnic origins, political opinions, religious beliefs, trade union membership, health, sexual life and any criminal history.

Equal opportunities legislation

The Equality Act 2010

The Equality Act 2010 aims to make the workplace a fair environment. Employers must not **discriminate** against employees on the grounds of the following 'protected characteristics':

- age
- disability
- gender reassignment
- marriage and civil partnership
- pregnancy and maternity
- race

> **Did you know?**
>
> An employer is legally required to have Employers' Liability Insurance as a safeguard against employee accident and illness while at work.

> **Key term**
>
> **Discriminate** – to treat someone differently, often unfavourably.

> **Did you know?**
>
> The Equality Act 2010 brings together previous equality legislation, including the Equality Act (2006), the Sex Discrimination Act 1995, the Race Relations Act 1976, the Disability Discrimination Act 1995, the Equal Pay Act 1970, the Employment Equality (Religion or Belief) Regulations 2003, the Employment Equality (Sexual Orientation) Regulations 2003, the Employment Equality (Age) Regulations 2006, and the Equality Act (Sexual Orientation) Regulations 2007.

- religion or belief
- sex
- sexual orientation.

Table 2.2 summarises the 'protected characteristics' and looks at how the Equality Act 2010 provides employees with protection from discrimination.

Protected characteristic	Definition of protected characteristic	What the Act says
Age	A person of a particular age group	• It is unlawful to treat an employee unfavourably because of their age. However, employers can discriminate where there is valid business reason for doing so. • From 6 April 2011, employers are not permitted to compulsorily retire employees at the age of 65.
Disability	A person who has a physical or mental impairment and the impairment has a major impact on their ability to carry out day-to-day activities	It is unlawful to treat an employee unfavourably because of their disability. Employers are not permitted to ask job candidates about their health at interview. Employers also have a duty to: • make reasonable adjustments to help disabled staff carry out their role • ensure that jobs do not have a requirement that puts disabled people at a disadvantage (unless there is a valid reason for this).
Gender reassignment	A person who is proposing to undergo, is undergoing or has undergone a process for the purpose of reassigning the person's sex by changing physiological or other attributes of sex	A person who is absent from work because of treatment for gender reassignment must not be treated less favourably than if they were absent because of illness, injury or other valid reason.
Marriage and civil partnership	A person is married or is a civil partner	A person must not be discriminated against because of their marital/civil partnership.
Pregnancy and maternity	A person who is expecting a baby or who has given birth	It is unlawful to treat a woman unfairly because of pregnancy, illness related to it or because she is on maternity leave.
Race	Includes colour, nationality, ethnic or national origins	It is unlawful to discriminate against a person because of their colour, nationality, ethnic or national origins.
Religion or belief	A person who has a particular religion or belief	It is unlawful to discriminate against a person because of their religion or belief, or lack of one.
Sex	Reference to a man or woman	• It is unlawful to treat a person differently because they are a man or a woman. • Men and women should receive equal pay for doing equal work. Employees who discuss their pay to find out if pay discrimination exists are protected from victimisation by their employer.
Sexual orientation	A person's sexual orientation towards some of the same sex, the opposite sex or either sex	It is unlawful to treat bisexual, gay, heterosexual and lesbian people unfavourably.

Table 2.2: The definition of 'protected characteristics' and how the Equality Act 2010 provides employees with protection from discrimination

Discrimination

The Equality Act defines different types of discrimination, as shown in Table 2.3. Discrimination covers all aspects of employment – recruitment and training, pay, benefits, terms and conditions, promotion, transfer, dismissal and redundancy, and retirement.

Figure 2.1: Men and women must be treated equally at work

Type of discrimination	Description
Direct discrimination	Occurs when a person is treated unfavourably because of a protected characteristic, e.g. not promoting someone because of their sex
Indirect discrimination	Occurs when a company has a working practice, policy or rule that unfairly puts someone with a protected characteristic at a disadvantage, e.g. a dress code that may affect an employee with a particular religion or belief. In certain circumstances, companies may be able to discriminate if it is necessary to do so for the business
Discrimination by association	Direct discrimination of a person because they associate with someone who has a protected characteristic
Perception discrimination	Direct discrimination of a person because it is believed that they have a protected characteristic, even if this is not the case
Harassment	Offensive behaviour targeted at a person's protected characteristic, e.g. through rude remarks, insults or jokes
Third-party harassment	Makes employers responsible for preventing employees from being harassed by non-employees such as suppliers, customers or clients
Victimisation	When a person is treated unfavourably because they have made a complaint under the Equality Act

Table 2.3: Types of discrimination under the Equality Act 2010

Remember

Where an employee believes that they have been unfavourably treated by their employer on the grounds of age, disability, and so on, they may make a claim to an employment tribunal. The claim must normally be made within three months of the employee leaving employment or the unfair treatment occurring. If the employer is found to have broken the law, the tribunal may award compensation to the employee or ask the employer to take action to stop the discrimination reoccurring.

Working life

Read the scenarios below and then answer the questions that follow.

1. Lauren is a female plumber and is often taunted and teased by her work colleagues as she is the only female plumber in the company. There are sexually explicit materials on display in the site cabin and offensive, suggestive remarks are often made to her. In confidence she tells you that she intends to leave the company due to the discrimination.

What should you do about Lauren's situation? Who, if anyone, should you inform?

2. Fahad is an apprentice plumber looking for work. At one interview he is told the company cannot employ him due to his colour and religious beliefs as this would have a potential effect on the company's clientele and would lead to a loss in business.

   • What should Fahad do?
   • Is the company in question within their rights to say this?

Health and safety legislation and regulations

The health and safety legislation that applies to the building services industry is covered in Unit 1 on pages 2–17. Many of the regulations that relate to the industry are on pages 5–16.

Employment legislation

Employment legislation is designed to protect your rights at work, and covers a wide range of aspects from pay and conditions, flexible working, holiday entitlement and trade union membership to disciplinary issues and unfair dismissal. The main laws are:

- **Employment Rights Act 1996**
- **Employment Relations Act 2004**
- **Employment Act 2002 and 2008.**

The Equality Act also protects your equal opportunities rights in the workplace by making discrimination illegal.

Employment Rights Act 1996

The Act covers the following areas:

- Written statement of employment terms – within two months of starting employment an employee should be given a written statement, outlining the main terms of their employment including job description, pay, and so on (see page 99).
- Pay and wages – an employee should be told when they will be paid, how much they will receive and the method of payment, e.g. cash or directly into their bank account. The employer should provide them with a payslip, showing their wage, deductions for tax and National Insurance and any other agreed deductions. It is unlawful for employers to make other deductions without the employee's consent.
- Time off work – an employee is entitled to time off work for antenatal care, training or public duties.
- Dismissal and **redundancy** – if an employee is dismissed without a good reason, they have a right to claim compensation from the employer. The Act outlines reasons for unfair dismissal and also guarantees payment in the event of redundancy.
- Resolving disputes – the Act lays down the process for resolving disputes between the employee and employer. The employee has the right to go to an employment tribunal if their rights have been ignored.

Employment Relations Act 2004

This Act, first introduced in 1999 and amended in 2004, established a number of new rights at work covering the following areas:

- trade union recognition
- ballots for industrial action
- unfair dismissal of striking workers
- maternity and parental leave
- time off work for dependants
- employment tribunal awards
- the right to be accompanied in disciplinary and grievance hearings
- part-time work – equality with people in full-time employment.

Did you know?

An employer must not make changes to a statement of employment without the employee's consent.

Key term

Redundancy – when a person is dismissed from their job because a company needs to downsize its workforce.

Did you know?

The Fixed-term Work Directive, a law which applies to all countries in the European Union, is designed to ensure that workers on fixed-term job contracts (e.g. 12 months' duration) have the same employment rights as permanent staff members.

Employment Act 2002 and 2008

These Acts cover key aspects of employment law including:

- paternity and maternity leave and pay
- reform of employment tribunal procedures
- resolution of disputes in the workplace
- provisions to introduce the Fixed-term Work Directive
- flexible working
- new rights to time off work for union learning representatives
- enforcement of the national minimum wage.

British Standards

You may be familiar with the BSI Kitemark. This symbol can be used only on materials and equipment that meet the standards of the British Standards Institution (BSI), the organisation for standards in the UK.

The BSI ensures standards of quality and also sets standard dimensions for such items as pipes and fittings.

British Standards all start with the letters BS followed by the number of the standard. For example, BS 6700 is one of the main standards for the plumbing industry. It's a specification for design, installation, testing and maintenance of services supplying water for domestic use within buildings and their **curtilages**.

Displaying a British Standard number (such as BS 6700) on a consumer product shows that the manufacturer has made the product in accordance with the British Standards. BS 6700 is often referred to throughout this book. The BS will sometimes also be accompanied by the letters EN and/or ISO. This also means that the standard was developed as a European (EN) or International (ISO) standard and then adopted for use in the UK as a British Standard.

Codes of practice

Codes of Practice (CoPs) make recommendations related to good practice but are not statutory. However, because of their nature, CoPs are widely used by clients and architects.

Manufacturer guidance and user instructions

Almost all equipment will have the manufacturer's fitting instructions and other technical data or information sheets. These should be read and understood before fitting the item concerned.

Often on a project, the manufacturer of certain products will have already been specified by the consulting engineer or client. For many of these products you will already be familiar with their use and any requirements for installation. However, you will also come across more complex products with many component parts that you are less familiar with. You will therefore need to access information

Did you know?

The BSI is an independent organisation and was set up in 1901 under a Royal Charter.

Key term

Curtilage – an area attached to a dwelling house and forming one enclosure within it.

Figure 2.2: The BSI Kitemark

Remember

Once installation of the item is complete, any information such as fitting or operating instructions should be kept safely in a central file, which is given to the client in a handover manual when the project is completed.

if they are to be installed, and subsequently, function correctly. Certain levels of information may be provided by the supplier but in some cases, perhaps more technical areas, you may need to refer to the manufacturer's catalogue, datasheets or website, or contact the manufacturer directly.

Information you may require from suppliers and manufacturers might include:

- the materials used in constructing the product
- the types of component used in constructing the product
- general equipment details such as operating characteristics, physical size and assembly instructions
- drawings and diagrams
- any special installation requirements
- delivery and storage details.

Installation instructions

Wherever possible, the manufacturer's installation instructions should be used and followed. If they are not, the item will not be covered by any warranty or guarantee and could possibly cause harm or damage. Most installation instructions will be linked to British Standards, Building Regulations and Water Regulations as well as to good working practice. This makes it very important to follow any installation instructions at all times.

Service and maintenance instructions

Remember

You have a duty of care to ensure that the client knows what level of servicing is required and when.

These instructions will be available from the manufacturer and are often supplied with the goods. Any service and maintenance instructions should be left with the client following the installation. This will allow the client to ensure that any repair work they carry out in the future meets the requirements of the product.

User instructions

User instructions are supplied to explain how to use an item safely, efficiently and correctly. After fitting a bathroom suite, for example you will need to explain to the customer how to use all the items. This could be done verbally and then by giving them the manufacturers' instructions, or you could provide the customer with a set of your own instructions.

Purpose of information used in the workplace

When you are in the workplace, there are certain key pieces of information you need to be familiar with. Some of these will affect you on a day-to-day basis. Others will have an impact on the project as a whole.

Job specifications

These usually accompany site and services drawings and details, and are mainly used on larger contracts. Job specifications provide details such as:

- type and quality of components, materials and fittings to be installed
- type of clips or brackets to be used
- system test specifications
- any specific installation requirements, e.g. who is to carry out any associated building work, such as cutting holes for pipework, drilling or notching joists and making good.

Plans and drawings

When major works are undertaken it is normal practice to have plans and drawings available showing details of the work required. You would not require these if simply replacing a toilet and cistern, but you would need a plan and drawing when replacing a bathroom suite in a new layout. The detail and requirement of the plans or drawing will depend on the scale of the works being carried out. On larger jobs each trade may have a set of plans for the work that needs to be carried out by them.

A technical diagram is simply a means of conveying information more easily or clearly than can be expressed in words. In the plumbing industry drawings and diagrams are used in different forms.

Block diagrams

A block diagram can be used to convey information about a circuit without giving details about components or how they are connected. It is typically used for a higher level, less detailed, description of overall concepts. It does not focus on understanding the finer details of a system.

In block diagrams the items are represented by a square or rectangle clearly labelled to show its purpose. This type of diagram shows the sequence of control for installations in its simplest form, as shown in Figure 2.3.

Figure 2.3: A block diagram

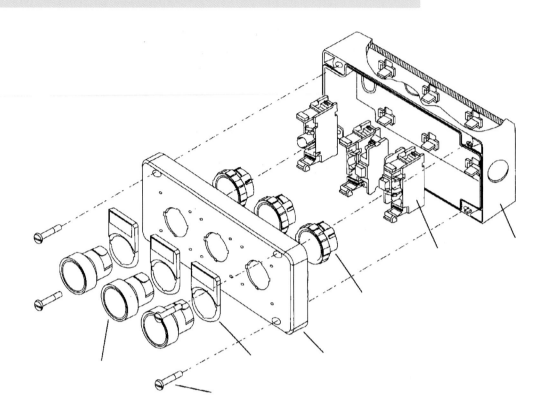

Figure 2.4: An assembly drawing

Assembly drawings

Assembly drawings show how the individual parts (or modules) of a product fit together. They normally contain scale drawings of all the components shown in their correct position relative to each other, with some overall dimensions.

Internal components are shown by sectioning. Each component is listed and described on the drawing. Figure 2.4 shows the assembly drawing for a push-button enclosure.

Scaled drawings

Layout and assembly drawings give information about physical objects, such as the floor layout in a building, or a mechanical object. If we were to make the drawing the same size as the object, the drawings would often be far too big to handle.

Figure 2.5 Using a scaled drawing

To make the drawing a sensible size, we use **scaled drawings**. You may, for example, have built model aeroplanes from a kit. Quite often these are described as 1/32 scale; in other words, every part in the model is 32 times smaller than the real thing.

This type of scale is known as a ratio scale, and it makes the drawings easy to use. To find a measurement on the actual object, you measure the distance on the drawing and multiply it by the scale. It does not matter what unit of measurement you choose, because you are simply going to multiply it by a number (the scale).

For example, on most construction projects, the scale used to show the floor layout of a building is 1:100. Therefore, 10 mm on the drawing represents something that is 100 times bigger in reality (i.e. 10 × 100 = 1000 mm (1 metre).

The drawing scale is chosen to make the drawing a reasonable size, according to its purpose. Although a scale of 1:100 may be fine for the layout of a building, it would be impractical for a road map, because you would only be able to get a few miles on each sheet. A scale of 1:500,000 (1 cm = 5 km) would be better.

In the same way, an assembly drawing for a wrist watch would be too small to read if we used 1:100; a better scale might be 20:1 (20 mm on the drawing represents 1 mm on the actual watch).

Work programmes

On a small job, you will have the work programme inside your head, based on agreed start and finish dates. On larger contracts, the approach is more scientific, and a work programme will have been provided with the contract. This could consist of an overall programme for all site trades held by the main contractor, as well as a separate programme existing for each individual trade.

The basic principle of the work programming is 'activity against time'. It will record exactly which activities need to be carried out at what time in the schedule. This will then allow the manager of the work site to keep track of the project, and quickly identify whether it is on schedule or falling behind.

Site documentation

Technical information is recorded on diagrams and drawings, so that everyone involved in a project during installation, and in the future, knows what to do and what has been done. There is plenty of other information people need to communicate to others in the workplace, and various types of documents are used for this.

Job sheets

Job sheets give detailed and accurate information about a job to be done. Plumbing contracting companies issue them to their plumbers. They will include:

- the customer's name and address
- a clear description of the work to be carried out
- any special instructions or special conditions (e.g. pick up special tools or materials).

Sometimes extra work is done, which is not included in the job sheet. In this case it is recorded on a day worksheet so that the customer can be charged for it.

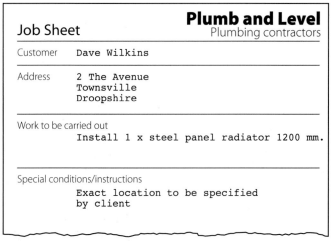

Figure 2.6: A typical job sheet

Variation orders

A variation order (VO) is issued ('raised') when the work done varies from the original work agreed in the contract or listed in the job sheet. If this situation arises, it is important for the site plumber to tell their supervisor immediately. A variation order can then be made out to enable the new work to be done without breaking any of the terms of the contract.

The purpose of the VO is to record the agreement of the client (or the consulting engineer representing the client) for the extra work to be done, as well as any alteration that this will make to the cost and completion date of the project.

Day worksheets

Work done outside the original scope of the contract, perhaps as a result of a VO initiated by the architect, engineer or main contractor, is known as day work. When the work is completed, the plumber or supervisor fills out a day worksheet and gets a signature of approval from the appropriate client representative. Day work is normally charged at higher rates than the work covered by the main contract, and these charges are usually quoted on the initial tender.

Typical day work charges are:

- Labour: normal rates plus 130%
- Materials: normal costs plus 25%
- Plant: normal rate plus 10%.

Disputes over day work can easily arise, so it is important that the installation team on site records any extra time, plant and materials used when doing day work.

> **Find out**
>
> When may you need to use a variation order?

Figure 2.7: A typical day worksheet

Time Sheet

Plumb and Level
Plumbing contractors

Employee Project/site

Date	Job No.	Start time	Finish time	Total time	Travel time	Expenses
Mon						
Tue						
Wed						
Thu						
Fri						
Sat						
Sun						
Totals						

Employee's signature Supervisor's signature

Date

Figure 2.8: A typical time sheet

Time sheets

Time sheets are very important to you and your company. They are a permanent record of the labour on a site, and include details of each job, travelling time, overtime and expenses. This information allows the company to track its costs on a project and also make up your wages. If you work on several sites during a week, you may need to fill in a separate time sheet for each job.

Remember

Accurate time sheets are essential to make sure you get paid correctly and that your customer is charged the right amount for the job.

Purchase orders

Before a supplier will despatch any materials or equipment, it will require a written purchase order. This will include details of the material, quantity required and sometimes the manufacturer. It may also specify a delivery date and place. In many cases the initial order is made over the telephone, via the website or using email, and a written confirmation is sent immediately afterwards. The company keeps a copy of the original order in case there are any problems.

Usually the purchasing department sends out these orders. Sometimes an order is raised directly from a site if immediate action is required.

Delivery notes

All materials delivered to the site by a third party will arrive with a delivery note. This is usually a form with several copies recording the delivery of materials and equipment. As the company representative on site, this is the form you are most likely to deal with.

The delivery note should give the following information:

- the name of the supplier
- the name of the person the materials are being sent to

Delivery note	S. BENDS *Plumbing supplies*
Order No.	Date
Delivery address 2 The Avenue Townsville Droopshire	Invoice address Plumb and Level Plumbing Contractors

Description	Quantity	Catalogue No.
Steel panel radiator 1200 mm	1	
Comments		
Date and time of receiving goods		
Name of recipient	Signed	

Figure 2.9: A typical delivery note

Remember

Prompt delivery – goods must be delivered without delay.

- a list of the type, quantity and description of materials that are being delivered to the site in this particular load
- the time period allowed for claims for damage.

When materials arrive on site you should ensure that they are unloaded and stored correctly. You should also check each item against the delivery note and look for obvious signs of damage.

If everything is fine, the delivery note can be signed. If not, you should note any missing or rejected items on the delivery note, and you and the delivery driver should both sign it. Store your copy of the note safely.

Check the materials thoroughly for damage within the time given stated in the delivery note (usually three days), and inform the supplier immediately if there are any problems.

A delivery will not always contain all the materials listed in the purchase order. Sometimes the material is not all needed on site at the same time, and it is delivered in several loads. This helps to reduce the need for on-site storage and minimises the risk of damage or loss.

Incomplete deliveries may also occur if the supplier is out of stock, or if some of the order is coming direct from the manufacturer. A completion order records that all the material on an original purchase order has been delivered.

Site reports, memos and minutes of meetings

The site foreman, supervisor or engineer in charge usually compiles reports for companies. Site reports contain details of work progress, defects, problems and delays. Sometimes other reports will be made about specific problems or incidents.

A memo is usually a short document sent to a relevant person about a single issue. For example, a problem installing a piece of equipment, or materials not being delivered on time.

No installation project will be complete without a site meeting. Normally chaired by the main contractor with representatives from the consultants and all related contractors present, these meetings establish whether a project is progressing as hoped, as well as attempting to solve any identified problems.

The minutes provide a record of each meeting. These are sent to all relevant parties. Meeting minutes should be checked for accuracy and retained as there are often action points to follow up on.

Policy documentation

Policy documentation is usually developed by the senior management of any organisation. Policies typically describe to staff where their organisation stands on a given subject such as customer relations or the environment. The finer details tend to be covered by procedures or work instructions.

Depending on the policy and size of an organisation, there may be many individuals involved in ensuring a policy is carried out. As an example, an extract from a typical health and safety policy is shown below.

Health and safety policy

It is the policy of this company to give prime importance to the health and safety of its employees while at work, which includes the risks that arise from driving on company business.

This is considered to be a management responsibility equal to that of any other function. In order to achieve and maintain the high standard required, both management and staff must be aware of, and accept, their respective responsibilities. The company is committed to using best endeavours in promoting staff training in a safe, healthy and supportive environment within the scope of health and safety legislation, and by adopting and encouraging industry methods of best practice.

The company will, through its organisational arrangements, ensure so far as is reasonably practicable the health, safety and welfare of its employees while at work and all persons likely to be affected by its operations, and in particular provide:

- a safe and healthy working environment and safe systems of work
- safe plant and equipment
- adequate information, instruction, training and supervision.

Purpose of information given to customers

Quotations, invoices and statements

On a smaller job, such as updating a customer's bathroom suite, it is unlikely that a detailed specification will be produced. Your employer, however, will have issued a detailed quotation that the customer has agreed. This in itself is a form of contract. Again, any revisions to the original quote must be passed on to your employer so that they can be agreed with the customer.

Key term

Invoice – a statement of the materials supplied, requiring payment

Quotations are also used by plumbers' merchants to supply a price for the materials required to carry out the job. This would be followed by an **invoice**. For example, a plumber may have an account with a merchant and during a month will purchase a number of items. Each purchase will be invoiced separately. At the end of the month the merchant will send a statement listing all the invoices for that month.

An example of the paperwork system

Some of the other important documentation can be illustrated by looking at how the paperwork system works for a bathroom installation:

- Customer asks plumber for a quote.
- Plumber visits customer, measures up, estimates materials.
- Plumber gets a quotation for the materials from the merchant, which enables plumber to finalise quote to customer.
- Customer accepts and quote is given to customer in writing.
- Plumber orders materials from merchant after the materials quotation is confirmed in writing. This is an order, though it may not be confirmed in writing if the plumber has an account with the merchant. Plumber receives delivery of materials, confirmed by a delivery note.
- Job is carried out.
- Plumber receives invoices for payment of materials and also invoices the customer for the work carried out. A remittance advice (record of payment) is sometimes issued. A receipt may be given to the customer.

On any job, large or small, if you are advised that the delivery of materials will be delayed, or a particular item is not the one ordered, notify the site manager or your immediate supervisor/employer so that they are aware of potential delays.

Estimates

An estimate states the likely cost of a job. The client explains the work they require and the plumber or plumbing company then prepares an estimate of the work. It is normal for client to obtain two or three estimates, especially for larger jobs. This process is called putting the job out to 'tender'. The client will invite several companies to produce an estimate for the job. This tender may also include the materials the company intends to use, stating how they intend to work and how long they believe the work will take. The client will then choose between these. For example, if the work is urgent they may take a more expensive tender which promises the work can be done in a quicker time.

Remember

An estimate is, as it states, an 'estimate'. Therefore the finished work can cost more or less than the original figure.

Statutory cancellation rights

After the work has been agreed between the client and the company, they will sign a contract stating the terms of the agreement, such as the cost and the timescale of the project.

A contract is a legal document, which makes it difficult to cancel unless there has been a **breach of contract**. Some contracts include a cooling-off period, which allows either party to cancel, under pre-agreed circumstances, before the work begins.

Key term

Breach of contract – failure by one of the parties to carry out something in the contract.

Handover information

On completion of any work carried out, you have a duty of care to ensure that the customer is handed instructions about how to use and maintain the item. If the item was serviced, you should issue a record or certificate.

Importance of company policies and procedures

Most companies have policies and procedures in place. The difference between 'policy' and 'procedure' can be confusing. Procedures follow on from policies. They can either be a separate document or a section of the same document. They can be defined as follows:

- **Policy** – the overall aims of the company. For example, a health and safety policy will outline the precautions the company wants to take and its guarantees for its workers' welfare.

- **Procedure** –the day-to-day ways of working that are designed to meet the policy. A worksite will have health and safety procedures, such as evacuation procedures, that help the workers to meet the policy of the company.

Remember

A policy change may or may not alter the procedure. However, a change in procedure should not be allowed to change the policy by default.

The procedure document should state clearly which policy or policies it relates to.

Policies and procedures cover items such as health and safety, employment issues, equality and communication. The larger the organisation, the more policies and procedures they are likely to have.

Contract of employment

A contract of employment is an agreement between employer and employee, which outlines the employee's rights, duties and responsibilities. It does not need to be a written contract, but all employees must by law be given a written statement of their employment terms within two months of starting work (see page 88). This should include the following information:

- names of employer and employee
- date employment started
- job title and summary of duties
- period of employment if not a permanent employee
- place of work
- amount of pay and when it will be paid
- hours of work

Did you know?

A staff handbook, explaining the company's procedures, e.g. disciplinary and grievance procedures, may be given to an employee as part of their terms of employment.

- holiday entitlement
- sick pay
- pension
- notice period
- disciplinary and grievance procedures.

Entitlement to paid leave

By law all workers are entitled to a minimum of 5.6 weeks of paid leave a year. Employers may include public holidays as part of this and set the times when employees take holidays, such as during a Christmas shutdown. When a person leaves employment they have a right to be paid for holiday not taken.

A person may also be given time off work for:

- dealing with emergencies involving dependants
- jury service
- trade union and public duties
- training.

Working hours

Requirements for working hours are laid down in the Working Time Regulations. Workers aged 18 or over cannot be forced to work for more than 48 hours a week on average (unless they volunteer to do so). For workers under 18 the maximum working week is 40 hours and each working day must be no longer than 8 hours.

Behaviour, timekeeping and dress code

All employers will expect their workers to behave well and to project a good image of the company to the public. They will also expect employees to work the hours they have been employed for, reporting for work at their contractually agreed start time and not leaving before their finish time. Some employers may also have a dress code. On a worksite this will largely be related to personal protective equipment. In an office, or for certain parts of your job role such as meeting clients, you may also be expected to dress smartly (e.g. no jeans).

Behaviour, timekeeping and dress code and the procedures to follow in the event of a breach of policy are covered by the contract of employment. An example of a policy is shown below.

> It is company policy for all employees to behave in a professional manner at all times and poor timekeeping will not be tolerated. When on company business, the company work wear should be worn.

Remember

- The disciplinary procedure lays down the steps that your employer must take to deal with issues related to your employment such as poor work performance or illegal activities.
- The grievance procedure lays down the steps that you must follow to raise issues of concern that you may have at work.

Did you know?

Your employer is not obliged to let you carry over paid leave from one year to another.

Did you know?

The standard working week for plumbing operatives as laid down by the JIB is 37.5 hours.

Figure 2.10 PPE is essential for many plumbing jobs

Limits to personal authority

As your career progresses, you will need to understand the level of responsibility that you have in your company, which will determine how you deal with communication and problem-solving issues. Table 2.4 gives some typical examples.

Job role	Level of responsibility	Responsibility for communication and problem-solving issues
Apprentice plumber	Works directly for a qualified member of staff who provides the necessary level of work instruction and supervision to undertake the work	Does not usually communicate directly with customers and co-contractors. Usually passes work issues to the supervisor
Level 2 (or equivalent) qualified plumber	Works under minimal supervision but does not take full responsibility for project	May respond to queries for information from customers and co-contractors. Usually forwards requests for additional work, or major changes, and any complaints to the supervisor
Level 3 (or equivalent) qualified plumber	Usually takes full responsibility for the project	Deals with queries from customers and co-contractors and requests for additional work, including confirmation of work, pricing information and customer complaints

Table 2.4: Examples of responsibility within the company

Supervisor and management responsibilities

A supervisor is normally an experienced, qualified tradesperson in a trade-related role. They are responsible for the day-to-day running of the project, for ensuring the works are carried out correctly and safely and they may also hire labour. They will be your main point of contact.

Remember

It is good practice to maintain regular communication with your employer about the extent of your job responsibilities as your career progresses.

Progress check

1. Explain what the Data Protection Act is designed to do.
2. With regard to discrimination, which Act protects our individual rights in employment?
3. List three areas which the Employment Rights Act 1996 covers.
4. List five statutory rights that an individual has with regard to the Employment Relations Act 2004.
5. Briefly explain what a Code of Practice (COP) is.

3. Know how to communicate with others in the building services industry

Effective communication is vital to running a successful business.

Communication methods at work

What is communication?

Communication involves the passage of information between the sender and one or more people (the receiver). It is a two-way process:

1. The sender must know what message they want to deliver. The message is sent to the receiver using an appropriate method.
2. When the receiver gets the message they respond to show the sender that the message has been received and understood. This is known as feedback.

Good communication should be:

- clear and easy to understand – presented in simple, direct language
- correct – contain accurate information
- concise and complete – contain all the information that is required
- constructive – enable the receiver to respond in a way that achieves the purpose of the message.

Poor communication, on the other hand, may lead to dispute and disagreement, so it is essential that you are able to communicate effectively with colleagues and customers.

Oral and written communications

There are two main types of communication:

- oral (verbal) – the communication is spoken and there is no written record, e.g. face-to-face and phone conversations, meetings
- written – used when a record is required, e.g. letters, memos, reports, faxes, text messages, email.

Communication may also be:

- visual (through the use of graphics, **body language**, etc.), e.g. signs, charts, graphs
- a combination of one or more of the above, e.g. video, video conferencing, DVD.

Formal and informal communication

It is important to choose the appropriate style of communication for the message – informal or formal. For example, a meeting with a client would involve informal (verbal) communication, whereas a letter to a client would be formal communication (written down).

Figure 2.11: Be clear when you talk to customers

Key term

Body language – messages conveyed through facial expressions, gestures, head nodding, posture, etc.

Remember

Remember to use the right type of communication for the particular purpose.

Many jobs will require a combination of formal and informal communication methods. For example, pricing for a job with a customer may include:

- arranging a site visit by phone
- meeting with the client to discuss their needs
- preparing a written list of information and tasks while on the job
- obtaining a figure for the cost of materials needed from the merchants
- producing a written quotation for the job.

Table 2.5 summarises the uses, advantages and disadvantages of verbal, written and graphic communication in the workplace.

> **Remember**
>
> Certain types of information may be considered private and confidential, such as profits made on a job or the purchase price of materials. You therefore need to be careful in identifying the types of information that can be passed on as part of your job.

Type of communication	Advantages	Disadvantages	Uses
Verbal	• More personal contact between sender and receiver. • Tone and body language can be used to help pass on the information, and to find out whether the receiver has understood it. • Feedback is usually instant.	• There is no formal record unless the conversation is recorded. This may be a problem later if a dispute arises as there is no record of what has been discussed or agreed. • Not appropriate for sending a lot of detailed or technical information. • The receiver can be easily distracted.	Used to pass on relatively 'low risk' information. It may include advice and guidance for a customer about how a component operates, arrangements for protecting the customer's furniture or confirmation of an appointment on site.
Written	• Provides a permanent record for future reference, e.g. legal documents, contracts, agreements including estimates, quotations and job orders. • Summarises key points from meetings. • A more reliable method for reporting complex or technical information.	• Producing written material takes time. • Feedback can be delayed, or not given at all. • There is a chance of misunderstandings, as a face-to-face explanation will not be provided. • Lacks the personal touch. • Can be used for negative reasons, such as delaying tactics.	Written communication tends to be used for 'higher risk' activities, which may have some financial risk, e.g. the agreement for the price for doing a job and the range of tasks to be carried out or the remittance advice confirming payment received for the work carried out.
Graphics	• Useful for explaining complex technical details. • Useful for people who may have difficulties understanding written language.	• No control over interpretation and feedback unless sender is on hand to explain. • No personal contact if graphics are forwarded on to the receiver.	Graphics are generally used to support other formal or written documents to enhance understanding of the subject.

Table 2.5: Advantages, disadvantages and uses of different communication options in the workplace

Use of information technology

Information technology enables written or visual communication to be forwarded almost instantly by email. Documents of all sizes can be forwarded as attachments and then downloaded by the receiver, either in hard copy or to be read on screen and stored on their computer system. Information technology is improving constantly, and mobile telephones have text facilities and Internet connections. Some phones have photo messaging, or software that will enable video messages to be forwarded. All these features are increasing the methods of effective communication.

Working life

You are waiting for a delivery of plumbing materials on a job. The delivery was due at 9.30 a.m. but, after several hours, the delivery has still not arrived.

- What should you do?
- Who should you inform or ask about this problem?
- Which form of communication would be best to use?

Remember

It is important not to patronise or offend people when communicating with them. Remember that one of the keys to effective communication is respecting the person you are communicating with.

Methods of effective communication for people with specific needs

Communicating with people with physical and learning disabilities

As with any communication, care should be taken to ensure that those giving and receiving fully understand each other. When communicating with someone who has a physical or learning disability you will need to decide how best to communicate. This may utilise a number of different methods so that they will understand. You may even have to consider using the help of others.

Communicating with people with language differences

Some of the people you speak with may understand and speak English but it may sound different due to their dialect or accent. Where people have learned English as a second language, they may sometimes have difficulty communicating and understanding. If you are talking to someone who speaks English as a second language, or whose accent you find unclear, you may need to speak a little slower and clearer to make sure that you are understood. Always confirm what you have said and ask them to confirm their understanding so you know that they have understood. In some cases it may be necessary to use the services of an interpreter.

Resolving conflict

Part of the work of a plumber will involve dealing with conflict or dispute situations or resolving customer complaints. How effectively you deal with these issues can have a major impact on any plumbing business. The extent to which you will be involved in resolving disputes depends on the extent of your level of responsibility in the business (see page xx). A few key points are covered below.

- Customer complaints should be dealt with as soon as possible, otherwise they may 'spiral out of control'. If you do not have the authority to deal with a complaint, make sure that you pass the information to your supervisor immediately. Remember to inform the client that you have done this.

- Plumbers sometimes witness serious problems between work colleagues and between supervisors and operatives on site. If this occurs, most businesses usually require their operatives to report the matter. Disputes such as this need urgent resolution as they have a tendency to affect the work carried out, including potential costs to the company, and may leave a bad impression of the company on its customers and co-contractors.

- Most companies encourage staff to raise personnel issues, such as problems with pay or working conditions, with them directly rather than by 'chatting with work colleagues'. Typically, such problems affect the work carried out if not properly addressed.

- Formal disputes between staff and management in larger companies may be dealt with via the trade union if key personnel involved in the dispute are members. Dispute resolution may involve formal meetings between management and affected staff. In smaller companies the owner will usually discuss the issue with affected operatives.

- In the event that disputes between the trade union and management of large companies cannot be resolved, the matter is usually handed over to independent **arbitration**. This is where a third party reviews the case and recommends solutions to overcome the problem.

Effects of poor communication on an organisation

Poor communication can lead to confusion between work colleagues and between workers, supervisors and senior management. It can also give a bad impression of the business to customers which, in turn, may have a serious impact on the future of the business. Effectively dealing with customers or co-contractors on site is essential to the smooth running of a job. Below are a few tips for effective communication:

- Always be polite – many problems on site arise from a failure to communicate effectively or in a proper manner, or because incorrect or wrong information has been given.

- Always deal with a query if it is raised – the vast majority of complaints raised by customers in the plumbing industry result from the failure to provide information to respond to a query while the work is being carried out.

> **Did you know?**
>
> Plumbing companies sometimes arrange social events outside work in order to assist with team building at work and the smooth running of jobs.

> **Key term**
>
> **Arbitration** – guidance on resolving employment disputes and the arbitration process, often provided by ACAS (Advisory, Conciliation and Arbitration Service).

Figure 2.12 Careful explanation can avoid many problems

Key term

Toolbox talk – a briefing given by management/ supervisory staff on subjects such as safety and customer care.

- If a query is raised, remember to take all the details. Only deal with queries that are regarded as your normal responsibilities; if necessary, arrange for the query to be passed on and try to set a timescale for the response.

- A good plumbing company will see good communication skills as a major part of their staff recruitment requirements. A company will want to ensure that effective communication takes place between its staff and customers/co-contractors, and between its management/ supervisory staff and its plumbers. For example, activities such as **toolbox talks** are designed to deal with better ways of carrying out the work.

Figure 2.13 Plumber in discussion with fellow worker

Working life

A delivery of plumbing materials is made to the premises where you are working. On checking the delivery note you see some of the items listed are not as indicated but have been produced by another manufacturer. The quality of the goods is similar and you know the contract will fall behind if you do not install the goods supplied.

- What should you do in this situation?
- Who, if anybody, should be informed?

Progress check

1. List and describe four forms of written communication.
2. List and describe four forms of verbal communication.
3. List and describe four forms of graphic communication.
4. Explain what is meant by a 'toolbox talk'.
5. Identify two advantages and two disadvantages of verbal communication.

Check your knowledge

1. Who is the building control officer employed by?
   a Local authority
   b Architect
   c Customer
   d Main contractor

2. Who is responsible for coordinating different trades on a large contract?
   a Clerk of works
   b Foreman
   c Site agent
   d Site manager

3. Which one of the following is true of a limited company?
   a Workforce may exceed five operatives
   b Size of contract must not exceed £15,000
   c Any trading loss is limited to company assets
   d Only local authority work can be done

4. Which is the correct statement in relation to day rate mark up?
   a 120% labour, 30% material, 10% Plant
   b 130% labour, 25% material, 10% plant
   c 125% labour, 25% material, 10% plant
   d 130% labour, 30% material, 5% plant

5. Which of the following statements describes the essentials of communication?
   a Clear and correct
   b Constructive
   c Concise and complete
   d All of the above

6. Which of the following would be able to deal with a formal dispute between staff and management in a large company?
   a Trade union
   b The managing director
   c ACAS
   d The problem should just be left to go away

7. In the plumbing sector 80% of businesses employ a similar number of employees. How many people are employed by each of these businesses?
   a 1 to 4
   b 5 to 10
   c 11 to 49
   d Over 50

8. Buyers are also referred to as:
   a hawkers
   b procurement officers
   c facilities managers
   d closers

9. What is the preferred method of communication when making a formal complaint about a product or service?
   a Letter
   b Fax
   c Telephone
   d Text message

10. Which document confirms that an invoice has been paid?
    a An advice note
    b A remittance note
    c A bill of quantities
    d A delivery note

Getting ready for assessment

The information contained in this unit of the book, as well as the continued practical assignments that you will carry out in your college or training centre, will help you prepare for both your end-of-unit test and the diploma multiple-choice test. It will also support you in preparing for the practical assignments you will be required to complete to demonstrate your understanding and your communication skills with others in building services engineering.

There are opportunities throughout the unit for you to test your progress and understanding of the required underpinning knowledge; this will enhance your preparation for the forthcoming assessment, so make good use of them.

With regard to the building services industry, you will need to be familiar with:

- the members of the construction team and their role within the industry, including, key roles of the site management team, key roles of the individuals who report to the site management team and key roles of site visitors

- how to apply information, including, types of statutory legislation and guidance information, that applies to working in the industry, the purpose of information that is used in the workplace and given to customers and the importance of company polices and procedures that affect working relationships

- how to communicate with others, including suitable communication methods for use in work situations, methods of effective communication, the actions to take to deal with conflicts and the effects that poor communication may have on an organisation.

Check it out – make sure you understand and know the roles of the following **key** words:

Architect, project manager/clerk of works, structural engineer, building surveyor, building services engineer, quantity surveyor, buyer, estimator, contracts manager, construction manager, sub contractors, site supervisor, trade supervisor, bricklayer, carpenter and joiner, plasterer, wall and floor tiler, electrician, floorlayer, H&V fitter, gas fitter, painter and decorator, groundworkers, building control officer, water inspector, HSE inspector and electrical services inspector.

Good luck!

UNIT 3

Understand how to apply environmental protection measures within building services engineering

Environmental awareness is the understanding of why it is important to conserve energy, dispose of waste properly, prevent wastage of materials and reduce our carbon emissions. The UK government is keen to promote initiatives that help to conserve energy and reduce waste. Building Regulations have been improved and updated to ensure they are more energy efficient and that building materials are used more effectively.

This unit will cover the following learning outcomes:

- Know the energy conversation legislation that applies to the building services industry

- Know the applications of energy sources used in the building services industry

- Know the importance of energy conservation when commissioning building services systems

- Know the methods of reducing waste and conserving energy while working in the building services industry

- Know how to safely dispose of materials used in the building services industry

- Know the methods of conserving and reducing wastage of water within the building services industry

1. Know the energy conservation legislation that applies to the building services industry

Energy conservation aims to use energy and water resources more efficiently, reduce waste and cut carbon emissions. The government has introduced legislation to improve the insulation of buildings and make heating appliances more efficient. It has also introduced measures to minimise waste, as this is harmful to the environment, requiring landfill sites, energy for transportation and treatment. We are all being encouraged to recycle as this helps to reduce waste by returning excess and used materials to the production cycle. Energy conservation also involves ensuring that we have sufficient fuel supplies for the future, and there is a drive towards the use of renewable energy sources such as wind and solar power.

Energy conservation legislation

Building Regulations

The Building Regulations exist to ensure the health and safety of people in and around all types of buildings (i.e. domestic, commercial and industrial). They also provide for energy conservation, access to buildings and the use of buildings.

The Building Regulations contain various sections dealing with definitions, procedures, and what is expected in terms of the technical performance of building work. They set out the notification procedures to follow when starting, carrying out, and completing building work. They also list the 'requirements' that the individual aspects of building design and construction must comply with in the interests of the health and safety of building users, of energy conservation and of access to and use of buildings. These requirements are contained in a schedule (Schedule 1) to the Building Regulations and are grouped under fourteen 'parts' as shown in Table 3.1.

Part		Part	
A	Structure	H	Drainage and waste disposal
B	Fire safety	J	Combustion appliances and fuel storage systems
C	Site preparation and resistance to contaminants and moisture	K	Protection from falling, collision and impact
D	Toxic substances	L	Conservation of fuel and power
E	Resistance to the passage of sound	M	Access to and use of buildings
F	Ventilation	N	Glazing – safety in relation to impact, opening and cleaning
G	Sanitation, hot water safety and water efficiency	P	Electrical safety

Table 3.1: The 14 parts of Schedule 1 of the Building Regulations

Each of the parts has a set of requirements attached. These lay out the broad objectives or functions that each aspect of the building design and construction must set out to achieve. Therefore, they are often referred to as functional requirements and are expressed in terms of what is reasonable, adequate, or appropriate.

Building work is defined in regulation 3 of the Building Regulations. The following types of project are categorised as building work:

- the erection or extension of a building
- the installation or extension of a service or fitting that is controlled under the regulations
- an alteration project involving work that will temporarily or permanently affect the ongoing compliance of the building, service or fitting with the requirements relating to structure, fire, or access to and use of buildings
- the insertion of insulating material into a cavity wall
- the underpinning of the foundations of a building.

Points 2 and 3 in the above list are of particular relevance to you, as plumbing and heating installation is classed as a service.

Approved Document L1: Conservation of fuel and power

Part L of the Building Regulations 2010 deals with the conservation of fuel and power. *Approved Document L1: Conservation of fuel and power* (2010 edition) lays down requirements relating to new dwellings (L1A) and existing dwellings (L1B). The regulations were originally introduced in 2000 to improve the energy efficiency of existing buildings. In particular, glazing and the upgrading of boilers and systems had to meet the same standards as new buildings. Inefficient boilers require higher levels of insulation for the building, and new boilers and heating systems must be certified to show that they have been correctly installed and commissioned, and operating and maintenance instructions provided for users.

For new dwellings, the main changes include:

- Big improvements to **U values**, meaning higher levels of insulation. The lower the U value, the lower the heat loss through the building fabric.
- Higher standards of insulation for dwellings.
- Certification of heating and hot water systems to show that they have been correctly installed and commissioned, and that operating instructions have been left for the user.
- Higher standards of design and workmanship to improve building performance, reduce gaps in insulation, and reduce thermal bridging and poor air tightness (e.g. gaps around windows and door frames). Air tightness testing is now mandatory for dwellings.
- All new buildings to have an SAP (Standard Assessment Procedure) rating.

Remember

Not all the functional requirements may apply to every project, but those that do must be complied with as part of the overall process of complying with the Building Regulations.

Did you know?

The *Domestic Building Services Compliance Guide*, published by the Department for Communities and Local Government, includes full technical details of the minimum energy efficiency requirements of heating and hot water systems.

Key term

U value – a measure of the thermal transmission of heat through the fabric of a building.

Find out

Find out more information about the Building Regulations 2010 by accessing the website for the Department for Communities and Local Government. Go to www.pearsonhotlinks.co.uk, search for this title and click on this unit.

Did you know?

Every new house has to have an SAP rating.

Did you know?

SAP ratings are similar to the 'miles per gallon/kilometres per litre fuel consumption' figures for cars.

- minimum levels of boiler efficiency (requirements that also now apply to existing systems)
- minimum requirements for central heating control systems (requirements that now apply to existing systems as well).

Standard Assessment Procedure (SAP)

The Standard Assessment Procedure (SAP) is the government's method for working out the energy rating of dwellings. It is a compulsory component of Part L1 of the Building Regulations.

SAP is used to:

- estimate the energy efficiency performance of a dwelling. SAP ratings are expressed on a scale of 1–100. The higher the number, the better the rating
- calculate the carbon index, which can be used to show that dwellings comply with Part L1
- predict heating and hot water costs – these depend on the insulation and air tightness of the house and the efficiency and control of the building's systems and controls.

The procedure for calculating the SAP rating is shown on SAP worksheets, although in practice most people calculate SAP using a specialist computer program. The SAP calculation is based on:

- size of house
- insulation levels
- ventilation levels
- heating and hot water systems.

Once calculated for new dwellings, or new extensions to dwellings, the SAP rating can be submitted for Building Regulations approval and is checked by the local building control department.

SAP rating of a building is undertaken by an individual who has passed a recognised SAP assessor's course. Plumbers will often use the details provided by the SAP assessment of a prospective new building in order to design an energy-conscious heating or hot water system.

Requirements of the *Domestic Building Services Compliance Guide*

Most of the requirements that heating and hot water systems need to comply with are included in the *Domestic Building Services Compliance Guide*.

Gas and oil systems

The guide lays down strict energy efficiency requirements for heating systems. This section covers the requirements for gas and oil systems in both new-build and existing properties.

Boilers

The minimum SEDBUK efficiency for gas boilers must be at least 86% in new properties. In existing properties, the appliance efficiency is usually expected to be 86%. However, in exceptional circumstances (as indicated on completing the condensing boiler assessment procedure), the efficiency can go down to 78%.

Oil boilers installed after 1 April 2007 must be of the condensing type and must have a SEDBUK efficiency of 86%.

System circulation

Systems installed in new properties must be fully pumped. In circumstances where a boiler is being replaced in an existing system, the system must be upgraded to a fully pumped system. If a bypass is required in the system, then it must be of the automatic type.

Hot water storage

New or replacement hot water storage cylinders must meet the following specifications relating to the insulation and the area of the heat exchange coil:

- vented copper cylinders – BS 1566
- unvented cylinders – BS 7206
- thermal primary stores – Hot Water Association (HWA).

In addition, all hot water storage vessels must carry proper labelling to show that they are properly approved. The compliance guide also lays down a minimum specification for the insulation of central heating and hot water pipework, and the circumstances in which pipework must be insulated.

System controls

There are different requirements for each component of a domestic heating system, as shown in Table 3.2.

System controls – component	Requirement
Boiler interlock	There must be a suitable wiring arrangement so that when there is no demand for either hot water or space heating, the boiler and pump are turned off.
Space heating zones	For new properties with a total usable floor area of up to 150 m², the area should be divided into two separate space heating zones with independent temperature control. One zone must be the living room, and the requirement could be met by installing a room thermostat in the hall and placing TRVs on all other radiators. A new system in an existing property must meet these requirements, although in the case of a replacement boiler in an existing system, it is acceptable to have one heating zone.
Water heating zone	All dwellings must have a separate water heating zone.

Table 3.2: Requirements for each component of a domestic heating system ▼

Did you know?

SEDBUK stands for Seasonal Efficiency of Domestic Boilers in the UK. The Boiler Efficiency Database enables users to compare the energy performance of different boilers. For more information, visit the SEDBUK website. Go to www.pearsonhotlinks.co.uk, search for this title and click on this unit.

Did you know?

The condensing boiler assessment procedure is a point-scoring system that can be used for existing properties. It determines if, in exceptional circumstances, it is permissible to install a non-condensing boiler, for example if a condensate pipe cannot be easily sited.

System controls – component	Requirement
Time control of space heating and hot water	A system in a new property or a full replacement system in an existing property must have independent time control of the space heating and hot water circuits, such as using a programmer. If a cylinder is replaced in an existing system, it is permissible to have one time clock controlling the operation of both circuits together if a separate time control is not provided to both circuits.
Temperature control of space heating	Separate temperature control of each zone can be provided by: • room thermostats or programmable room thermostats in all zones • a room thermostat or programmable room thermostat in the main zone and individual TRVs on radiators in the other zones • a combination of both of the above. For existing systems in which the cylinder is being installed and there is a semi-gravity system, it is acceptable for a thermo-mechanical thermostat to be installed.
Temperature control of hot water	A cylinder must have a thermostat controlling a motorised valve. For new properties with a total usable floor area over 150 m², there must be two water heating circuits, each with their own time and temperature control. For existing systems in which the cylinder is being installed and there is a semi-gravity system, it is acceptable for a thermo-mechanical thermostat to be installed.

Table 3.2: Requirements for each component of a domestic heating system (cont.)

Figure 3.1: Thermo-mechanical hot water valve

Solid fuel systems

Boilers and heating appliances must meet the minimum efficiency requirements as defined in the guide.

- There must be thermostatic control of the boiler for all appliances, excluding those that will have an open fire.
- Simple appliances such as open fireback boilers and room heaters will usually be fed from a semi-gravity system so that they work safely; this means that boiler interlock and temperature control of the hot water circuit cannot be provided. However, the requirements for control of the space heating circuit are the same as for gas and oil systems in terms of zoning requirements, time control and temperature control.
- Less sophisticated independent boilers, such as gravity feed boilers, are subject to the same requirements as room heaters.
- Fully automatic solid fuel boilers, such as pelleted woodchip boilers burning biomass type fuel, are available if the manufacturer indicates that it may be possible to fully pump these, incorporating full boiler interlock and controls as per gas and oil systems.
- The system should be commissioned according to the same requirements as gas and oil systems.

The level of sophistication of the control system is identified by the manufacturer as a control system that is safe to use in conjunction with the appliance.

Planning and Energy Act 2008

The Act enables local planning authorities (LPAs) to set 'reasonable requirements' for energy use and energy efficiency in local plans, such as ensuring that new developments contain:

- a proportion of renewable energy sourced from the locality of the development
- a proportion of low carbon energy sourced from the locality of the development.

LPAs can also require new developments to meet energy efficiency standards exceeding those of the Building Regulations, for example by requiring a particular Code for Sustainable Homes rating (see page 122).

(see page 122)

> **Did you know?**
>
> The Code for Sustainable Homes is a method of environmental assessment which rates the performance of new homes.

Responsibilities of members of the construction team under energy conservation legislation

Clients

The client is the person wanting construction work to be done. While they have specific responsibilities under health and safety legislation, they are likely to require professional expertise to guide them through environmental legal requirements.

Designers

Designers are responsible for converting a client's wishes into reality. Their design solution should satisfy the client and comply with all rules and regulations such as the Planning and Energy Act 2008 and Part L of the Building Regulations.

Employers

Where developers are the main contractors, such as house builders, the Code for Sustainable Homes introduces a national standard for new-build residential developments and may be used by local authorities to ensure planning applications meet certain environmental standards.

Nominated subcontractors, such as plumbing installation companies, are specified in the contract by the architect or client to carry out certain work. As such this work will have been obtained by submission of tenders against specifications of work and materials prepared by the design team. Environmentally, this means that employers are required to comply with these specifications and work in line with industry best practices, most of which will be covered through health and safety legislation (e.g. waste reduction).

Employees

As for employers, employees are required to comply with installation specifications and work in line with industry best practices, mainly covered through health and safety legislation (e.g. only running energy efficient generators when needed). Some other common practices are listed below:

- Order materials that have a recycled content.

- Order materials appropriately – this means calculating how much of each material is needed, including likely wastage.
- Knowing when and where materials are required.
- Asking suppliers to minimise packaging and to guarantee a take-back service.
- Reduce waste.

Working life

Beth is a Level 2 plumbing apprentice who is planning to get a Level 3 qualification and become a registered Gas Safe engineer. Beth has been asked to work on a property, being modernised by Graham, a builder friend. Graham has collected an array of plumbing and heating appliances and components.

To maximise his profits he asks Beth to install a gas-fired boiler, which is as new, having never been fitted but using his components, some of which are over ten years old. The existing one-pipe heating system in the property is even older. It has an indirect hot water cylinder that is leaking and requires replacement with another one of Graham's components. Beth has only ever seen a cylinder like this in old installations.

Graham also does not want any special controls to be fitted. This would make the installation too complicated and may put prospective buyers off.

Discuss the following points:

- What should Beth do?
- What advice could Beth give to Graham regarding his proposals?
- What are the minimum requirements that Graham could expect regarding the hot water and central heating systems?
- Should anyone be informed of the proposed works and, if so, who?
- Beth has quite a lot of experience working on gas but never unsupervised. Would she be competent to work on gas installations?

Progress check

1. What is the aim of energy conservation?
2. Where has energy conservation directly affected the building services engineering industry?
3. Why do the Building Regulations exist?
4. What is the purpose of the Domestic Building Services Compliance Guide ?
5. What is the SAP calculation based on?
6. What is the minimum efficiency percentage for gas boilers fitted in new buildings?
7. What is meant by the space heating zone in a central heating system?
8. Other than complying with legislation, how else can the plumbing industry become more environmentally aware?

2. Know the applications of energy sources used in the building services industry

Types of energy used in properties

Using fuel creates energy, for example burning coal to generate electricity. There are three categories of energy source, as shown in Table 3.3.

Energy source and fuel	Description
High carbon (non-renewable) energy source, e.g. coal, fuel oils, solid fuels (coal and peat), natural gas and LPG, electricity from fossil fuels such as coal	Natural resources that cannot be re-made, re-grown or regenerated. In the long term it will not be possible to rely on these because they are becoming depleted.
Low carbon (renewable) energy source, e.g. solar thermal, biomass (solid fuel), hydrogen fuel cells, heat pumps, combined heat and power, combined cooling, heat and power	Natural resources such as sunlight, wind, rain, waves and tides, and geothermal (underground) heat are naturally replenished at a rate that is **sustainable** with our rate of consumption. Biomass wood fuels are classified as renewable only when the wood is produced sustainably.
Zero carbon energy source, e.g. wind power, tidal power, hydroelectric power, solar photovoltaic power	Energy comes from the environment and creates no carbon waste. Includes wind and wave/tidal power to produce electricity.

Table 3.3: Types of energy source and resources used to generate them

Basic operating principles of installations containing environmental energy sources

Energy generation (by power stations) is a major source of carbon emissions. There are a number of initiatives aimed at producing 'clean' energy.

Solar thermal

Solar water heating systems use heat from the sun to work alongside a conventional water heater. They collect heat from the sun's radiation via a flat plate system or an evacuated glass tube system before being stored in a hot water cylinder. The system is normally not pressurised

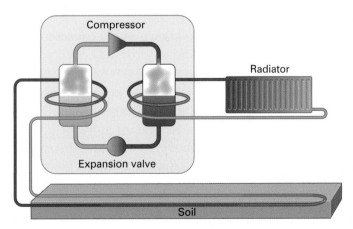

Figure 3.2: Water heating system

and often uses 'drain back' technology. This system uses pipework that slopes between components and does not have water inside it when not in use. Therefore, there is no need for insulation.

When a solar water heating and hot-water central heating system are used in conjunction, solar heat will either be concentrated inside a pre-heating tank that feeds into the tank heated by the central heating, or the solar heat exchanger will replace the lower heating element. The upper element will remain in place to provide for any heating that solar cannot provide.

However, central heating is usually used at night and in winter, when solar gain is lower. Therefore, solar water heating for washing and bathing can often be a better application than central heating because supply and demand is better matched.

Solid fuel (biomass)

Biomass, also known as solid fuel, is biological material produced from living, or recently living, organisms. It can be converted into electricity or clean-burning fuels in an environmentally friendly and sustainable way, and can be used instead of fossil fuels. Energy produced from biomass is carbon neutral – it takes carbon out of the atmosphere while it is growing and then returns it as it is burned.

Biomass energy can be generated from virgin wood, energy crops and agricultural, food and industrial waste. As a plumber, wood is the most common type of biomass that you are likely to come into contact with. Wood logs or chips can be burnt in a stove to provide heat and domestic hot water.

Water, air and ground source heat pumps

A ground source heat pump works by absorbing heat from the ground and then raising its temperature. It is normally used in domestic properties. An air source heat pump does the same from the air. Both work using the operating principles of a fridge.

The operating principle of the heat pump can be demonstrated using the ground source version (see also Figure 3.3):

- Brine circulates in a closed loop known as the collector coil, which is buried in the ground and absorbs heat energy from the ground.
- Inside the heat pump, the collector coil is wound around a heat exchanger (the evaporator) and the lukewarm brine in the collector coil begins to warm the ice-cold refrigerant in the heat exchanger. This refrigerant has a very low boiling point, and therefore the brine in the collector causes its temperature to rise by a few degrees. In turn this is enough to cause it to boil and evaporate. Think of a boiling kettle which is moving liquid into vapour as it boils.
- The evaporated refrigerant now moves into a compressor that compresses it. This compression raises the pressure, which causes the

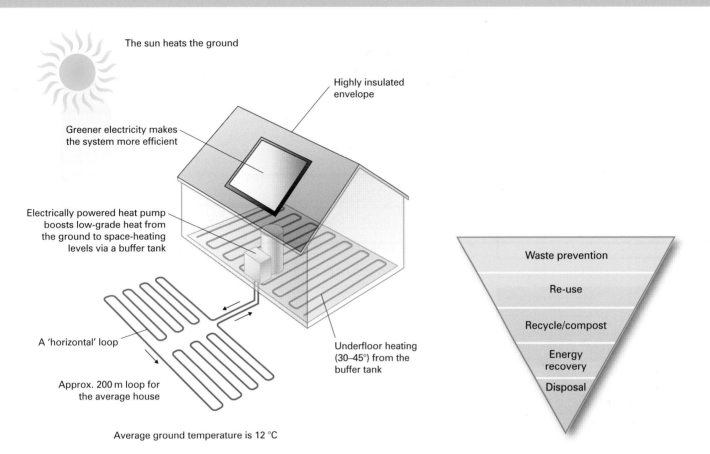

The sun heats the ground

Highly insulated envelope

Greener electricity makes the system more efficient

Electrically powered heat pump boosts low-grade heat from the ground to space-heating levels via a buffer tank

A 'horizontal' loop

Approx. 200 m loop for the average house

Underfloor heating (30–45°) from the buffer tank

Average ground temperature is 12 °C

Figure 3.3: Ground source heat pump

Waste prevention

Re-use

Recycle/compost

Energy recovery

Disposal

Figure 3.4: The waste management triangle

refrigerant to rise in temperature (to about 50 °C). It then passes into another heat exchanger (the condenser) and condenses. Once again think of the kettle: the boiling steam vapour condenses back to liquid.

- Another closed loop (the distribution system out to the radiators) is wound around the condenser, causing the generated heat in the condenser to transfer from it to the closed loop feeding the radiators. From there it is sent out hot to the radiators, cools as it is passes through them and returns to be heated again by the condenser.

- The condensing refrigerant circulates from the heat exchanger and into an expansion valve that lowers the pressure and the refrigerant becomes cold once again. The process then begins again when the circulating refrigerant meets the warm brine in the collector coil.

Combined heat and power (CHP)

Micro-combined heat and power (Micro-CHP) uses heat produced via a heat generator (such as a sophisticated boiler) to generate electricity as a by-product of the heating process.

Although not renewable, this system is sometimes called cogeneration. It is essentially a domestic boiler that contains a condensing boiler to heat the home and provide hot water, but also includes a Stirling engine to generate electricity.

Did you know?

The ground source heat system makes use of the fact that in most places only a few metres under the ground surface, the temperature remains at a fairly constant 12 °C, even in winter.

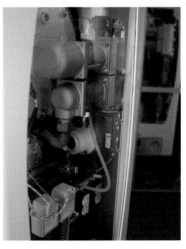

Figure 3.5: A micro combined heat and power (Micro-CHP) unit

Figure 3.6: Wind power is one of many alternative power sources which may increase in use over the next few years

Wind turbine

Wind energy generation relies on wind to turn a shaft linked to a generator that in turn produces electricity. The greater the wind speed, the more power that is produced. Wind speed increases with height so the large rotating blades are normally seen at the top of a tall supporting mast or tower.

The two main types of wind turbine used in a domestic setting in the UK are:

- mast mounted (2.5 kW to 6 kW)
- roof mounted (1 kW to 2 kW).

Any surplus electricity generated by a domestic wind turbine can be sold back to the electricity supplier.

Wind speed and direction is variable. Therefore, wind power may not be the best option for domestic properties unless they are in remote locations.

Solar photovoltaic

This system consists of photoelectric cells set in panels, which convert the sun's rays (solar radiation) into electricity. The power generated is first used within the property to reduce electricity consumption and any surplus is sold to the electricity provider.

For houses solar panels are usually located on the roof, but they may also be located on the ground. The panels must face south towards the sun, and for maximum efficiency, turn in order to face the sun throughout the day.

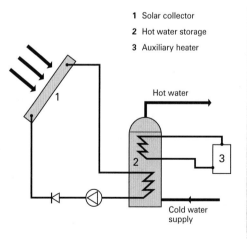

1 Solar collector
2 Hot water storage
3 Auxiliary heater

Hot water

Cold water supply

Figure 3.7: Solar heating system with indirect twin coil cylinder

Figure 3.8: Solar panels are increasingly being used as alternative energy sources

Table 3.4 summarises the advantages and disadvantages of alternative energy sources.

Energy source	Advantages	Disadvantages
Solar thermal	• Environmentally friendly • No fuel deliveries required • Low maintenance	• High set-up cost
Wind power	• Plentiful • Renewable • Does not release harmful carbon emissions or other pollutants • As equipment is fairly low maintenance and wind is free, once the initial installation has been paid for, electricity costs are reduced • Useful in remote locations away from the electricity grid • Electricity generated can be stored in batteries for use on windless days • Surplus electricity can be sold to electricity supplier	• Wind is required to turn the turbine and wind speed is variable • Small domestic turbines require an exposed location without nearby obstacles that cause turbulence and affect air speed • Planning permission is required
Biomass	• Carbon neutral • Excellent green use for waste materials that would otherwise be sent to landfill • Biomass energy can produce up to 80% of a home's energy needs	• Although biomass energy (like most alternative fuels) is more cost-effective than fossil fuels, there are initial set-up costs and ongoing running costs • Storage space is required, which requires ventilation and a dry environment
Solar photovoltaic	• Sunlight is free • Solar power is pollution-free • Installations can operate for many years with little maintenance after the initial set-up • Useful for areas where connection to the electricity grid or fuel transport is difficult, costly or impossible • Grid-connected solar electricity can be used locally reducing transmission/distribution costs • Electricity costs reduced • Surplus electricity can be sold to electricity supplier • Planning permission not usually required	• Costly to install. While the modules are often guaranteed for 20 years, the personal investment in a domestic system can be lost if the owner moves • Requires space (roof or ground) for solar panels • Solar panels must face south towards the sun • The amount of electricity generated depends on sunshine. Cloudy days reduce the quantity of electricity produced
Air and ground source heat pumps	• Low carbon 'footprint' • No fuel deliveries required • Provide space heating and hot water • Lower fuel bills, especially when measured against electric heating • Low maintenance	• High set-up costs • Ground source heat pumps produce a lower temperature heat than traditional boilers. Building needs to be well insulated and draught proofed for the heating system to be effective • Air source pumps tend to produce lower temperatures than ground source pumps

Table 3.4: Advantages and disadvantages of alternative energy sources

Organisations providing guidance and advice on energy saving

Various organisations and initiatives offer guidance on energy saving. Some of the main ones are covered on page 123.

Code for Sustainable Homes

The Code was introduced by the government to encourage the building of sustainable homes. The minimum standards for compliance with the Code, which is voluntary, have been set above the requirements of Building Regulations.

The Code uses a sustainability rating system – from one star (minimum standard to six stars) to show the overall sustainable design of a home. Table 3.5 summarises the minimum standards required under the Code.

Minimum standards		
Code Level	**Category**	**Minimum standard**
1(*)	Energy CO_2 Percentage improvement over Target Emission Rate (TER) as determined by the 2006 Building Regulation Standards	10%
2(**)		18%
3(***)		25%
4(****)		44%
5(*****)		100%
6(******)		A 'zero carbon home' (heating, lighting, hot water and all other energy uses in the home).
1(*)	Water Internal potable water consumption measured in litres per person per day (l/p/d)	120 l/p/d
2(**)		120 l/p/d
3(***)		105 l/p/d
4(****)		105 l/p/d
5(*****)		80 l/p/d
6(******)		80 l/p/d
1(*)	Materials Environmental impact of materials	At least three of the following five key elements of construction are specified to achieve a BRE Green Guide 2006 rating of at least D: 1. Roof structure and finishes 2. External walls 3. Upper floor 4. Internal walls 5. Windows and doors
1(*)	Surface water run-off Surface water management	Ensure that peak run-off rates and annual volumes of run-off will be no greater than the previous conditions of the development site
1(*)	Waste Site waste management	Ensure that there is a site waste management plan in operation, which requires the monitoring of waste on site and the setting of targets to promote resource efficiency
	Waste Household waste storage	Where there is adequate space for the containment of waste storage for each dwelling. This should allow for the greater (by volume) of the following: **either** accommodation of all external containers provided under the relevant Local Authority refuse collection / recycling scheme. Containers should not be stacked to facilitate ease of use. They should also be accessible to disabled people, particularly wheelchair users and those with a mobility impairment **or** at least 0.8 m² per dwelling for waste management as required by BS 5906 (Code of Practice for Storage and On-site Treatment for Solid Waste from Buildings)

Department for Communities and Local Government, *Code for Sustainable Homes: Technical Guide* (2010).

Table 3.5: Code for Sustainable Homes minimum standards star system

The Energy Saving Trust

The Energy Saving Trust offers a free advice and information service to help people save energy, conserve water and reduce waste. It also provides advice, support and services to local authorities to help them reduce carbon emissions.

Energy rating tables and their effect on component selection

Buildings produce nearly half of the UK's carbon emissions. The way a building is constructed, insulated, heated and ventilated and the type of fuel used, all contribute to its carbon emissions. Government measures to improve the energy performance of buildings include:

- Energy Performance Certificates (EPCs) for homes when they are bought, sold or rented
- Display Energy Certificates to be clearly displayed in all large public buildings
- inspections for air conditioning systems
- advice and guidance for boiler users.

These measures will impact on the selection of construction materials, air conditioning systems and boilers, which will need to be energy efficient with carbon emissions kept to a minimum.

Where to find information on alternative energy sources

Alternative energy sources can be classified as:

- micro-generation – small scale as used in individual households
- macro-generation – large scale commercial renewable energy sources.

Did you know?

Energy Performance Certificates provide 'A' to 'G' ratings for buildings, with 'A' being the most energy efficient and 'G' being the least.

Remember

Although a plumber may not design such systems, they may be involved in a systems installation or maintenance, and every visit to site, along with the plumbing systems used, has an environmental effect. Plumbers therefore have a responsibility to work in an environmentally friendly way.

Find out

Use the Internet to find further information on alternative energy sources. For some useful websites (listed below), go to www.pearsonhotlinks.co.uk, search for this title and click on this unit.

- RenewableUK – wind, wave and tidal power
- Biomass Energy Centre – biomass fuels
- The Micropower Council – photovoltaic and solar thermal systems, ground source heat pumps
- Energy saving Trust.

Progress check

1. Energy is created by the use of fuel. There are several types of fuel, but these fall into three main categories. What are these categories?
2. Which energy sources take energy from the environment and create no carbon waste?
3. What are the sources of biomass material?
4. Would planning permission be required for a domestic wind turbine?
5. Biomass production is carbon neutral. What does this mean?
6. Ground source and air source heat pumps work on the same principle as which domestic appliance?
7. What is the fairly constant temperature (even in winter) only a few metres below the Earth's surface, which may be used to harness energy with a ground source heat pump?

3. Know the importance of energy conservation when commissioning building services systems

Role of the commissioning process in conserving energy usage

Commissioning is the process by which we verify and record, through inspection and functional testing, whether the installed systems within a building are working to the intended criteria and specification.

Most often, this task is seen as something that occurs at the end of the job. It is through the commissioning process that we can:

- prove energy is being used as efficiently as the design process intended
- identify any problems within a system that require rectification.

However, if the designers provide the installers with clear information from an early stage, giving them a greater understanding of the design criteria, many systems may be set up at their most efficient operating levels from the beginning.

System handover procedure and conserving energy usage

The process of handing over systems should begin with inspection checks of the completed installation. This will ensure that systems are safe for use. Manufacturer's data will be needed to complete this stage.

The next important part of the commissioning process is the functional testing of any systems. Faults within a system can arise from design, installation or set-up, and functional testing allows these to be detected.

Specific guidelines for commissioning a system

The various tests involved will depend upon the system that has been installed and are too numerous to cover in this book. However, there are some general guidelines for commissioning a system:

- The system should be thoroughly flushed on completion of the installation.
- A corrosion inhibitor must be added to the system.
- The system must be commissioned in line with manufacturer requirements.
- The property owner must be fully instructed in the effective operation of the system and all user operating instructions must be left on site.
- A commissioning record must be left on site.

Additionally, the installation of a boiler, central heating or hot water system is identified under the Building Regulations as work on a controlled service or fitting. Therefore, such an installation requires **one** of the following:

- A notice must be given to the local authority before the work begins.
- A compliance certificate must be issued on completion of the work by an installer who is a member of a competent persons' scheme.

4. Know the methods of reducing waste and conserving energy while working in the building services industry

Working practices to conserve energy and protect the environment

The following procedures are particularly relevant to efficiency:

- All systems pipework should be insulated and installed to comply with the Water Regulations.
- Systems and components should be serviced and maintained regularly to ensure that they are working to design specification and not wasting water.
- Systems should be designed to keep dead legs as short as possible, and avoid over capacity, i.e. heating high volumes of hot water storage that may not be required.

Working life

Jordan is a second-year apprentice working under the supervision of Anthony, who has been with the company for a few years. They have nearly finished installing a heating system in an occupied terraced house. Jordan decides to use some initiative and begins to insulate the hot water primaries in the airing cupboard. Anthony tells Jordan not to bother doing this as it is a waste of time and effort. Jordan is about to tell Anthony about the new regulations and the importance of energy efficiency he learned about at college when Anthony tells him to get on with something worthwhile.

The job is completed (minus insulation). While Jordan is putting the tools in the van the customer thanks them and gives them a £10 tip each for doing such a good job. Jordan feels guilty because the job is not as complete as it should be.

- What should Jordan do now?
- Should Jordan and Anthony accept a tip anyway?
- Is the insulation important?
- Have Jordan and Anthony contravened any legislation by not insulating the pipework and, if so, which?
- What should Jordan have done, when challenged by Anthony?

Methods of reducing and conserving material usage

You should get into the habit of keeping waste material to a minimum when you are carrying out plumbing work. A checklist is provided below.

Checklist

Waste reduction:

1. If it is economically possible, try to repair an appliance or component, rather than replace it. New components have to be made and the manufacturing process uses energy. Old appliances have to be transported and disposed of, although replacing an inefficient boiler would be advisable.

2. Take time to carefully measure and set out pipework for bends. This will reduce the amount of wasted pipe when cutting it to length.

3. When using integral solder capillary ring fittings, do not use additional wire solder on the joint.

4. Treat screws and other fixings carefully – too often they are left lying around.

5. Be extra careful when fitting sanitary ware: a broken item has to be disposed of and replaced.

6. Take care when storing sanitary ware and other damageable goods.

7. Take care of your tools. Defective tools have to be replaced – and so do stolen ones.

8. Think about how you use water when you are working on a system. Check all fittings thoroughly to make sure they have been soldered. You do not want to be filling and draining systems unnecessarily.

9. Do not overuse jointing compounds, fluxes and other materials.

10. Take care when taking up floorboards; avoid having to replace damaged boards with new ones.

Planning work methods

A carefully planned installation can reduce the environmental impact of the work in terms of the number of visits to site, consequent travel and welfare arrangements, material and equipment delivery arrangements, selection of material and equipment both in terms of installation and maintenance factors and the removal or disposal of waste.

Approved Document L2A: Conservation of fuel and power

As stated earlier, Part L of the Building Regulations deals with energy conservation. It states: 'Responsible provision shall be made for the conservation of fuel and power in buildings by … providing and commissioning energy-efficient fixed building services with effective controls.'

Part L also states that the owner of a building must be provided with sufficient information about the building, its fixed building services and any maintenance requirements, to enable the building to use no more fuel and power than is reasonable under the circumstances.

Saving energy in an eco-friendly building

Table 3.6 looks at ways to save energy in an eco-friendly building.

Materials	Energy	Water
• Use natural, sustainably managed renewable sources (e.g. there are timber building companies in Norway that have planting policies, which mean they are growing more trees than they cut down to make their products) • Source materials near to the point of use, therefore reducing transportation effects • Use minimal processing and minimise added content such as chemicals • Make use of natural insulation properties • Materials should be non-toxic and not hazardous to users or building occupants (e.g. paint fumes) • Materials should be durable with low maintenance • Materials should have the capability to be recycled • 'Measure twice, cut once' is a good message to follow to avoid wastage	• Use natural light wherever possible • Use low-energy appliances wherever possible (e.g. 'A' rated washing machines) • Use local expertise and labour wherever possible, thus reducing the effects of transportation • Use renewable sources where possible	• Use rainwater harvesting (using rainwater for irrigation, vehicle washing or toilet flushing) • Use grey water recycling (using water from baths, showers etc. for toilet flushing) • Use low volume flush toilets • Use aerated taps (these make the water spray) • Use instantaneous water heaters over sinks instead of heating large volumes of water • Lag hot water pipes to avoid losing heat • Only heat as much as you need in a kettle or boiler

Table 3.6: Methods of saving energy in an eco-friendly building

Progress check

Within an eco-friendly building, what measures can be taken to reduce the use of water?

5. Know how to safely dispose of materials used in the building services industry

Waste management legislation

Waste (England and Wales) Regulations 2011 and List of Waste Regulations 2005

The Hazardous Waste Regulations 2005 and the List of Waste Regulations 2005 introduced the requirements of the European Hazardous Waste Directive into England and Wales. The main aim of the European Directive is to define hazardous waste and to make sure it is properly managed and regulated. The Waste Regulations 2011 came into force on 29 March 2011 and amend the Hazardous Waste Regulations 2005.

The Hazardous Waste Regulations:

- control waste that can harm human health or the environment
- control waste that is difficult to handle

Did you know?

If you leave materials on site when your work is complete, you may be discarding them. If they are discarded they will be 'waste' and, as the producer of the waste, you will be responsible for it.

- make sure that hazardous waste is properly managed at all times
- define hazardous waste in England and Wales
- require producers or consignors (the company delivering the goods) of hazardous waste to notify (register) their premises
- restrict mixing of wastes and require separation of wastes where appropriate
- make sure that companies document the movement of hazardous waste
- require consignees receiving hazardous waste to keep thorough records and provide the Environment Agency with information on the disposal and recovery of hazardous waste every three months.

The List of Waste Regulations:

- introduced the List of Wastes, also known as the European Waste Catalogue
- explain the list, giving help on choosing the code for a specific waste
- show how waste is classified as either hazardous or non-hazardous
- identify limits for certain hazardous properties.

The List of Wastes

The Waste Regulations set out the rules for assessing whether or not a waste is hazardous. As part of the assessment of waste, the regulations refer to the list of wastes, which categorises each type of hazardous and non-hazardous waste by a six-digit code.

Guidelines for using the List of Wastes:

- Wastes categorised as hazardous are marked with an asterisk.
- Some waste types are classed as hazardous waste outright and are known as absolute entries. In the Environment Agency consolidated guidance they are highlighted in red and marked with an 'A'.

Figure 3.9: Are these hazardous goods?

European Regulation on Classification, Labelling and Packaging of Substances and Mixtures

Countries in the United Nations, including those in the European Union (EU), have been working together with industry representatives and others to agree a classification and labelling system that can be used worldwide. The outcome is the Globally Harmonised System of Classification and Labelling of Chemicals (GHS).

The GHS provides a single system to identify hazards and to communicate them in transporting and supplying chemicals across the world. This is not law but an international agreement. To make the GHS legally binding, each country must implement the GHS through legislation.

The EU agreed to adopt the GHS through the European Regulation on Classification, Labelling and Packaging of Substances and Mixtures. This is known as the CLP Regulations or CLP.

Remember

Waste needs to be assessed to see if it contains dangerous substances. You must no longer refer to the Health and Safety Executive's Approved Supply List to classify hazardous wastes; instead you must use Table 3.2 of Annex VI of the CLP Regulations.

Examples of hazardous materials in waste include lead acid batteries, solvents, asbestos, chemicals and some IT equipment. There are specialist provisions for explosives and radioactive wastes.

Methods for safely disposing of waste materials

There are special arrangements for the disposal of some types of asbestos. On larger sites, skips are usually provided for waste. These, once full, are taken to licensed sites for disposal. Do not overfill the skip and do not put in items that are dangerous, such as flammable material.

On smaller jobs you might still use a skip, for the disposal of an old bathroom suite for example. If you try to take 'industrial waste' to your local tip, you are likely to be turned away or asked to pay for its disposal. Remember that the disposal of refrigerators is covered by legislation and must be dealt with at specialist sites. Do not be tempted to let the customer just dump them in the skip.

Dealing with waste

Current government policy regarding this could be summarised as 'protection of human health and the environment by producing less waste and by using it as a resource wherever possible'. Through more sustainable waste management such as reduction, reuse, recycling, composting and using waste as a source of energy, the government aims to reduce the environmental impact of waste.

This approach introduces the hierarchy of waste management as shown below, where waste prevention is the preferred option and disposal the least favourite, as shown in Figure 3.11.

> **Did you know?**
>
> Across the world, countries have different rules on classification and labelling. For example, a chemical could be classified as 'toxic' or 'explosive' in one country but not in another. Different symbols are also used to indicate the same hazards.

Figure 3.10: Dispose of construction waste responsibly

> **Did you know?**
>
> Plumbing work often involves transporting waste items such as building rubble, offcuts of pipe and old sanitary ware. By law, plumbing companies are required to hold a Waste Carriers' Licence.

> **Remember**
>
> 'Fly tipping' – the disposal of waste in places other than a registered commercial or private site – is illegal. Recycling is an energy efficient and cost-effective way of getting rid of scrap waste.

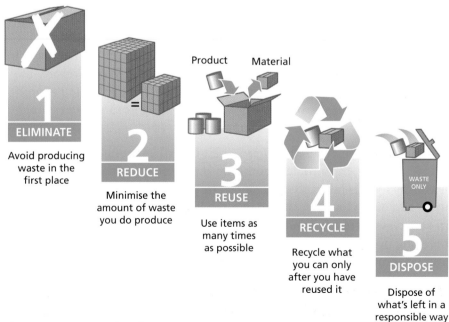

1 ELIMINATE	2 REDUCE	3 REUSE	4 RECYCLE	5 DISPOSE
Avoid producing waste in the first place	Minimise the amount of waste you do produce	Use items as many times as possible	Recycle what you can only after you have reused it	Dispose of what's left in a responsible way

Figure 3.11: Hierarchy of waste management

Waste prevention

Examples of waste prevention include:

● using less product packaging
● using reusable shopping bags
● sustainability in building design.

Energy recovery

Energy recovery (sometimes known as Waste to Energy) systems create energy in the form of electricity or heat from the incineration of waste materials.

Incineration has previously introduced its own problems, through the release of acid gases and ash left over by the burning process. However, modern plants now use lime filters in chimneys to achieve acceptable control of the problem. Use of these plants can be linked to the development of 'district heating' in cities such as Newcastle upon Tyne, Nottingham and Sheffield. In such systems, the incinerator acts as a huge central boiler that then supplies heat to a large area such as a housing estate.

There are also a number of developing technologies that can produce energy from waste without using incineration. Some are able to efficiently convert the energy into liquid or gaseous fuels.

Disposal

The most common disposal methods, particularly in the UK, are landfill and, to a lesser extent, incineration. Each year millions of tonnes of controlled waste (household, commercial and industrial waste) are disposed of in landfill sites in the UK. Some waste from sewage sludge is also placed in landfill sites, along with waste from mining and quarrying.

There are over 4000 landfill sites in the UK and as landfill waste decomposes, methane is released in considerable quantities. Methane is a greenhouse gas and contributes to global warming. Furthermore, the leachate fluids formed from decomposing waste can permeate through the underlying and surrounding ground, polluting groundwater, which may be used for drinking water supplies.

With incineration covered earlier (albeit not all incinerators are linked to district heating), a third, less common but more sustainable method of waste disposal is anaerobic digestion. In this process waste decomposes in an enclosed chamber, unlike in a landfill site. Digestion takes place in an oxygen-free environment where bacteria decompose waste by breaking down the molecules to form gaseous by-products (methane) and small quantities of solid residue. Anaerobic sewage plants produce significant quantities of methane, which can be burnt to generate electricity. Liquid and solid organic fertilisers are also formed with this method.

Figure 3.12: Landfills should be the last resort when disposing of waste

Waste transport

A business can transport most of the waste it produces directly to an authorised waste management site or recycling facility. However, it must register with the environmental regulator as a 'waste carrier' if it transports:

- construction and demolition waste produced by its own business
- any waste produced by another business.

As a waste carrier, the business must also ensure that it holds the correct registration for the type of waste being transported. It must complete a Waste Transfer Note (WTN) for every load of waste passed on or accepted. Copies of all WTNs must be held for at least two years.

Here are some guidelines for good waste disposal practice:

- Do not burn waste on site; find another method of disposal.
- Before allowing a waste hauler or contractor to remove waste material from the site, ask where the material will be taken and for a copy of the waste management licence or evidence of exemption for that facility.
- Separate out the different types of waste created on site. This will make it easier to supply an accurate description of the waste for waste transfer purposes.
- Minimise the quantity of waste produced to save money on raw materials and disposal costs.
- Label all waste skips – make it clear to everyone which waste types should be disposed of in each skip.
- Check if waste is hazardous/or special waste before transporting it.

If a business transports hazardous or special waste it must:

- keep it separate from other wastes
- use sealed and clearly labelled containers
- check that it is transferred to a facility that is authorised to receive it.

Items such as asbestos and refrigerants need to be disposed of at an approved centre that can deal with this type of waste. Not all recycling centres will be able to deal with these.

Burning materials on site

It is common practice to burn waste material on larger sites. All sorts of things get thrown into a site fire: plastics, paint cans and so on. This activity is not environmentally friendly and is usually against site rules. In particular, plastics and polystyrene should never be burned as these give off dangerous fumes, and oils and fuels must never be poured down drainage systems. All these things can damage the environment and as such should be reported to the relevant responsible persons on site.

Did you know?

In general the burning of waste in the open is an environmentally unsound practice, and you should use less damaging options of waste disposal.

Remember

A consignment note must be completed whenever hazardous waste is transferred. Copies of consignment notes must be kept for at least three years.

Find out

Find out what sort of recycling facilities are available in your local area.

Key term

Pulping – where the products are dropped into a huge tank and soaked with water, which separates the paper from any other impurities.

Remember

Emulsified cutting oils are highly polluting in water and great care should be taken in their disposal. Contact a specialist contractor for advice.

Remember

Food and drink cans are made up of up to 50% steel. You may only be able to drink the cola or eat the beans once but, by recycling the can, the same steel can be used again and again. Steel in Europe contains 56% recycled steel and is 100% recyclable.

Approved processes for recycling materials

There are two general aspects to recycling:

- Turning something old into something new.
- Reusing a product we no longer want, even if it contains a hazardous material.

For example, paint is not turned into something new, but if a half-full tin of paint is taken to a recycling plant, it could be collected and carefully distributed for use by others in community projects, where there may be little money to purchase new tins of paint.

Equally, a school or youth club may be grateful to receive the computer monitor that you were going to throw away. You can give unwanted items to a charity shop or community project. Items made out of materials that can be turned into something new provide the basic concept for recycling.

Most paper-based products such as magazines, newspapers, fliers, and Christmas cards can be processed via **pulping** and turned into new paper-based products.

Recycling dangerous substances

Industrial oil is also a useful substance that, when recovered, can be used as a fuel and save resources. It should be treated as such. Larger quantities of used oil, such as hydraulic fluid or lubricants from lorries, buses or mechanical plant should be stored securely to await collection by a registered waste carrier. There are specialist companies that will collect used cutting oils, and then treat and recover the oil.

Cooking oils from commercial users, such as caterers and fish and chip shops, must not be disposed of with in general waste. Although not hazardous they can be collected by specialist contractors and even be reused as fuel for diesel engine cars.

Recyclable materials at work

Plumbers work with:

- copper
- brass
- lead
- cast iron
- low carbon steel.

All of these have a value at the scrap merchant. Think about recycling before you throw anything away.

Even waste plastics and packaging such as cardboard and wood can be taken to an approved centre for recycling. However, this will be classed as trade waste if it is from plumbing work and some council recycling centres may not take trade items. Some private recycling centres may take it without a charge.

By adopting measures that contribute to the basic '3 Rs' – Reduce, Reuse and Recycle your waste – you are helping to improve the environment locally and globally.

Working life

Aaron and John are working on a house conversion. They remove the old bathroom suite and complete a lot of the preparation work required. They leave the old suite in the front garden, having saved money by not having a skip.

The next day, the owner of the property complains about the suite being left in the front garden and demands that they load it into the van immediately and dispose of it. When they take the suite to the waste disposal point, they are turned away as the site sees their van and their dress and classes them as delivering industrial waste.

Instead they unload their van at their yard. Their boss is furious, saying they should have lied at the waste disposal point and threatens to take the cost of the disposal from their wages.

When more waste is removed the next day, Aaron and John load it into the van and decide to dump the rubbish in a country lane where they have seen other rubbish left. They are spotted and the van registration number, reported. Aaron, John and their boss are currently awaiting a court appearance and a maximum fine of £20,000.

Discuss the following points:

- What should Aaron and John have done with the waste materials?
- Who is responsible for the offence: Aaron and John, the employer, or all three?
- Should they have a licence for carrying redundant bathroom suites in their van?
- What would have been the simplest solution in this scenario?
- Was the employer right in threatening to withhold their wages?

Progress check

1. Which regulations deal with waste management?
2. What is the List of Wastes?
3. Would a plumbing company require a waste carriers' licence?
4. How long should a business keep a Waste Transfer Note?
5. What is Energy Recovery?
6. Where should materials such as asbestos and refrigerants be disposed of?
7. What metals do plumbers use, which can be recycled?

6. Know the methods of conserving and reducing wastage of water within the building services industry

Plumbers see water wastage in the form of dripping taps, leaking joints and overflows on both hot and cold water systems. A dripping hot tap means that the hot water lost is replaced by cold water, which then has to be re-heated, wasting boiler energy. There are around 18 million homes in the UK: if a leaking hot tap lost an average of half a litre of water a week in each dwelling, 9 million extra litres of water would have to be heated overall.

In the case of cold taps and cistern overflows, wasted water has to be replaced, which on a wider scale means additional treatment and distribution by water companies, using extra electrical and mechanical energy.

Water can also be wasted through burst pipes, brought about by freezing, due to a lack of insulation or because pipe runs are in exposed conditions.

Undue consumption

Excessive consumption can be caused by bad design and in particular by dead legs. This results in high volumes of cold water being run off before hot water arrives at the tap. The Water Regulations also make recommendations about the insulation of hot water pipes and storage vessels, which is another way to reduce heat loss and decrease the load on the boiler.

Water wastage and misuse legislation

Water Supply (Water Fittings) Regulations 1999

These regulations apply only to England and Wales, and have replaced Water Bye-laws. The regulations are national regulations made by the government's Department of the Environment, Food and Rural Affairs (DEFRA). The bye-laws that they replace were made locally and applied in that area.

The purpose of the Water Regulations is to prevent:

- contamination of a water supply
- the waste of water
- the misuse of a water supply
- undue consumption of water
- erroneous measurement (fiddling the meter).

Two of these areas have a direct impact on energy conservation.

Water Regulations Advisory Scheme (WRAS)

Formally known as the Water Bye-laws Scheme, the WRAS has been carrying out fitting testing for many years and will continue to advise on Water Regulations in the future. The scheme produces a *Water Fittings and Materials Directory*, which lists all approved fittings and is an important guide for all who aim to comply with or enforce the Water Regulations. Products approved by WRAS carry the symbol shown in Figure 3.14.

Water efficiency calculations for new dwellings

The government has introduced a national water efficiency calculator for new dwellings. It is used to assess compliance against the water performance targets in Regulation 17K (Part G of the Building Regulations) and the Code for Sustainable Homes.

Figure 3.13: A dripping tap represents a waste of water and energy

Figure 3.14: WRAS-approved product symbol

It is not a design tool for water supply and drainage systems, nor is it able to calculate the actual **potable water** consumption of a new dwelling. Changing behaviour can also have an effect on the amount of potable water used throughout a home.

The calculation method requires the use of water consumption figures provided by manufacturers' product details. Before the assessment can be carried out, figures will need to be collected from manufacturers' product information to determine the consumption of each terminal fitting and this includes taps, bidets, toilets, baths, washing machines and dishwashers.

Table 3.8 shows some figures from manufacturer product details used to calculate the consumption of each fitting in litres per person per day.

Performance target	Maximum consumption of potable water (litres/person/day)
Building Regulations 17K Compliance	125
Code for Sustainable Homes (Level 1/2)	120
Code for Sustainable Homes (Level 3/4)	105
Code for Sustainable Homes (Level 5/6)	80

Table 3.7: Recommended consumption of potable water

Installation type	Unit of measure	Capacity/ flow Rate (1)	Use factor (2)	Fixed use (litres/ person/day) (3)	Litres/ person/ day = [(1) × (2)] + (3) (4)
WC (single flush)	Flush volume (litres)		4.42	0.00	
WC (double flush)	Full flush volume (litres)		1.46	0.00	
	Part flush volume (litres)		2.96	0.00	
WCs (multiple fittings)	Part flush volume (litres)		4.42	0.00	
Taps (excluding kitchen/utility rooms)	Flow rate (litres/ minute)		1.58	1.58	
Kitchen/utility room sink taps			0.44	10.36	
Bath (where shower also present)	Capacity to overflow (litres)		0.11	0.00	
Bath only			0.50	0.00	
Shower (where bath also present)	Flow rate (litres/ minute)		4.37	0.00	
Shower only			5.60	0.00	
Washing machine	Litres/kg dry load		2.1	0.00	
Dishwasher	Litres/place setting		3.6	0.00	
Waste disposal unit	Litres/use	If present = 1 If absent = 0	3.08	0.00	
Water softener	Litres/person/day		1.00	0.00	
		Total calculated use (litres/ person/day) = (Sum column 40)			

Table 3.8: Figures from manufacturer product details used to calculate the consumption of each fitting

Where there are multiple fittings of the same type that have various flow rates or capacities (e.g. hot and cold taps with different flow rates), then Table 3.9 should be used as a template to determine the average flow rate or capacity of such fittings.

Taps (excluding kitchen sink taps)			
Tap fitting type	Flow rate (litres/min) **(a)**	Flow rate (litres/min) **(b)**	Flow rate (litres/min) **(c)**
1			
2			
3			
4			
Total (sum of all quantities) **(d)**			
Total (sum of all totals per fitting types) **(e)**			
Average flow rate (litres/min) **(e/d)**=			
Maximum flow rate (litres/min) **(f)**			
Proportionate flow rate (litres/min) [**(f)** × 0.7]=			

Table 3.9: Template for determining the average flow rate or capacity of multiple fittings of the same type that have various flow rates or capacities

All values throughout the water efficiency calculator for new dwellings should be rounded to two decimal places with the exception of the total water consumption figures for Regulation 17K and the Code for Sustainable Homes, which should be rounded to one decimal place.

The total calculated use, resulting from Table 3.9, is the total consumption of all water consuming fittings per person. To calculate the litres of water consumed per person per day (l/p/d), any savings from grey or rainwater need to be deducted from the total calculated use.

The resulting figure is used to determine compliance with the Code for Sustainable Homes and 17K water targets. To calculate the total water consumption for 17K, an additional allowance for external water use is added on to the total water consumption. This figure is set at 5 litres per person per day. The allowance for external water use is only applied to 17K as external water use is assessed separately in the Code for Sustainable Homes.

Methods for reducing water wastage

Flow-reducing valves

A flow-reducing valve is designed to reduce the effect of high pressure by restricting the flow of water. Often seen in the cold water feed to a sink, where an 'explosion' of high pressure water going everywhere is not required and is wasteful.

Spray taps

Spray taps use less water by reducing droplet size or by **aeration** of the water flow. The increased pressure and velocity compensate for reduced volume of the flow rate. These taps are commonly used in kitchens and bathrooms.

Low-volume flush WC

WC cisterns today have to be dual flush to avoid wasting water. A full flush will allow 6 litres of water to pass and 4 litres on a short flush compared to 7.5 litres in older cisterns. The user will have the option of using either the full flush or a half flush. By avoiding always using a full flush, water can be conserved.

> **Key term**
>
> **Aeration** – a process where air is circulated through, or mixed with, water.

Figure 3.15: Low-volume flush WC

Capturing surface water and recycling used water

Micro (small scale) hydro

This concept has been used for hundreds of years. The micro hydro process uses running water to turn something. Years ago water was used to turn wheels in a mill, which in turn was used to grind flour. Now the system uses running or falling water to turn a turbine to produce electricity.

Useful power may be produced from even a small stream. Therefore, for houses with no mains connection but with access to a micro hydro site, a good hydro system can generate a steady, more reliable electricity supply than other renewable technologies.

Turbine technology is now available that can utilise quite small spring-fed streams for power generation if the fall is sufficient.

Advantages include:

- It is environmentally friendly.
- No fuel deliveries are required.
- It works well in rural areas.
- It is very low maintenance.

Disadvantages include:

- Total system costs can be high initially (£20,000) but often less than the cost of a grid connection.
- It requires a sufficient flow of water.

Basic working principles of captured and recycled water systems

Grey water recycling

Grey water is waste water generated by activities in the home such as bathing, laundry and dishwashing. Up to half of the water used in a house can end up as grey water.

As grey water is less contaminated than waste water, common recycling sees it treated and then used for flushing toilets or for watering gardens. Some toilets now include the sink built into them to feed grey water straight to them without the need for additional pipework.

Advantages include:

- It can bring a reduction in water consumption without changing consumer behaviour.
- It is easy to install and maintenance free.
- It removes the need for complex water treatment.
- If properly designed a grey water system can lower sewage costs.
- It can reduce ground water usage for irrigation.
- Less water will enter a city's sewage systems. This saves building new, or extending old, treatment plants.

The main disadvantage is that the water is not suitable for drinking.

Rainwater harvesting

Rainwater harvesting is the gathering and storage of rainwater. Once stored, it can be used for any normal use of water, including drinking, flushing, washing and gardening – these processes equal half the water used in an average home. Collecting rainwater can make an important contribution to drinking water and, in some situations rainwater may be the only available, or economical, water source.

Most rainwater systems are simple and inexpensive to build, as they will use the existing guttering and downpipes on a house. Water then flows via a filter to an underwater storage tank.

Water can be of a good quality that may not need any form of treatment before being of drinking standard (potable).

Did you know?

Grey water does not include waste water from toilets but, with water reserves dwindling, it would seem sensible to put it to better use wherever possible.

Remember

Water gathered from a roof can pick up anything that was on the roof. This could include bird droppings, dead animals or substances in the roof material.

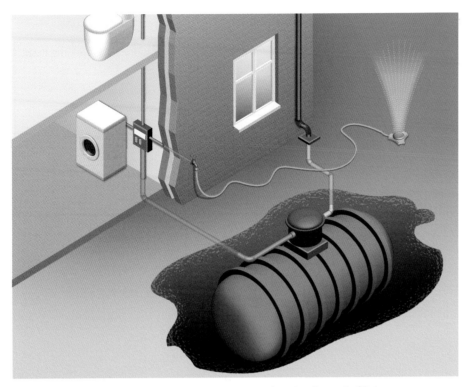

Figure 3.16: Rainwater harvesting systems can be simple and efficient

Advantages include:

- It is usually free.
- Large volumes of water are kept out of the storm-water management system, thereby helping to reduce flooding risks.
- It is low maintenance.

However, treatment is required to make the water drinkable.

Working life

Congratulations! You have won the lottery. You decide to invest some of your money in the future. You design a state-of-the-art, five-bedroomed house standing in 20 acres of ground, including a fishing lake and a fast-flowing stream running through it.

Discuss the following points:

- Which type of renewable energy and/or environmentally friendly systems could be used to reduce/replace conventional methods used with regard to building services input and removal from buildings?

- Which renewable energy resources would be suitable which would create a zero-carbon rating?
- Would the selected zero-carbon resources be adequate to satisfy the building needs on their own?
- Which forms of non-renewable energy would you select to enhance zero-rated resources if they are at times inadequate or temporarily unavailable?
- Which regulations need to be applied for zero-rated resources?

Progress check

1. What can cause undue or excessive consumption of water?

2. What is the purpose and general aims of the Water Regulations?

3. What is aeration?

4. Flow-reducing valves, spray taps and low-volume flush WCs are methods of what?

5. How many litres of water are used in a low-volume flush WC, on a full flush?

6. How many litres of water are used in a low-volume flush WC, on a short flush?

7. What is grey water?

8. What is the main disadvantage with using grey water in domestic properties?

9. Excluding drinking (which is possible in certain cases), how can harvested rainwater be used in a dwelling?

Check your knowledge

1. Which of the following is one of the key aims of the Water Regulations?
   a To consume water
   b To waste water
   c To prevent the use of water
   d To prevent the waste of water

2. Which of the following will produce carbon dioxide (CO_2) when it is used?
   a LPG equipment
   b A threading machine
   c Solvent cement
   d Flux

3. Which of the following does not produce emissions that are hazardous to the environment?
   a Gas central heating boiler
   b Solar power
   c Gas multi-point boiler
   d Oil-fired boiler

4. Fuels fall into two categories. What are these two categories?
   a Non-renewable and usable
   b Renewable and waste
   c Expensive and cheap
   d Renewable and non-renewable

5. If asbestos is found in a skip, what should be done about it?
   a Ask for a face mask
   b Nothing
   c Report to a supervisor
   d Set fire to the skip

6. Which of the following will produce toxic fumes if burnt?
   a Cardboard
   b Wood shavings
   c Paper
   d Polystyrene packing

7. In order to reduce carbon emissions, what must a new building have?
   a GHSP
   b CWSC
   c GHS certificate
   d SAP rating

8. Which of the following appliances requires specialist disposal?
   a Washing machine
   b Dishwasher
   c Gas cooker
   d Refrigerator

9. Which Building Regulation covers the conservation of fuel and power?
   a J1
   b K1
   c L1
   d M1

10. It is deemed good practice to minimise consumption of water. How can this be achieved in a hot water system?
    a Fit more bends
    b Fit extra gate valves
    c Remove dead legs
    d Fit service valves

11. What does SAP stand for?
    a Self-assessment procedure
    b Standard assessment procedure
    c Service and plumbing
    d Signed assessment product

12. Which of the following is the most energy-efficient boiler?
    a Economy 7
    b Combination boiler
    c Condensing boiler
    d Oil-fired boiler

13. What does a boiler's SEDBUK rating represent?
    a Maximum output from a boiler
    b Amount of emissions from a boiler
    c The type of fuel used by the boiler
    d Seasonal efficiency of a boiler

14. How would a plumber help the customer to maximise the economy of the heating/hot water system that they had installed?
    a Explain the servicing
    b Tell them about the insulation
    c Set the controls and commission the system correctly
    d Set the pump to the highest setting

Getting ready for assessment

The information contained in this unit of the book, as well as the continued practical assignments that you will carry out in your college or training centre, will help you prepare for both your end-of-unit test and the diploma multiple-choice test. It will also support you in preparing for the work for the practical assignments you need to complete to demonstrate your understanding and use of environmental protection measures within the building services engineering industry.

There are opportunities throughout the unit for you to test your progress in and understanding of the required underpinning knowledge; this will enhance your preparation for the forthcoming assessment/s, so make good use of them.

Make sure you know about:

- the energy conservation legislation that applies to the building services industry, including, the aims of energy conservation legislation and the responsibilities of members of the construction team under energy conservation legislation

- the applications of energy sources used in the building services industry, including the types of energy used in properties, the basic operating principles of installations containing environmental energy sources, organisations giving guidance and advice on energy saving and conservation, energy rating tables and the location of information on alternative energy sources

- the importance of energy conservation when commissioning building services systems, including, the commissioning process in conserving energy usage and actions to be covered during the system handover procedure to the customer

- the methods of reducing waste and conserving energy while working in the building services industry, including, working practices that can be employed to conserve and protect the environment and conservation of material usage

- how to dispose safely of materials used in the building services industry, including legislation for waste management, safely disposing of waste materials, approved processes for recycling materials, disposal requirements of potentially hazardous materials and the action to be taken when work activities endanger the environment

- the methods of conserving and reducing the wastage of water within the building services industry, including, statutory legislation for water wastage/ misuse, water efficiency calculations, methods of reducing water wastage, methods of capturing surface water and recycling used water, the uses of captured and recycled water in properties and the basic working principles of captured and recycled water systems.

Good luck!

UNIT 4

Understand how to apply scientific principles within mechanical services engineering

A plumber's job involves the installation, maintenance and servicing of many different systems using a large range of materials for system pipework, fittings and components. You need to have a good understanding of the properties of these materials and their suitability for the type of work for which they will be used.

This unit covers the science of plumbing and acts as an introduction to the basic materials, theories and concepts that you will encounter and work with on a daily basis.

This unit will cover the following learning outcomes:

- Know the standard units of measurement used in the mechanical services industry
- Know the properties of materials used in the mechanical services industry
- Know the relationship between energy, heat and power in the mechanical services industry
- Know the principles of force and pressure and their application in the mechanical services industry
- Know simple mechanical principles and their application in the mechanical services industry
- Know the principles of electricity as they relate to the mechanical services industry

1. Know the standard units of measurement used in the mechanical services industry

In the UK there are two principal systems of measurement: metric and imperial. The standard units of measurement (SI units) in the plumbing trade are usually metric (metres, kilograms, etc.), although imperial measurements such as feet, inches, pounds and ounces may also be used. See Figure 4.1 for the basic SI units.

Attribute	SI unit	Abbreviation	Conversion	Imperial unit(s)	Imperial abbreviation	Conversion
Length	millimetre metre centimetre	mm m cm		inch, feet	in, ft	1 in = 25.4 mm 1 in = 2.54 cm 1 ft = 0.3048 m
Mass	kilogram	kg		ounce, pound	oz, lb	1 oz = 28.35 g 1 lb = 0.4536 kg
Time	second	s	60 s = 1 min 60 mins = 1 hour			
Electric current	ampere	A				
Temperature	kelvin	K	degrees Centigrade (°C) $K = °C + 273.15$	degrees Fahrenheit	°F	$°C = \frac{5}{9}(°F - 32)$
Angle	radian	rad	1 rad = 57 degrees (°) 1 degree = 60 mins (°) 1 min = 60 seconds (°)			
Area	square metre	m^2	1 hectare = 10,000 m^2	square inches, acre		
Volume	cubic metre	m^3	cc = cm^3	cubic inches		
Capacity	litre	l	1 ml = 1 cc or cm^3	pint, gallon	1 pint = 0.5663 l	
Speed	metres per second	m/s		miles per second feet per second	mph fps	
Acceleration	metres per second per second	m/s^2 or ms^2				
Force	newton	N		pounds per square inch	lb/in^2	

Figure 4.1: SI units with appropriate metric and imperial equivalents

2. Know the properties of materials used in the mechanical services industry

Density of common materials

Density of solids

Solid materials with the same size and shape can frequently have a completely different mass. This relative lightness or heaviness is referred to as density. In practical terms the density of an object or material is a measure of its mass (for example grams) compared to its volume (for example m³) and can be worked out using the following formula:

$$\text{density} = \frac{\text{mass}}{\text{volume}}$$

The densities of the common materials you will come into contact with during your plumbing career are all known. Lists of these comparative densities can be found at a variety of reference sources.

Density of liquids

Liquids and gases also have differing densities depending on the number of molecules that are present within a particular volume of the substance. As a plumber you will need to understand the density of water and how this changes with the water's temperature. Water is less dense when it is heated:

- 1 m³ of water at 4 °C has a mass of 1000 kg
- 1 m³ of water at 82 °C has a mass of 967 kg.

This is because heat energy excites the molecules so that they move further apart and the water becomes less dense. This explains why hot water floats on cold water in plumbing systems.

Relative density

Relative density (occasionally known as specific gravity) is an effective way of measuring the density of a substance or object by comparing its weight per volume to an equal volume of water. Water has a relative density (specific gravity) of 1.0. For example:

- 1 m³ of water has a mass of 1000 kg
- 1 m³ of mild steel has a mass of 7700 kg.

The mild steel is 7.7 times heavier than water and therefore has a relative density of 7.7.

Did you know?

SI stands for 'Système Internationale d'Unités' (international system of units).

Find out

Find a list of comparative densities. You could use your learning centre or local library.

Did you know?

Water is at its maximum density at 4 °C.

Remember

The effect of density changes in relation to temperature changes can be put to good use in making plumbing systems work.

Density of gases

Just as water is said to have a relative density of 1.0 to enable comparisons of relative density between solids and liquids, air has a relative density of 1.0 to enable comparisons between gases, depending on whether they are lighter or heavier than air. Think about helium-filled balloons, for instance. Do you think helium is lighter or heavier than air?

Principal applications of solid materials

Several materials which have a range of different uses are commonly used in the mechanical services industry. There is no perfect pipework material that is suitable for all applications; different materials perform better in relation to different factors and conditions such as pressure, type of water, cost, bending and jointing method, corrosion resistance, expansion and appearance. The most common types of these are explained below.

Metal

Metals rarely occur in their pure form. More often they occur as ores, which are **compounds** of the metal and have unwanted impurities. To produce the required metal, a process of smelting is necessary. Metals commonly used in the manufacturing industry include iron, copper, lead, tin, zinc and aluminium.

The most common method of producing metals is by removing the oxygen from the ore in a process known as reduction.

Figure 4.2: Copper and plastic pipes with push-fit fittings

> **The industrial production of iron and steel:**
>
> 1. Iron ore (haematite – iron oxide) is loaded into a blast furnace along with coke and limestone.
> 2. Hot air is blasted into the base of the furnace and carbon from the coke reacts with oxygen in the air to form carbon monoxide.
> 3. Carbon monoxide reacts with oxygen from the haematite (iron oxide) to form carbon dioxide and iron.
> 4. Limestone combines with impurities in the ore (mainly silicates) to form slag.
> 5. The molten iron is tapped from the base of the furnace and solidifies into billets known as 'pigs' – hence the term 'pig iron'.
>
> At this point the iron is impure. To form steel, which is an alloy of iron and carbon, it is necessary to reheat the iron to drive off the impurities, and then to add up to 1.5% of carbon. Other metals can give the steel particular properties – the addition of chromium will produce stainless steel, for example. Alloys can be produced either by mixing different metals or by mixing metals with non-metallic elements, such as carbon.

Alloys

An alloy is a type of metal made from two or more other metals. Table 4.1 shows some commonly used alloys.

Name	Alloy	Common uses
Brass	Copper and zinc	Used for electrical contacts and corrosion-resistant fixings (e.g. screws, bolts) and pipe fittings
Bronze	Copper and tin	Used for decorative or artistic purposes and corrosion-resistant pumps
Solder	Lead and tin, tin and copper	Used for electrical connections and as a jointing material
Duralumin	Aluminium, magnesium, copper and manganese	Used in aircraft production
Gunmetal	Copper, tin and zinc	Used for underground corrosion-resistant fittings

Table 4.1: Some commonly used alloys

Copper

Copper tube has been used as a material for pipework for over 100 years. It is a malleable and ductile material, which you will use frequently throughout your plumbing career.

There are four main types of copper tube used in the plumbing industry:

- R250 half hard (also called Table X) is the copper tube most commonly used above ground for plumbing and heating installations. It is fairly rigid and will usually need to be bent using a bending machine.
- R290 hard (also called Table Z) is a more rigid copper tube. Its increased hardness means that the walls of the pipe don't need to be as thick, so the internal diameter of the tube or bore can be wider than that of the R250 type of tube; the tube cannot easily be bent.
- R220 soft coil (also called Table W) is a copper tube used for micro-bore pipework, typically on central heating systems.
- R220 soft coil (also called Table Y) is a softer copper tube, which is most commonly used underground for the supply of water.

The outside diameter is the same for each type of pipe, but there are differences in the internal bore due to variations in the pipe wall thickness. R250 half hard lengths and R220 soft coils are available with a plastic coating. This type of tube can be used in potentially corrosive environments as the plastic coating protects the copper from corroding.

> **Remember**
>
> Table X and Table Z types of copper tube are not suitable for underground use.

Steel

Low carbon steel

Low carbon steel (LCS) or mild steel is an alloy made from iron and carbon. It is frequently used in the plumbing and heating industry and is manufactured to BS 1387. The tube comes in three grades of weight: light, medium and heavy (see Table 4.2). As with copper tube, the outer diameter is similar, but the internal bore and wall thickness vary.

> **Did you know?**
>
> Galvanised tubes have an outer and inner layer of zinc, which prevents oxidisation or rusting.

Grade	Wall thickness	Bore	Colour code
Light LCS tube	Thin walls	Larger bore	Brown
Medium LCS tube	Medium walls	Medium bore	Blue
Heavy LCS tube	Thick walls	Smaller bore	Red

Table 4.2: Grades of LCS or mild steel

Light LCS is usually used for conduits. As a plumber, you will come across it on occasion, but you will work far more frequently with medium and heavy LCS. Medium and heavy LCS tubes are used for water-supply services, as they are capable of sustaining the pressures involved. When LCS tube is used for domestic water supplies, it must be galvanised.

Stainless steel

Stainless steel is the most recently developed pipe material used for water services. It is a complex alloy made up of a number of elements, as shown in the Table 4.3.

The tube has a shiny appearance due to the chromium and nickel content and is protected from corrosion by a microscopic layer of chromium oxide, which quickly forms around the metal and prevents further oxidisation. This tube is produced with bores of 6 mm to 35 mm and has an average wall thickness of 0.7 mm. The outside diameters are similar to those of R250 copper tubes.

Stainless steel is commonly used where exposed pipework and sanitary appliances are needed, as it is a very strong metal (much stronger than copper) and is easy to clean. Stainless steel is also commonly used for:

- sink units
- urinal units and supply pipework
- commercial kitchen discharge pipework.

Element	% of element in stainless steel
Iron	70 (approx)
Chromium	18
Nickel	10
Manganese	1.25
Silicon	0.6
Carbon	0.08
Sulphur	Trace
Phosphorus	Trace

Table 4.3: Composition of stainless steel

Lead

Lead is a very heavy, valuable metal that requires specialist handling. It is one of the oldest known metals and is highly ductile, malleable and corrosion-resistant. During your career, you will come into contact with lead in sheet form, which is used for weatherings on buildings. It was used in the past for mains, sanitary and rainwater pipework, but this practice stopped in 1986 with the Model Water Bye-laws, due to the possibility of lead poisoning. Its use has now been superseded by the use of materials such as plastics.

Did you know?

The main use of lead in plumbing today is for the weathering of buildings.

Cast iron

Cast iron is an alloy of iron and is approximately 3% carbon. It is very heavy but quite brittle, although it can stand years of wear and tear. It has been used in the plumbing industry for many years for above- and below-ground sanitary pipework; mainly guttering, soil stacks and baths. You will probably come into contact with it on older properties and new industrial or commercial properties.

Plastics commonly used in the plumbing industry

Plastics (polymers) are products of the oil industry.

Ethene, a product of crude oil, is a building block of plastics. It is made up of carbon, hydrogen and oxygen atoms. Molecules of ethene (monomers) can link together into long chains (polymers) to make polythene (poly + ethene) when they are heated under pressure with a **catalyst**. If the ethene monomer is modified by the replacement of one of the hydrogen atoms with another atom or molecule, further monomers result, producing other plastics. This process is called polymerisation.

There are two main categories of plastics used in the plumbing industry:

- thermosetting plastics
- thermoplastics.

> **Key term**
>
> **Catalyst** – a substance that increases the rate of a chemical reaction while itself remaining unchanged.

Thermosetting plastics

These are generally used for mouldings. They soften when first heated, which enables them to be moulded, but when they cool they set hard and their shape is fixed; it cannot be altered by further heating. WC cisterns can be made of thermosetting plastic.

Thermoplastics

Thermoplastics can be re-softened by heating. Most of the pipework materials you will come into contact with fall into this category. The different types of thermoplastics (see Table 4.4) share many of the same characteristics:

- strong resistance to acids and alkalis
- low specific heat (i.e. they do not absorb as much heat as metallic materials)
- poor conductors of heat
- affected by sunlight, which makes the plastic brittle (also called degradation).

Material	Max. usage temp (°C)	Main plumbing industry purpose
Polythene – low density	80	Flexible pipe material used to channel chemical waste
Polythene – high density	104	More rigid, again used for chemical or laboratory waste
Polypropylene	120	Tough plastic with a relatively high melting temperature, can be used to channel boiling water for short periods of time (e.g. traps)
Polyvinyl chloride (PVC)	40–65	One of the most common pipework materials, used for discharge and drainage pipework
Unplasticised polyvinyl chloride (UPVC)	65	More rigid than PVC, used for cold-water supply pipework
Acrylonitrile butadiene styrene (ABS)	90	Able to withstand higher temperatures than PVC. Used for small-diameter waste, discharge and overflow pipework

Table 4.4: Types of thermoplastics

Other materials relevant to the plumbing industry

Ceramics include products that are made by baking or firing mixtures of clay, sand and other minerals: bricks, tiles, earthenware, pottery and china. The kiln firing process fuses the individual ingredients of the product into a tough matrix. The main constituent of all these products is the element silicon (Si). Clay is aluminium silicate and sand is silicon dioxide (silica). Ceramics also include those products made by 'curing' mixtures of sand, gravel and water with a setting agent (usually cement) to form concrete, or mortar, using a sand, water and cement mixture. You may come across ceramics in the form of discs in modern taps as an alternative to rubber washers.

Glass is also produced by melting minerals together. The basic ingredients are sand (silicon dioxide), calcium carbonate ($CaCO_3$) and sodium carbonate ($NaCO_3$). The resulting mixture of calcium and sodium silicates cools to form glass. Again, additives can change the character of the product: boron will produce heat-resistant 'Pyrex'-type glass, and lead will produce hard 'crystal' glass. Glass is very resistant to chemical attack and was used for waste pipework in laboratories.

Properties of solid materials

The properties of materials relate to things such as how strong they are, how well they conduct heat or electricity or how flexible they are. We will examine these different properties before looking at common plumbing materials in more detail.

Materials are classified according to a variety of properties and characteristics. Properties can be measured by the way materials react to a variety of influences.

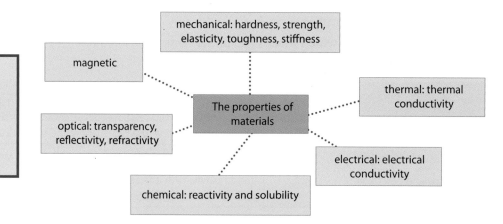

Figure 4.3: The properties of materials

Hardness

Hardness is a measure of a material's resistance to permanent or plastic deformation by scratching or indentation. It is an important property in materials that have to resist wear or abrasion – teeth on wrenches, for example – and frequently needs to be considered along with the strength of materials. Some tools in plumbing (e.g. a chisel) require a hard point and a tough head to prevent shattering when struck with a hammer. Hardness is measured on a scale of 1 to 10 based on the hardness of ten naturally occurring minerals.

Strength

The strength of a material is the extent to which it can withstand an applied force or load (stress) without breaking. The load is expressed in terms of force per unit area (newtons per square metre, N/m^2), and can be in the form of:

- compression force, as applied to the piers of a bridge or a roof support
- tensile or stretching force, as applied to a guitar string, tow rope or crane cable
- shear force as applied by scissors or when materials are torn (see Figure 4.4).

Materials are therefore described as having compressive, tensile or shear strength.

Materials that can withstand a high compression loading include cast iron, stone and brick, hence the common use of these materials for building purposes. However, these are brittle and will break if subjected to high tension. If a building needs to resist tensile strain – in an earthquake-prone area, for example – steel, which has high tensile strength, would be a more suitable building material.

Did you know?

The hardness scale is called Mohs' scale (see Table 4.5). It is named after the German mineralogist Friedrich Mohs (1773–1839), who developed the scale in 1812.

Scale (1 is the softest material; 10 is the hardest)	Material
1	Talc
2	Gypsum
3	Calcite
4	Fluorite
5	Apatite
6	Feldspar
7	Quartz
8	Topaz
9	Corundum
10	Diamond

Table 4.5: Mohs' hardness scale

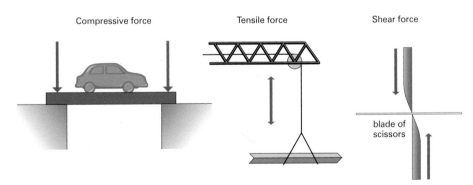

Figure 4.4: Compressive, tensile and shear forces

Elasticity and deformation

Almost all materials will stretch to some extent when a tensile force is applied to them. Table 4.6 illustrates some of the properties of metals. The increase in length on loading, compared to the original length of the material, is known as strain. As loading continues, a point is reached when the material will no longer return to its original shape and size when the material is removed. Permanent **deformation** occurs; the material is said to have exceeded its elastic limit or yield stress, and is suffering plastic deformation: it has been stretched irreversibly. Eventually, at maximum stress, the material reaches its breaking point – its ultimate tensile strength – and failure or fracture rapidly follows. This sequence is illustrated for four different materials below:

- Mild steel has little elasticity but has the highest yield stress of all the samples. It is fairly ductile, i.e. it has a large range over which it can sustain plastic deformation, and it has the highest ultimate tensile strength.

- Cast iron is brittle. It has the least elasticity of the four samples, and no ability to sustain plastic deformation, although its tensile strength is higher than that of concrete.

- Copper has little elasticity but is the most ductile of the four samples. It has an ultimate tensile strength less than half that of mild steel.

- Concrete has little elasticity, and the lowest tensile strength of the four samples.

Find out

What can we learn about the materials from the stress/strain diagram?

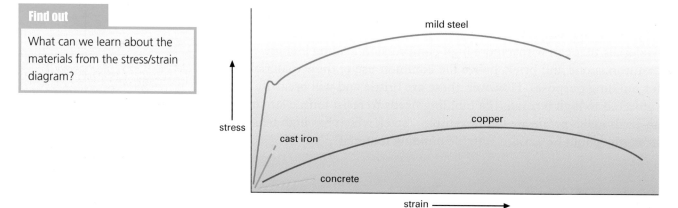

Figure 4.5: The effects of stress and strain on materials

Property	Description
Elasticity	The ability of a material to resume its normal shape after being stretched or squeezed
Plasticity	The exact opposite of elasticity – a material that is not able to return to its original shape when deformed
Ductility	The ability of a material to withstand distortion without fracture, such as metal that can be drawn out into a fine wire (plastic deformation under tension)
Durability	A material's ability to resist wear and tear
Fusibility	The melting point of a material, i.e. when a solid changes to a liquid
Malleability	The ability of a metal to be worked without fracture
Temper	The degree of hardness in a metal
Tenacity	A material's ability to resist being pulled apart
Thermal expansion	The amount a material expands when heated

Table 4.6: Properties of metals

Conduction

Conduction is the transfer of heat energy through a material. It takes place as a result of the increased vibration of molecules, which occurs when materials are heated. The vibrations from the heated material are passed on to adjoining material, which then heats up in turn. Some materials are better at conducting heat than others. For example:

- metals tend to be good conductors of heat
- wood is a poor conductor of heat.

Gases and liquids also conduct heat, but poorly. In the plumbing industry, conduction is principally between solids. Of the metals commonly used by plumbers, copper has a higher conductivity than steel, iron or lead. Wooden, ceramic and plastic materials, which are poorer conductors of heat, are known as **thermal insulators**. Pipework and cylinders must be insulated to prevent heat loss, for efficiency and environmental reasons.

Conduction also refers to the ability of a material to transfer electricity. Most metals are good conductors of electricity, but plastics are not. This will be covered in greater detail on page 185.

Find out
How good are the following materials as heat conductors? Plastic, rubber, ceramic, wool and carbon.

Why solid materials break down

If materials are not kept under the correct conditions they can decay and break down. Once these materials break down they can sometimes be unusable or less effective and in need of replacement. Some of the main reasons why solid materials break down are explained below.

Corrosion

Both acids and alkalis can cause **corrosion** and thereby damage plumbing materials. Metals are at particular risk.

The main causes of **corrosion** are:

- the effects of air
- the effects of water
- the direct effects of acids, alkalis and chemicals, e.g. from environmental sources
- electrolytic action.

Atmospheric corrosion

Pure air and pure water have little corrosive effect, but together in the form of moist air (oxygen + water vapour) they can attack **ferrous** metals such as steel and iron very quickly to form iron oxide or rust. The corrosive effects of rusting can completely destroy metal. This process is called **oxidisation**.

Various other gases (carbon dioxide, sulphur dioxide, sulphur trioxide), which are present in our atmosphere, can also increase the corrosive

The effects of corrosion on brass

Key term
Ferrous – metals that contain iron.

effect of air on particular metals; especially iron, steel and zinc. These gases tend to be more abundant in industrial areas as they are often waste products from various industrial processes.

Coastal areas also suffer from increased atmospheric corrosion due to the amount of sodium chloride (salt) from the sea, which becomes dissolved into the local atmosphere.

Non-ferrous metals, such as copper, aluminium and lead, have significant protection against atmospheric corrosion. Protective barriers (usually sulphates) form on these metals to prevent further corrosion. This protection is also known as **patina**.

Corrosion by water

Ferrous metals are particularly vulnerable to the effects of corrosion caused by water. These are commonly seen in central heating systems as black ferrous oxide and red rust build-up in radiators. A by-product of this process is hydrogen gas, which accumulates in the radiator, leading to the need for radiator to be 'bled' (see Unit 8).

In certain areas of highly acidic water, copper may become slightly discoloured. This won't affect the quality or safety of the drinking water. However, in areas where lead pipework is still in use, very soft (acidic) water may dissolve minute quantities of lead, thus contaminating the water. This can have potentially toxic effects, especially for children.

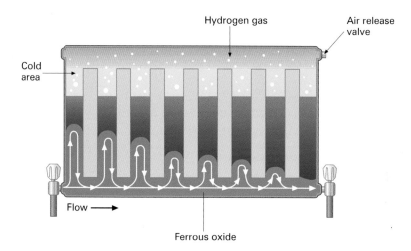

Figure 4.6: Corrosion effects in steel panel radiators

Corrosive effects of building materials and underground conditions

Some types of wood (such as oak) have a corrosive effect on lead, and latex cement and foamed concrete will adversely affect copper. Certain types of soil can damage underground pipework. Heavy clay soils may contain sulphates, which can corrode lead, steel and copper. Ground containing ash and cinders is also very corrosive as these are strongly alkaline; if pipes are to be laid in such ground they should be wrapped in protective material.

Electrolytic action and corrosion

Electrolytic action describes a flow of electrically charged ions from an **anode** to a **cathode** through a medium known as the **electrolyte** (usually water) as shown in Figure 4.7.

In chemistry, an electrolyte is any substance containing free ions that make the substance electrically conductive. The most typical electrolyte is an ionic solution, but molten electrolytes and solid electrolytes are also possible.

Electrolytic corrosion takes place when the process of electrolysis leads to the destruction of the anode. The length of time it takes for the anode to be destroyed will depend on:

- the properties of the water that acts as the electrolyte: if the water is hot or acidic the rate of corrosion will be increased
- the position of the metals that make up the anode and the cathode in the electromotive series.

The electromotive series

The list below shows the common elements used in the plumbing industry; the order in which they appear indicates their electromotive properties:

- copper
- tin
- lead
- nickel
- cadmium
- iron
- chromium
- zinc
- aluminium
- magnesium.

The elements higher up in the list will destroy those lower down through the process of electrolytic corrosion. The further apart in the list that the materials appear, the faster the corrosion will take place. For example, copper will destroy magnesium at a faster rate than lead will destroy chromium.

> **Key terms**
>
> **Anode** – negative electrode of every electrochemical current source while being discharged.
>
> **Cathode** – designation for the positive electrode of every electrochemical current source while being discharged.

> **Did you know?**
>
> A simple battery uses electrolysis to produce electricity.

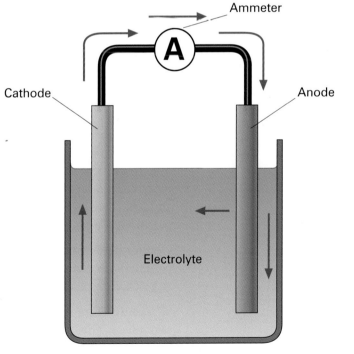

Figure 4.7: The process of electrolysis

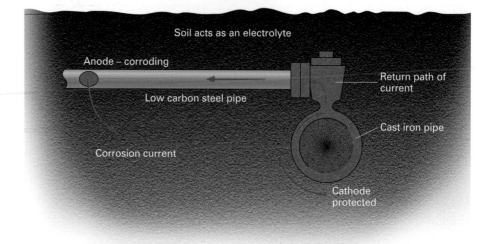

Figure 4.8: Example of possible electrolytic corrosion in pipework systems

Find out

Try to think of some practical examples of electrolytic corrosion.

As a plumber you will need to be aware of the potential for electrolytic corrosion when two very dissimilar metals such as a galvanised tube and a copper fitting are in direct metallic contact. If these metallic elements are then surrounded by water (of a certain type) or damp ground, a basic electrical cell is effectively created and electrolytic corrosion can take place.

Methods of preventing corrosion

The location and type of pipe used will determine the type of protection required from corrosion.

Plastic pipes are generally not affected by soil and water but if brass fittings were used they would need to be wrapped to prevent them corroding.

Low carbon steel, which is not galvanised, is not suitable for hot- and cold-water supplies and left unpainted would corrode both inside and out. To prevent this, you can use a galvanised pipe. Low carbon steel pipes run underground would need to be protected from the damp and soil there by wrapping it with a non-corrosive covering such as petrolatum tape.

Copper pipes and fittings need to be installed so they do not come into contact with cement. The use of a sleeve where pipes pass through holes in brickwork is recommended.

Where different materials are used in a heating system, corrosion can take place. This can be prevented by adding a corrosion inhibitor when you complete an installation.

UV damage to plastics

UV degradation is the decay of many natural and synthetic polymers from exposure to ultra-violet radiation. Products made from these materials may crack or break. Continuous exposure to sunlight is a more serious problem than intermittent exposure, since the problem depends on the extent and degree of exposure. The degradation can be prevented by adding anti-UV chemicals to the polymer when mixing the ingredients prior to shaping.

Heat damage to plastics

It may seem obvious that heat melts plastic. However, to prevent decomposition of the polymer during processing, where temperatures could reach 180 °C, manufacturers will add heat stabilisers to the compound. Once on site however, operations such as welding or soldering can damage plastic components and care should be taken, for example, by using a heat shield gel.

Principal applications and properties of liquids

Refrigerant

A refrigerant is a substance used in a heat cycle usually including, for greater efficiency, a reversible phase change from a gas to a liquid, as in a heat pump.

The ideal refrigerant has thermodynamic properties of a boiling point below the target temperature, a high heat of vaporisation, a moderate density in liquid form, a relatively high density in gaseous form, and a high critical temperature. Since boiling point and gas density are affected by pressure, refrigerants may be made more suitable for a particular application by choice of operating pressure.

Traditionally, fluorocarbons were used as refrigerants. However, these are no longer used because they affect the planet's ozone layer.

Natural refrigerants such as ammonia and non-halogenated hydrocarbons such as methane help preserve the ozone layer and are used in air-conditioning systems.

Antifreeze/glycol mixes

Antifreeze, much as the name suggests, is a freeze preventative used in internal combustion engines and other applications such as HVAC chillers.

The purpose of the antifreeze is to prevent enclosures from damage due to expansion when water turns to ice.

All antifreezes are glycol based, with both ethylene glycol (EG) and propylene glycol (PG) used as the antifreeze base. Both perform to similar levels. However, the major difference between EG and PG is toxicity levels, with the acute toxicity of PG, especially in humans, being substantially lower than that of EG.

Propylene glycol, like alcohol, is not toxic at low levels.

When water turns to ice, damage can occur

Safety tip

In applications where ingestion is a possibility, PG-based antifreeze is the sensible choice.

Fuel oils

Fuel oil is gained from petroleum distillation, either as a distillate or a residue. Fuel oil is generally said to be any liquid petroleum product that is burned in a furnace or boiler for the generation of heat or used in an engine for the generation of power. In this sense, diesel is a type of fuel oil.

Fuel oil is classified by six classes according to its boiling point, composition and purpose. Viscosity also increases with number, and the heaviest oil has to be heated to get it to flow. As examples, Number 1 is similar to paraffin and Number 2 is diesel and, consequently, is the same as heating oil.

Lubricants/greases

A **lubricant** is a liquid substance that we introduce between two moving surfaces to reduce the friction between them; thus improving efficiency and reducing wear. Some lubricants can also help remove foreign particles.

Typically lubricants contain 90% base oil derived from petroleum and less than 10% additives, where the additives deliver reduced friction and wear, increased viscosity and resistance to corrosion. Consequently, the most common use for lubricants is protecting the internal combustion engines in cars and equipment.

Grease is used to describe semi-solid lubricants and typically applies to a material consisting of a soap emulsified with mineral oil. The characteristic feature of greases is that they possess a high initial viscosity, which upon the application of shear, drops to give the effect of an oil-lubricated bearing of approximately the same viscosity as the base oil used in the grease.

Greases are used where a mechanism can only be lubricated infrequently and where a lubricating oil would not stay in position. They also act as sealants to prevent the ingress of water and incompressible materials. Grease-lubricated bearings have greater frictional characteristics due to their high viscosity.

Properties of water

Water is a chemical compound of two gases: hydrogen and oxygen (H_2O). It is formed when hydrogen gas is burned.

One of the most important properties of water is its solvent power. It can dissolve numerous gases and solids to form **solutions**. The purest natural water is rainwater collected in the open countryside. It contains dissolved gases such as nitrogen, oxygen and carbon dioxide, but this does not affect its **potability**.

Water may be classified as having varying degrees of hardness or softness.

Key term

Potability – suitability for drinking.

Behaviour of water at different temperatures

Matter, such as water, can exist in what is known as three different states:

Solid (ice) ⟶ Liquid ⟶ Gas (steam)

In moving from one state to another a **change of state** must take place, during which the water is either cooled or heated.

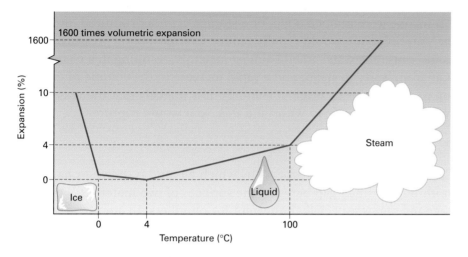

Figure 4.9: Change of state diagram – atmospheric pressure

Look at Figure 4.9, which shows the behaviour of water at atmospheric pressure. In its natural state between 0 °C to 100 °C water is in the form of a **liquid.** When heated during its liquid state the water will increase in volume by up to 4% (i.e. it will expand) and an increase in volume will result in a decrease in density.

In other words, the density of water will depend on its temperature. When cooled from room temperature, liquid water becomes increasingly dense, as do many other substances. When it reaches about 4 °C, pure water reaches its maximum density. However, as it is cooled further, it expands and actually becomes less dense and, upon freezing, the density of water has decreased by about 10%.

We can demonstrate this very easily, as the solid form of most substances is denser than the liquid phase. For example, if we take a block of most solids, they will sink in the liquid. However, do the same with a block of ice and it floats in liquid water because the ice is less dense.

The **freezing point** of water is 0 °C. At temperatures of 0 °C and below, water changes state, into ice, and expands. This expansion or increase in volume in an enclosed space can result in components rupturing, such as a burst pipe.

The **boiling point** of water is 100 °C. At temperatures of 100 °C and above, water under atmospheric conditions changes to steam. This causes a rapid increase in volume (up to 1600 times), which can have explosive effects if the water is stored in an enclosed space.

Did you know?

Hot water temperature must be very effectively controlled and maintained below 100 °C in order to avoid potentially serious explosions.

Water stored at above atmospheric pressure

The pressure at which water is stored is also linked to water temperature and water volume. If water is stored in an enclosed space (at a constant volume – such as in a storage cylinder) at above atmospheric pressure, the temperature at which it boils will rise above 100 °C. For example, the boiling point of water at 1 bar pressure is approximately 120 °C.

So why is water stored at above atmospheric pressure and above temperatures of 100 °C so dangerous if it is not boiling? Put simply, if someone were to open a tap, or a storage cylinder ruptured, then the water pressure would rapidly reduce to atmospheric pressure. This would cause the water to boil almost instantly, resulting in a change of state from water to steam and a rapid increase in the volume of the gas (up to 1600 times its original volume). It is unlikely that the vessel or system would be able to withstand this, so it would rupture.

Capillary action

Capillary action is the process by which a liquid is drawn or hauled up through a small gap between the surfaces of two materials. This phenomenon is especially important for plumbers. In a positive way, it helps when soldering, but it can also affect the way in which water can get into buildings, which can be damaging.

Forces of attraction

Surface tension

Surface tension describes the way in which water molecules 'cling' together to form what is effectively a very thin 'skin'. This can be demonstrated by filling a glass beaker right up to the top and examining the top of the glass. The water will appear above the upper limit of the glass – why doesn't it spill down the side?

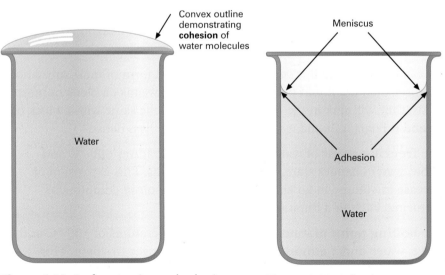

Figure 4.10: Surface tension and cohesion

Figure 4.11: Adhesion

The answer is because of the **cohesion** (water molecules 'sticking together') as a result of surface tension.

Adhesion

Adhesion is the force of attraction between water molecules and the sides of the vessel the water is contained in. This leads to the slightly curved 'skin' that appears when water is held in a vessel as shown in Figure 4.11. The correct name for this skin is a meniscus. The processes of adhesion and cohesion cause capillary action.

Practical examples of capillary action

Plumbers must consider the possible effects of capillary action when planning lead roofing work. Water can find its way into a building's lapped roof if materials are close together. Figure 4.12 shows how this occurs and how it can be remedied.

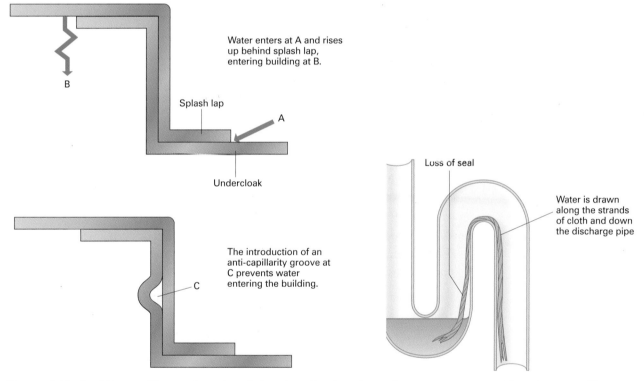

Water enters at A and rises up behind splash lap, entering building at B.

Splash lap

A

Undercloak

The introduction of an anti-capillarity groove at C prevents water entering the building.

C

Loss of seal

Water is drawn along the strands of cloth and down the discharge pipe

Figure 4.12: Avoiding capillary action when fixing sheet lead work

Figure 4.13: Trap seal lost due to capillary action

Capillary action can occur in 'S bend' drainage traps, which are found under sinks. If a length of waste material, such as a piece of dishcloth, becomes lodged in the 'S bend', capillary action could take place. This can lead to the loss of seal at the bottom of the trap, allowing bad smells from the drainage pipe to filter back into the home.

pH value and corrosion

The term **pH value** refers to the level of acidity or alkalinity of a substance. As a plumber, you will need to be particularly aware of the

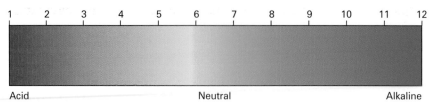

Figure 4.14: The pH scale of acidity

potential effects of water acidity or alkalinity upon materials, appliances and components.

Both acids and alkalis can cause **corrosion** and thereby damage plumbing materials. Metals are at particular risk.

All water has a 'pH value'. Rainwater is naturally slightly acidic, due to small amounts of carbon dioxide and sulphur dioxide in the atmosphere being dissolved into it, forming very weak carbonic and sulphuric acids. The pH value of groundwater is affected by the different rock types it passes though. For instance, water with dissolved carbonate from chalk or limestone is alkaline.

> **Did you know?**
>
> Pure water has a neutral pH of around 7.

> **Did you know?**
>
> Rainwater collected in towns contains higher percentages of dissolved substances such as soot and other pollutants from the atmosphere.

Water hardness

Hard water

Water is classified as hard if it is difficult to obtain a lather with soap. Hard water is created when it falls on ground containing calcium carbonates or sulphates (chalk, limestone and gypsum), which it has dissolved and taken into solution.

Soft water

Water is said to be soft when it is easy to produce a lather with soap. This is because of the absence of dissolved salts such as calcium carbonates and calcium sulphates.

Soft water can cause corrosion in plumbing components because it is relatively acidic.

Water hardness

Hard water is undesirable in domestic installations as it can produce limescale in pipework, heating equipment and sanitary appliances. This can lead to high maintenance costs. Hard water also requires the use of much more soap and detergent for washing purposes, as the 'hardness' makes it far more difficult to produce a lather.

Water hardness can be described as temporary or permanent:

- **Permanent hardness** is a result of ions of nitrates and sulphates. It makes it difficult to form a lather and cannot be removed by boiling.
- **Temporary hardness** is a result of the amount of carbonate ions in the water. Temporary hardness can be removed from the water by boiling, which results in the carbonate being precipitated out as

Limescale can build up in pipes and taps

limescale. This hard scale accumulates inside boilers and circulating pipes, restricting the flow of water, reducing the efficiency of appliances and components and ultimately causing damage and system failure.

Principal applications of gases used in the mechanical services industry

Air and steam

Compressed air is used mainly for cleaning components or testing pipework. It is available in canisters. Steam can be used for hot water heating and central heating but is only normally found in commercial and industrial premises.

LPG

LPG (liquefied petroleum gas) is mainly used for heating equipment in the plumbing industry. It is the fuel used in most blowlamps for soldering copper pipe and fittings. It is supplied in small cylinders or canisters. It can also be used in hot water and central heating boilers, fires and stoves. Gas for these is normally supplied in large cylinders or storage tanks.

Natural gas

Natural gas is used as a fuel for heating and cooking appliances and is normally supplied through the network of gas mains throughout the country. Natural gas is made up of 94% methane, 3% ethane and another 3% of a mixture of other constituents. It is also known by its chemical symbol CH_4.

Did you know?

Natural gas is odourless so an additive is used to enable us to smell it. This means it is detectable when it leaks. Otherwise, the gas could build up to dangerous levels without detection.

Carbon dioxide

Carbon dioxide is used for the freezing of water in pipes. It is supplied in a canister (which is normally disposable) or in re-chargeable cylinders. Caution needs to be taken when using carbon dioxide. It is also available in liquid form but this tends to be more for industrial and commercial pipework freezing. Carbon dioxide is a colourless gas with a chemical symbol CO_2.

Refrigerant gases

Probably the most widely used current application of refrigerants is the air-conditioning of private homes and public buildings, and refrigerators. It is unlikely that you will encounter these during your work as a domestic plumber. With the invention of synthetic refrigerants based mostly on a chlorofluorocarbon chemical (CFC), safer refrigerators are possible for home and consumer use.

Principal applications of gases used in the mechanical services industry

Pressure exerted by a gas

Solids exert pressure only in downward direction and liquids exert pressure downward as well as to the sides. However, gases exert pressure in all directions (a good example is a balloon). This pressure is due to bombardment of the particles against the walls of the container.

Charles's Law and Boyle's Law

Charles's Law describes the direct relationship between temperature and volume of a gas. Assuming that pressure does not change, a doubling in absolute temperature of a gas causes a doubling of the volume of that gas. A drop of absolute temperature sees a proportional drop in volume. The volume of a gas increases by $\frac{1}{273}$ of its volume at 0 °C for every degree Celsius that the temperature rises.

To explain why this happens, we need to explore temperature and volume in terms of gases. Temperature is an average of molecular motion. This means that, while all of the gas molecules are moving around their container in different directions at different speeds, they will have an average amount of energy that is the temperature of the gas. The volume of the gas is the size of its container because the molecules will move in a straight line until they impact something (another molecule or the container). However, to move as they do, the molecules require kinetic energy, which is measured by temperature.

Volume and temperature are very closely related. If the temperature was not sufficient, the molecules would not be able to overcome the weak forces of attraction among them and would not be able to fill the container.

Charles's Law must be used with the Kelvin temperature scale. This scale is an absolute temperature scale. At 0 °K, there is no kinetic energy (Absolute Zero). According to Charles's Law, there would also be no volume at that temperature. This condition cannot be fulfilled because all known gases will liquify or solidify before reaching 0 °K. The Kelvin temperature scale is Celsius minus 273.15°. Therefore, zero Kelvin would be –273.15 °C and any Celsius temperature can be converted to Kelvin by adding 273.15 (273 is often used).

Boyle's Law (sometimes referred to as the **Boyle-Mariotte Law**) is one of many gas laws and a special case of the ideal gas law. Boyle's Law describes the inversely proportional relationship between the absolute pressure and volume of a gas, if the temperature is kept constant within a closed system. The law was named after chemist and physicist Robert Boyle, who published the original law in 1662.

Remember

Any unit of volume will work with Charles's Law, but the most common are litres (dm³) and millilitres (cm³).

The equation for Boyle's law is

$$pV = k$$

where p = pressure
V = volume
k = constant

Therefore, for a fixed amount of an *ideal gas* kept at a fixed temperature, p (pressure) and V (volume) are inversely proportional (when one doubles, the other halves).

Heat pump and refrigeration cycle

You will typically come across two types of heat pump: ground and air.

A ground source heat pump works by absorbing heat from the ground and then raising its temperature; normally for use in the home and an air source heat pump does the same from the air. The operating principle using the ground source version can be demonstrated as follows.

Brine circulates in a closed loop known as the collector coil, which is buried in the ground and absorbs the heat energy from the ground. Inside the heat pump, the collector coil is wound around a heat exchanger (the evaporator) and the lukewarm brine in the collector coil begins to warm the ice-cold refrigerant in the heat exchanger.

This refrigerant has a very low boiling point, and therefore the brine in the collector causes its temperature to rise by a few degrees, which in turn is enough to cause it to boil and evaporate. If it helps, think of a boiling kettle which, as it boils, is moving liquid into vapour.

The evaporated refrigerant now moves into a compressor that compresses it. This compression raises pressure, which causes the refrigerant to rise in temperature (to about 50 °C) and it then passes into another heat exchanger (the condenser) and condenses. Think of the kettle again; the boiling steam vapour condenses back to liquid.

Another closed loop (the distribution system out to the radiators) is wound around the condenser causing the generated heat in the condenser to transfer from it to the closed loop feeding the radiators. From here it is sent out hot to the radiators, cools as it is passes through them and returns to be heated by the condenser again.

The condensing refrigerant circulates from the heat exchanger and into an expansion valve that lowers the pressure and the refrigerant becomes cold once again. The process then restarts when the circulating refrigerant meets the warm brine in the collector coil.

Progress check

1. At what temperature is water at its maximum density?
2. What is the relative density of water and of air?
3. How many types of copper tube are commonly available for plumbing and heating work? Which types of tube are not suitable for underground use?
4. How many types of low carbon steel (LCS) are commonly available for plumbing and heating work?
5. Which colours apply to which weights, when referring to LCS tubes?
6. What are the two main categories of plastic used in the plumbing industry?
7. According to Mohs' scale, which natural material has the greatest hardness?
8. What are the most common forms of corrosion found to affect plumbing materials and systems?
9. Is copper higher or lower than iron in the electromotive series?
10. When water is heated in its liquid state, by approximately how much does it expand?
11. What is the pH value of pure water?
12. What effect does boiling have on the permanent hardness of water?

3. Know the relationship between energy, heat and power in the mechanical services industry

Difference between heat and temperature

The main difference between heat and temperature is that heat is recognised as a unit of energy, measured in joules (J). Therefore:

- Temperature is the degree of hotness of a substance.
- Heat is the amount of heat energy (J) that is contained within a substance.

For example, imagine an intensely heated short length of wire and a bucket of hot water:

- The wire has a temperature of 350 °C.
- The water has a temperature of 70 °C.

The wire is far hotter, but contains less heat energy. The water would obviously take longer to heat, although to a lower temperature than the wire. Therefore, more heat energy would be required.

Measuring temperature

The SI unit of temperature measurement is the degree Kelvin (°K) but the unit you will deal with most frequently is degrees Celsius (or centigrade), written as °C.

Temperature is measured using thermometers. There are many types, but the most common depend upon the expansion of either a liquid or a **bi-metallic strip**.

The two most common liquids used in thermometers are alcohol and mercury. These liquids are used because they expand at a uniform rate when exposed to heat.

Bi-metallic strips work on the principle of thermal expansion and contraction (covered in more detail later) and the fact that some metals expand and contract at a faster rate than others. Bi-metallic strips are frequently used in thermostats, where they will bend (see Figure 4.15) when a particular temperature is reached. This process will break an electrical circuit and turn off the heating.

> **Did you know?**
>
> Examples of the bi-metallic strip may be found in many thermostats, such as those used in ovens.

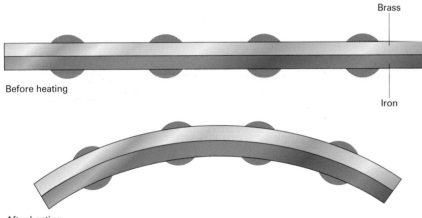

Brass

Before heating

Iron

After heating

Figure 4.15: A bi-metallic strip. The brass expands more than the iron on heating, thus bending the strip

Simple temperature measuring devices

Simple thermometer

In its simplest form a thermometer is a glass tube containing a substance such as alcohol or mercury that expands or contracts at a constant rate in response to temperature changes. A metal bulb is sited at one end of the thermometer for taking the temperature or immersing in water.

The most common simple thermometer used in the plumbing industry is the waterproof pen type. The pen-like probe is placed in the water and the LCD screen will show the temperature. This type of thermometer can be used for checking the water temperature at taps or outlets.

Clip-on thermometer

The clip-on, or clamp-on, thermometer, often called a pipe thermometer, is used for measuring the surface temperature of pipes. The thermometer uses the bi-metallic strip principle as its method of operation. This type of thermometer can be used when commissioning central heating systems: two clip-on thermometers are used to measure the temperature of both the flow and return pipework in order to establish the correct water flow through each.

Differential digital thermometer

This device comes with a range of attachments that can be used to measure simple water temperatures by immersing a probe in the water stream, or by connecting a device known as a **thermistor** to pipe surfaces. When commissioning central heating systems two thermistors are attached by means of a strap to the flow and return pipework. The digital thermometer can measure the temperature of each pipe and the difference in temperature between each pipe.

Figure 4.16: A type of simple thermometer

> **Key term**
>
> **Thermistor** – a device that changes electrical resistance in response to a temperature change.

Changes of state

Molecules are always in a state of rapid motion, but when they are densely packed together, this movement is restricted and the substance formed by these molecules is solid. When the molecules of a substance are less tightly bound, there is a great deal of free movement and the substance is a liquid. Finally when the molecule movement is almost unrestricted, the substance can expand and contract in any direction and is a gas.

Latent heat and sensible heat

Latent heat is the amount of energy released by a substance during a change of state. The name latent means the energy is hidden. In other words it is 'hidden' in the substance and is only realised when heat or pressure causes a change of state.

Figure 4.17: Differential digital thermometer

Sensible heat is the increase in temperature (or energy released) by a substance before it changes state. The rise in temperature is directly related to the amount of heat or pressure the substance is subjected to.

Specific heat capacity

To size various plumbing components, such as boilers and radiators, plumbers need to be able to understand the concept of heat. Heat is different from temperature and is a measure of the amount of energy in a substance. The standard unit of measurement of heat is the joule.

In order to work out the amount of heat required to heat a substance we need to be able to measure the amount of heat required over time or the **power** required. This is a measure of the energy divided by the time taken to heat the substance and is expressed in in kW/h.

$$1 \text{ kW/hr} = \frac{1000 \text{ joules}}{1 \text{ second}} \times 3600 \text{ seconds (number of seconds in one hour)}$$

In order to be able to undertake plumbing calculations involving heat we usually need to work out the amount of energy required to raise a quantity of a substance such as water from one particular temperature to another. To do this we need to know the substance's **specific heat capacity**.

The specific heat capacity of a substance is the amount of heat required to raise the temperature of 1 kg of a substance by 1 °C. The specific heat capacity of water is 4.186 kJ/kg °C. Table 4.7 shows the specific heat values for a range of materials.

Example:

Calculate the heat energy and power required to raise 200 litres of water from 10 °C to 60 °C, assuming 1 litre of water weighs roughly 1 kg.

$$\text{Heat energy} = 200 \text{ litres} \times 4.186 \text{ kJ/kg °C} \times (60 \text{ °C} - 10 \text{ °C})$$
$$= 41.860 \text{ kJ}$$

Power required to heat the water in 1 hour (assuming no energy is lost):

$$\frac{41,860}{3600} = 11.63 \text{ kW}$$

The power calculation is essential for determining factors such as the amount of energy required to re-heat a hot-water storage cylinder within a specific period of time. For example, if our previous power calculation had required the water to be reheated in a 30-minute time period, then the power required would be doubled as the re-heating period has been halved (from 3600 seconds to 1800 seconds).

$$\frac{41,860}{1800} = 23.26 \text{ kW}$$

Thermal expansion and contraction

Most materials expand when heated. This is because all substances are made up of molecules (groups of atoms), which move about more vigorously when heated. This causes the molecules to move further apart from each other – resulting in the material taking up more volume.

As the material cools, the molecules slow down and move closer together; thus the material gets smaller or contracts. The amount that the material expands in length when heated can be calculated using the following formula:

length (m) × temperature rise (°C) × coefficient of linear expansion

Table 4.8 shows the coefficient values for some of the most common materials used in the plumbing industry.

Material	kJ/kg°C
Water	4.186
Aluminium	0.887
Cast iron	0.554
Zinc	0.397
Copper	0.385
Lead	0.125
Mercury	0.125

Table 4.7: Specific heat values

Find out

What do you think happens when materials are cooled down?

Material	Coefficient °C
Plastic	0.00018
Zinc	0.000029
Lead	0.000029
Aluminium	0.000026
Tin	0.000021
Copper	0.000016
Cast iron	0.000011
Mild steel	0.000011
Invar	0.0000009

Table 4.8: Coefficients of expansion

Safety tip

Plastics tend to expand the most. We often need to leave an expansion gap in the pipework system when working with plastics to prevent the material from failing.

Example:

Calculate the increase in length of a 6 m long plastic discharge stack due to a temperature rise of 19 °C.

Use the following formula:

length (m) × temperature rise (°C) × coeffcient of linear expansion

$$6 \times 19 \times 0.00018 = 0.02052 \text{ m or } 20.52 \text{ mm}$$

We need to take this expansion and contraction into account in plumbing systems that are constantly being subjected to heating and cooling processes. If we don't, the system or component may break down, causing leakage or other problems.

Heat transfer

As a plumber you will need a good understanding of the methods of heat transfer because you will be dealing with the effects of this process on a daily basis. There are three methods of heat transfer:

1. Conduction (covered on page 153)
2. Convection
3. Radiation.

Convection

Convection is the transfer of heat by means of the movement of a locally heated **fluid** substance (usually air or water). As a fluid is heated, it expands, which causes a lowering of density. The less dense warm fluid begins to rise, and is replaced by cooler, denser fluid from below. Eventually, convection currents are set up, which allow a continuous flow of heat upwards from the source.

Examples of systems that use convection currents for heat transfer are:

- convector heaters, which warm the air at one place in a room; the resulting convection currents transport the heat around the room
- domestic hot-water systems, which depend on convection currents to transfer heat from an immersion heater (similar to the 'element' in an electric kettle) to the rest of the water in the hot tank (cylinder).

It is easy to demonstrate the 'updraught' part of a convection current by hanging a piece of light material above a convector heater. The movement of the hanging material will clearly show the presence of rising currents of warm air. Figure 4.18 shows a practical example of the use of convection currents in domestic hot-water systems.

Radiation

Radiation is the transfer of heat from a hot body to a cooler one without the presence of a material medium (other than air), by means of 'heat' waves. Heat radiation can be felt as the 'glow' from a fire or the heat from the sun.

Heat radiation is better absorbed by some materials than others; in these instances colour can be an important factor. Dull, matt surfaces will absorb radiated heat more efficiently than shiny, polished surfaces.

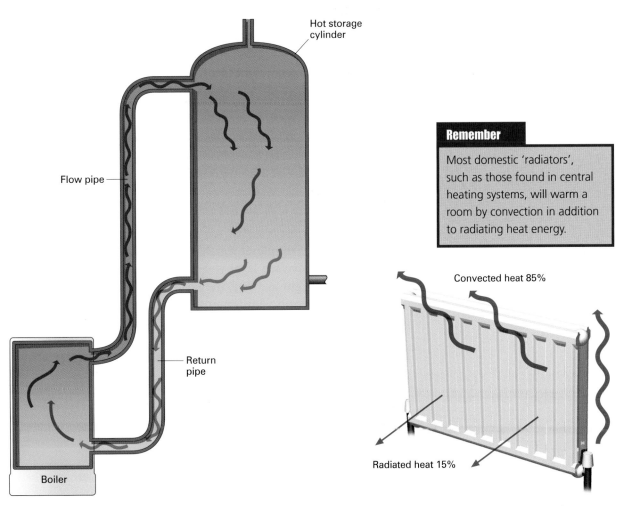

Figure 4.18: Gravity circulation taking place owing to the effects of convection

Figure 4.19: Heat production from panel radiator

Remember

Most domestic 'radiators', such as those found in central heating systems, will warm a room by convection in addition to radiating heat energy.

Progress check

1. Why are alcohol and mercury used in thermometers?
2. With regard to a central heating system, what would a differential digital thermometer be used for?
3. What is the specific heat capacity of water?
4. Calculate the increase in length of a 9-metre long plastic gutter, due to a temperature rise of 17 °C.
5. What are the three modes of heat transfer?

4. Know the principles of force and pressure and their application in the mechanical services industry

Force and pressure

The difference between weight (a force) and mass is a very important concept.

Mass

Mass is simply the amount of matter contained in an object. Assuming we do not cut or change this object, its mass will stay the same wherever we are. The unit of mass is the kilogram (kg).

Weight

The weight of an object is defined as the force exerted by its mass and gravitational pull. On planet Earth this can vary according to where we are (the higher above sea level you go, the less you weigh), and although the changes in weight are tiny, they can be measured with very sensitive and expensive scientific equipment. The unit of weight, and force in general, is the newton (N).

The difference between mass and weight

To give an example of the difference between mass and weight we need to go to the moon.

On Earth:

- The gravitational pull of the Earth is 9.8 m/s^2.
- An object with a mass of 1 kg would weigh 9.8 N on Earth.

On the moon:

- The gravitational pull of the moon is approximately 1.633 m/s^2.
- An object with a mass of 1 kg would weigh 1.633 N on the moon.

The mass of the object does not change, whether it is on the Earth or the moon, but the weight of the object changes considerably because of the reduced gravitational pull of the moon.

Expressed as a formula, we can say: Weight = Mass × Gravitational Pull.

Back on Earth, if we disregard the effect of height above sea level, the weight acting on 1 kg of mass is equal to 9.81 newtons (N). So, we can say that 1 kg weighs 9.81 N.

Force

Using this knowledge, we can see that force is a push or pull that acts on an object. If the force is greater than the opposing force, the object will change motion or shape. Obvious examples of forces are gravity and the wind. Force is measured in newtons.

Did you know?

In many situations 9.81 N can be rounded up to 10 N, but only do this in an exam if the question tells you to.

The presence of a force is measured by its effect on a body. For example, a heavy wind can cause a stationary football to start rolling.

Equally, gravitational force will cause objects to fall towards the Earth. Therefore, a spring will extend if we attach a weight to it, because gravity is acting on the weight.

As the force of gravity acts on any mass, such a mass tends to accelerate and exert a force that depends upon the mass and the acceleration due to gravity. This acceleration due to gravity is agreed worldwide as being 9.81 m/s^2 at sea level and, therefore, a mass of 1 kg will exert a force of 9.81 N.

Linking our knowledge together and expressed as a formula, this is:

$$\text{Force (N)} = \text{Mass} \times \text{Acceleration}$$

Work

If an object is moved, then work is said to have been done. The unit of work done is the joule. Work done is the relationship between the effort (force) used to move an object and the distance that the object is moved. Expressed as a formula, this is:

$$\text{Work done (J)} = \text{Force (N)} \times \text{Distance (m)}$$

Example:

A combination boiler has a mass of 50 kg. How much work is done to move it 10 m?

$$\begin{aligned}
\text{Work} &= \text{Force} \times \text{Distance} \\
&= (50 \times 9.81) \times 10 \\
&= 490.5 \times 10 \\
&= 4905 \text{ J}
\end{aligned}$$

Energy

Energy, measured in joules, is the ability to do work, or to cause something to move or the ability to cause change. Machines cannot work without energy and we are unable to get more work out of a machine than the energy we put into it. This is due mainly to **friction**. Friction occurs when two substances rub together. Try rubbing your hands together. Did you feel them get warmer?

The work produced (output) is usually less than the energy used (input). Energy can be transferred from one form to another, but energy cannot be created or destroyed. The loss of energy by friction usually ends up as heat.

There are two types of energy:

1. Potential Energy (Energy of Position or Stored Energy)
2. Kinetic Energy (Energy due to the motion of an object).

There are many forms of energy, such as: solar, electrical, heat, light, chemical, mechanical, wind, water, muscles and nuclear.

Remember

In calculations, it is often assumed for ease that the value of acceleration is taken as 10 m/s^2

Remember

Do not assume that the acceleration due to gravity is equal to 10 m/s^2. Only take it as 10 m/s^2 if you are told to do so in a question, or if you are doing a rough calculation for your own purposes.

Key term

Friction – force that opposes motion.

Pressure

Pressure is defined as force applied per unit area and is measured in newtons per square metre (N/m^2), a unit also known as a pascal (Pa).

You will probably come across other terms used to identify pressure such as the 'bar' or 'pounds per square inch' (lbs/in^2). These can be expressed as:

- 1 bar = 100,000 N/m^2
- 1 lbs/in^2 = 6,894 N/m^2.

Pressure is therefore a measurement of a concentration of force. The effect of a concentration of pressure can be seen if water flowing through a pipe is forced through a smaller gap by reducing the diameter of the pipe. (Think about a hosepipe, and how to maximise the force of the jet of water.)

Pressure can also be lowered by 'spreading' the applied force over a wider area. For example, rescue teams will often spread themselves over fragile roofs or on thin ice to minimise the chance of the surface giving way.

As a plumber you will need to have a basic understanding of the effects pressure has on the pipes and fittings you will install. The internal pressure in a pipe or vessel will be affected by what is being transported (water or gas), and must be considered when deciding which material and which size of pipe or vessel should be used.

Pressure in liquids

The pressure in a liquid increases with depth, so in a plumbing system water pressure is higher at the lowest points of the system and lower at the highest points of the system.

Water pressure is measured using a number of different units:

- Metres head (m)
- The pascal (Pa), also the newton, per metre squared (N/m^2)

Did you know?

Pressure is exerted by solid objects in a downwards direction only; liquids, however, exert pressure downwards and outwards.

Figure 4.20: The effects of pressure in solids and liquids

- Bar pressure (bar)
- psi (pounds per square inch).

1 metre head = approx. 10,000 Pa (10 kPa) = approx. 0.1 bar

An understanding of the **pressure head** created in systems is important in determining component sizes (for example, pipe sizes) and confirming that components will be able to withstand the pressure of water created within them.

Intensity of pressure and total pressure

Intensity of pressure is the force created (kPa) by the weight of a given mass of water acting on a unit area (m²).

Total pressure is the intensity of pressure multiplied by the area acted on.

Example 1:

Calculate the intensity of pressure and total pressure acting on the base of a lower boiler as shown in Figure 4.21.

$$\text{Intensity of pressure} = \text{head} \times 9.81 \text{ kPa}$$
$$= 3 \times 9.81$$
$$= 29.43 \text{ kPa or approximately } 0.3 \text{ bar pressure.}$$

An alternative method of calculating this is to multiply the head by 0.1 bar (0.1 bar = 1m head):

$$3 \times 0.1 \text{ bar} = 0.3 \text{ bar pressure}$$

$$\text{Force acting on boiler base} = \text{intensity of pressure} \times \text{area of base}$$
$$= 28.43 \times (2 \times 15)$$
$$= 88.29 \text{ kN}$$

Example 2:

If a tap is sited 5 metres below a plumbing cistern feeding it, the pressure created at the tap will be:

$$5.0 \text{ metres head} \times 0.1 \text{ bar pressure} = 0.5 \text{ bar}$$

Atmospheric pressure

The pressure exerted by the weight of the Earth's atmosphere pressing down on the ground varies depending on height above sea level. The pressure at the top of Mount Everest is not as high as the pressure in the bottom of a valley below sea level (such as the Great Rift Valley in Africa). The pressure at sea level is 101 325 N/m² (approximately 1 bar).

The siphon

The siphon uses atmospheric pressure to operate. A quantity of water is forced through a short leg of pipe (Column A) and over the crown of the siphon. The weight of water in Column B (the longer leg) is greater than in Column A. The water continues to flow owing to the difference in the weight of water with positive (atmospheric) pressure in Column B and negative (suction) pressure in Column A.

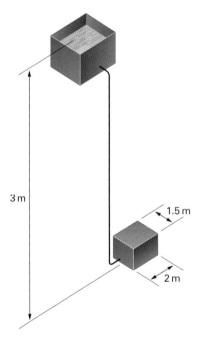

3 m

1.5 m

2 m

Figure 4.21: Pressure calculation

> **Remember**
>
> Plumbers must be aware of the effects of atmospheric pressure to ensure that they avoid creating 'negative' pressure or vacuums within pipework systems. Negative pressure can damage components.

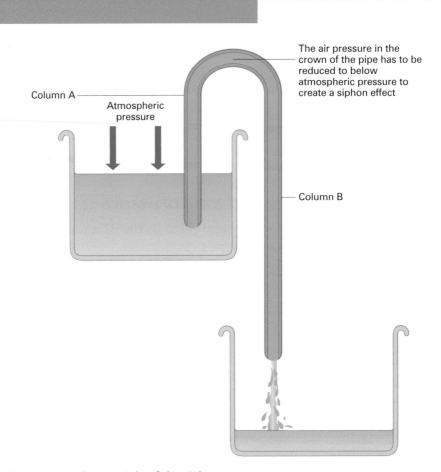

Column A

Atmospheric pressure

The air pressure in the crown of the pipe has to be reduced to below atmospheric pressure to create a siphon effect

Column B

Figure 4.22: The principle of the siphon

The principle of the siphon is used to good effect in plumbing applications.

Examples include:

- siphonic WC pan
- WC-flushing siphon
- siphoning the contents of a hot-water storage cylinder using a hose pipe.

Flow rate

Flow rate means how much of something is passing by a certain point in a given time. Both gas and liquid flow can be measured in volumetric or mass flow rate. There are various measures used to express this. The most common volumetric ones being cubic metres per second (m^3/s), litres per second (l/s) or, for mass, we tend to use kilograms per second (kg/s).

The rate of water flow through pipework is affected by friction. This is an important consideration when designing pipework installations. To get an idea of how friction works, compare the speed of a car driving freely through a tunnel with the speed of a car that is in contact with the tunnel walls. As you can imagine, the result of rubbing on the wall will cause a considerable reduction in speed.

Pipe walls have the same effect on water flow: the water is 'rubbing' on the pipe. This friction is further increased if the inside surface of the pipe wall is rough. This effect is known as frictional resistance or frictional loss.

Comparison of flow in smooth and rough pipework

Water flowing through a 10 m length of 25 mm diameter pipe would flow at approximately 22 litres/minute for copper and plastic pipes, which have a smooth internal surface.

Water flowing through a 10 m length of 25 mm diameter pipe would flow at approximately 18 litres/minute for galvanised LCS pipe, which has a rougher inner surface.

The size of pipes will obviously have an effect on the rate of flow. The use of knuckle bends, or elbows, provides a greater resistance to the flow of water than the use of manufactured machine bends. The use of certain types of valves, such as stop valves, creates resistance to the flow of water in a pipeline. The valve features reduced dimensions/minor obstructions within its design.

Remember

Frictional resistance is also increased by the number of fittings in a pipework run.

Relationship between velocity, pressure and flow rate

'The amount that comes out depends on how hard I push and what gets in the way.' Bear this in mind when considering the following formula describing the flow of a fluid in a pipe:

$Q = A \times v$

where:

Q is flow rate

A is the cross-sectional area (CSA) of the pipe

v is the average fluid velocity in the pipe.

To clarify, if we have a fluid travelling at an average velocity of 1 metre per second through a pipe with a csa of $1m^2$, then we can say that the flow rate of that fluid will be 1 cubic metre per second. Bear in mind that as Q is giving a flow rate in m^3/s it is therefore expressing volumetric flow rate, in other words the volume of fluid passing by per second.

Mass flow rate represents the amount of mass that passes by per second and we use the following formula to calculate this:

$W = Q \times r$

where:

W = the mass flow rate

Q = the volumetric flow rate

r = liquid density (*rho*)

So, if we take a pipe with water in it and it is lying horizontally on the ground, although the water is not coming out of one end of the pipe, there is still pressure acting on the water. We call this static pressure (no movement).

If we now lift one end of the pipe into the air, we have a difference in height between the top and bottom, and we refer to this vertical distance as the head. This can be measured in either height or pressure. This is called dynamic pressure (movement) acting on the water.

The water will now flow down the pipe at a velocity determined by the head and by the CSA of the pipe. However, the water will be slightly restricted in its ability to come out of the other end by any obstacles or resistances in its way.

One such obstacle would be the fittings on the pipework and one such resistance would be the internal surface of the pipe. If the inside of the pipe is extremely rough, then there is more friction between the water and the pipe sides and this restricts flow rate.

We can therefore say that if we increase the pressure then we are pushing harder and therefore more water can come out. This is classed as being directly proportional, i.e. one factor increasing causes the other factor to increase.

However, if we increase the resistance to flow (by having greater friction, more fittings or reducing pipe size) then less water comes out. This is classed as inversely proportional, i.e. one factor increasing causes the other to decrease.

In terms of a relationship with velocity, all things being equal, an increase in head (lifting one end of the pipe higher) will result in an increase in velocity. In other words, the two are directly proportional.

Working life

A copper tube carries high-temperature hot water from one part of a factory to the other. The pipe is well insulated, but there is evidence of damage to the surrounding brickwork where it passes through a wall. The pipe is 50 metres in one straight length; it runs at high level and is adequately clipped throughout. The pipe varies in temperature by as much as 50 °C at any time during the process.

Matt is a Level 3 apprentice carrying out some plumbing maintenance at the factory and is asked by the site factory manager what the problem could be.

- How is the damage to the brickwork being caused?
- What should be done to the wall to alleviate the problem in future?
- Calculate how much the pipe would increase in length on each temperature change.
- What steps should Matt take to remedy the problem?
- Is copper tube the best option for this particular usage, or which alternative could be offered?

Progress check

1. What units is force measured in?
2. Define the term 'pressure'.
3. In a plumbing system where is the greatest head of pressure: higher at the lowest points or higher at the highest points?
4. Which actions, components and appliances use the principle of the siphon in plumbing systems?
5. The rate of water flow through pipework is affected by what?
6. Why is it good practice to use manufactured bends in pipework rather than knuckle bends or elbows?

5. Know simple mechanical principles and their application in the mechanical services industry

Principles behind simple machines

A simple machine is a device that helps us to perform our work more easily when a force is applied to it. A screw, wheel and axle and lever are all simple machines.

A machine also allows us to use a smaller force to overcome a larger force. Machines can help us change the direction of the force and work with a faster speed. The most common simple machines are covered below.

Levers

Levers allow us to use a small force to apply a larger force to an object. They are grouped into three classes, depending on the position of the fulcrum (the pivot).

Class 1

The fulcrum is between the force and the load, like a seesaw.

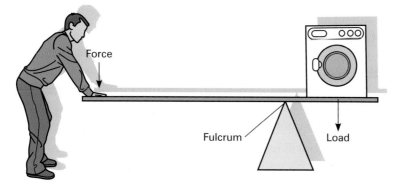

Figure 4.23: Class 1 lever

Class 2

The fulcrum is at one end, the force at the other end, and the load is in the middle. A wheelbarrow is a good example.

Class 3

The fulcrum is at one end, the load at the other end and the force in the middle, like a human forearm.

A small force at a long distance will produce a larger force close to the pivot:

$$10 \times 2 = F \times 0.5$$
$$F = \frac{10 \times 2}{0.5} \quad \frac{20}{0.5} = 40 \text{ N}$$

> **Remember**
>
> To make any simple machine work for us, we need to apply a force on it.

> **Did you know?**
>
> In medieval times, siege engines were used to hurl rocks at enemy castles. A siege engine is simply a Class 1 lever.

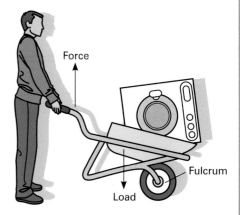

Figure 4.24: Class 2 lever

Figure 4.25: Class 3 lever

Gears

Gears are wheels with teeth; the teeth of one gear fit snugly into those around it. You can use gears to slow things down or speed them up, to change direction, or to control several things at once. The gears, when placed together, have to be fitted into the teeth. Each gear in a series changes the direction of rotation of the previous gear. The smaller gear will always turn faster than the larger gear and, in doing so, turns more times.

Figure 4.26: Gears

The inclined plane

The inclined plane is the simplest machine of all, as it is basically a ramp or sloping surface. The shortest distance between two points is generally taken as a straight line, but it is easier to move a heavy object to a higher point by using stairs or a ramp. If you think of the height of a mountain, the shortest distance is straight up from the bottom to the top. However, we always build a road on a mountain as a slowly winding inclined plane from bottom to top.

As a plumber, you will use the inclined plane most days in the form of a screw, which is simply an inclined plane wound around a central cylinder.

The inclined plane works by saving effort but you must move things a greater distance.

Pulleys

A pulley is made with a rope, belt or chain wrapped around a wheel and can be used to lift a heavy object (load). A pulley changes the direction of the force, making it easier to lift things. There are two main types of pulleys: the single fixed pulley and the moveable pulley.

A **single fixed pulley** is the only pulley that uses more effort than the load to lift the load from the ground. The fixed pulley, when attached to an unmoveable object, e.g. a ceiling or wall, acts as a first-class lever with the fulcrum being located at the axis but with a minor change: the bar becomes a rope. The advantage of the fixed pulley is that you do not have to pull or push the pulley up and down. The disadvantage is that you have to apply more effort than the load.

A **moveable pulley** is one that moves with the load. The moveable pulley allows the effort to be less than the weight of the load. The moveable pulley also acts as a second-class lever. The load is between the fulcrum and the effort.

There are many combinations of pulleys. The most common is the block and tackle, which use the two main types as its principle of operation.

We will look at examples of the two main types of pulley to understand their operating principles. Imagine that you have the arrangement of a 20 newtons (N) weight suspended from a rope, but actually resting on the ground as shown Figure 4.27.

Example 1:

In this example, if we want to have the load suspended in the air above the ground, then we have to apply an upward force of 20 N to the rope in the direction of the arrow. If the rope was 3 m long and we wanted to lift the weight up 3 m above the ground, we would have to pull in 3 m of rope to do it.

Now imagine that we add a single fixed pulley to the scenario, as shown in Figure 4.27. We have not really changed anything in our favour. The only thing that has changed is the direction of the force we have to apply to lift the load. We would still have to apply 20 N of force to suspend the load above the ground, and would still have to reel in 3 m of rope in order to lift the weight 3 m above the ground. This type of system gives us the convenience of pulling downwards instead of lifting.

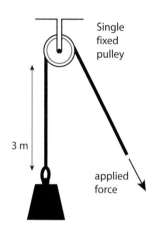

Figure 4.27: A single fixed pulley

Example 2:

Figure 4.28 shows the arrangement if we add a second, moveable pulley. This new arrangement now changes things in our favour because effectively the load is now suspended by two ropes rather than one. That means the weight is split equally between the two ropes, so each one holds only half the weight, or 10 N. This means that if you want to hold the weight suspended in the air, you only have to apply 10 N of force (the ceiling exerting the other 10 N of force on the other end of the rope). However, if you want to lift the weight 3 m above the ground, then you have to reel in twice as much rope – i.e. 6 m of rope must be pulled in.

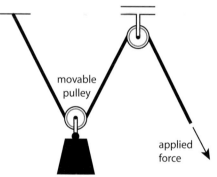

Figure 4.28: A moveable pulley

The more pulleys we have, the easier it is to lift heavy objects. As rope is pulled from the top pulley wheel, the load and the bottom pulley wheel are lifted. If 2 m of rope are pulled through, the load will only rise 1 m (there are two ropes holding the load and both have to shorten by the same amount).

With pulley systems, to calculate the effort required to lift the load, we divide the load by the number of ropes (excluding the rope connected to the effort). Figure 4.30 shows a four pulley system, where the person lifting the 200 kg mass or 2000 N load (remember 1 kg weighs 10 N) has to exert a pull equal to only 500 N (i.e. 2000 N divided by four ropes).

Mechanical advantage

All these machines enable us to increase our ability and gain an advantage over nature. This is a relationship between the effort needed to lift something (input) and the load itself (output) and we call this ratio the **mechanical advantage**. Consequently, when a machine can put out more force than is put in, the machine is said to give a good mechanical advantage. Mechanical advantage can be calculated by dividing the load by the effort. There are no units for mechanical advantage; it is just a number.

$$\text{Mechanical Advantage (MA)} = \frac{\text{Load}}{\text{Effort}}$$

Using Figure 4.29, what is the mechanical advantage of the pulley system?

$$\text{MA} = \frac{\text{Load}}{\text{Effort}} = \frac{2000 \text{ N}}{500 \text{ N}} = 4$$

In a lever, an effort of 10 N is used to move a load of 50 N. What is the mechanical advantage of the lever?

$$\text{MA} = \frac{\text{Load}}{\text{Effort}} = \frac{50}{10} = 5$$

This effectively means that, for this lever, any effort will move a load that is five times larger. To summarise:

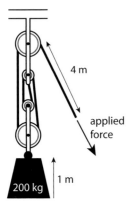

Figure 4.29: A four pulley system

- **where MA is greater than 1:** the machine is used to magnify the effort force (e.g. a class 1 lever)
- **where MA is equal to 1:** the machine is normally used to change the direction of the effort force (e.g. a fixed pulley)
- **where MA is less than 1:** the machine is used to increase the distance an object moves or the speed at which it moves (e.g. the siege machine).

Velocity ratio

Sometimes machines translate a small amount of movement into a larger amount (or vice versa). For example, in Figure 4.30, a small movement of the piston causes the load to move a much greater distance. This property is known as the velocity ratio, and is found by dividing the distance moved by the effort by the distance moved by the load in the same period of time. There are no units for velocity ratio; it is just a number.

$$\text{Velocity Ratio (VR)} = \frac{\text{Distance effort moves}}{\text{Distance load moves}}$$

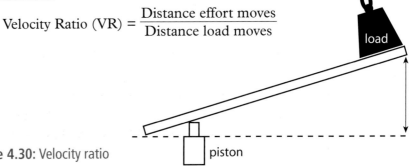

Figure 4.30: Velocity ratio

In Figure 4.32, the piston moves 1 m to move the load by 5 m. The velocity ratio is:

$$VR = \frac{\text{Distance effort moves}}{\text{Distance load moves}} = \frac{1}{5} = 0.2$$

Principles of basic mechanics

In physics, mechanics is typically categorised into two groups:

1. Classical mechanics
2. Quantum mechanics.

Linking some of our previous topics together puts them under the heading of 'classical mechanics'; in other words, a set of physical laws that describe the motion of bodies under the action of a system of forces. Velocity and acceleration were two such examples that we looked at.

Theory of moments

The theory of moments, more commonly referred to as the principle of moments, refers to the fact that the application of a force on an object can distort that object.

If an object is balanced, then the sum of the clockwise moments around the pivot point must be equal to the sum of the anticlockwise moments about the same pivot point. In other words, if the push and the pull factors are the same then nothing is happening.

Actions and reactions

This leads onto Isaac Newton's Third Law, which famously states 'for every action there is always an equal and opposite reaction'. As an example, if we look at a car on a road, the road itself is exerting a frictional force on the car tyres but the tyres themselves are exerting a pressure on the road.

The concept here is that there is no such thing as force that acts on only one body. Instead a force acts between a pair of objects, in this case the tyres and the road, and not on a single object. Therefore, each force effectively has two ends. Each of the two ends is the same, except for being opposite in direction.

Centre of gravity

These points lead to another point, namely the centre of gravity.

Keeping things simple, we refer to the centre of gravity as being the point in an object where we believe the total weight of the object may be concentrated. We use this idea when designing cars, buildings and bridges, or when predicting the behaviour of a moving object when it is acted on by gravity.

As an example, if a tower block office building has its centre of gravity near its base, it will be more stable than an office block that has its centre of gravity near the top of the building.

Equilibrium

Definitions of equilibrium include 'a condition in which all acting influences are cancelled by others, thus resulting in a stable or unchanging system'. Another is 'the state of a body or physical system at rest or in un-accelerated motion in which the resultant of all forces acting on it is zero and the sum of all torques about any axis is zero'.

In other words, a rigid object body in mechanical equilibrium is not undergoing any linear or rotational acceleration. It is balanced.

We refer to mechanical equilibrium of a stationary object as static equilibrium; for example, a television sitting on a table would be in static equilibrium.

However, a better example of mechanical equilibrium would be watching someone trying to compress a spring. They can only compress the spring up to a certain point, after which the force of the person trying to compress it and the force of resistance from the spring are equal. At this point we have mechanical equilibrium, as when the person stops compressing the spring, it will return to its original state.

6. Know the principles of electricity as they relate to the mechanical services industry

Try to imagine a world without electricity. It's hard because everything, from computers to MP3 players, washing machines, lighting and heating, relies on it.

So what precisely is it? You cannot see it with the naked eye; you cannot smell it or hear it. But if you misuse or touch it, it can hurt or even kill. Because of its possible dangers, you need to have a deeper understanding of the subject.

What are electrons?

Molecules and atoms

Every known substance is made up of molecules. A molecule is a very tiny part of matter, which can only be seen using special microscopes. Molecules are always in a state of rapid motion, and the ease with which they move around determines the form of the substance they make up.

- When molecules are densely packed together, their movement is restricted and the substance formed by them is called a **solid**.
- When the molecules are less densely bound together, and where they can move more freely, the substance is called a **liquid**.

- A substance that allows the molecules almost unrestricted movement is known as a **gas**.

These three conditions, solid, liquid and gas, are sometimes known as the three states of matter.

Molecules are made up of **atoms,** and we need to take an imaginary look inside the atom to begin to understand what electricity is. Atoms are the smallest parts of matter that can be subdivided, and they are not solid. They have a **nucleus** at their centre, made up of very tiny particles, known as **protons** and **neutrons**. Protons are said to have a positive charge (+) and neutrons are electrically neutral (that is, they have no charge). We can think of neutrons as the 'glue' that holds the nucleus together. Around the nucleus orbits a third type of particle, the **electron**, which has a negative charge (–).

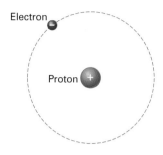

Electron

Proton

Figure 4.31: A hydrogen atom

You may remember from your school science studies that like charges repel each other (+ and +, or – and –) and unlike charges attract each other (+ and –).

All atoms contain equal numbers of protons and electrons, and, in this unaltered state, the matter is said to be electrically neutral (in other words, no electricity is flowing). In some cases, it is possible to add or remove electrons, leaving the atom with a positive (+) or negative (–) charge.

Consider our own solar system, where the planets (including Earth) rotate around, or 'orbit', the sun. In each tiny atom, there is the nucleus containing protons and neutrons (the 'sun') and electrons (planets), which orbit the nucleus. Sometimes these electrons 'break free' and flow to a neighbouring atom. It is these wandering, or free, electrons, moving through the material structure, that give rise to electricity.

What are conductors?

In some materials, it is very easy to get the electrons to move; these materials are called **conductors**. In other materials, it is very hard to get the electrons to move; these materials are called **insulators**. Examples of good conductors of electrons are copper and aluminium (gold is probably the best conductor, but it is too expensive to use in everyday installations). Typical insulators are wood, plastic and rubber.

Measuring electricity

For practical purposes, three things need to be present to create an electrical circuit:

- current
- voltage
- resistance.

Current

If electricity is the movement of electrons through a conductor, we need to be able to measure how much electricity will flow in any given circuit

Key terms
Electricity – the flow of electrons through a conductor. **Conductors** – materials through which electrons flow easily. **Insulators** – materials through which electrons cannot move easily.

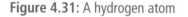

Did you know?

A coulomb consists of approximately 6,240,000,000,000,000,000 electrons

Key terms

Ampere – a flow of one coulomb in one second. In other words, the quantity or amount of electricity that flows every second.

Direct current – the current that is produced by connecting a battery to a circuit.

Alternating current – the electricity produced by power stations that we use in our homes.

Remember

The amount of current flowing in amps is measured using an instrument called an ammeter, which is connected into a circuit in series (see page 190).

Remember

An instrument known as a voltmeter is used to measure the voltage of a circuit. Voltmeters are connected into a circuit in parallel (see page 192).

in a given time and to control how much electricity is flowing to create a practical electric circuit. A single electron, because it is only a tiny part of an atom, is much too small a quantity to have any practical use or for us to measure usefully. Instead, we 'group' millions and millions of electrons together into useful amounts and then measure the groups. These groups of electrons are known as **coulombs**; a coulomb is an extremely large number of electrons.

If we consider the electron as a single drop of water, ask yourself whether it is practical to measure water flow in terms of drops of water flowing in a second. It is more practical to measure water current flow in litres per second.

Similarly, an electrician would consider the flow of electricity through a conductor in terms of coulombs per second. This brings us nicely to one of the main properties of an electric circuit: the electric **current** flow (given the symbol I) is measured in **amperes** (usually abbreviated to amps).

Voltage

The next essential ingredient is **electro-motive force** (emf) measured in terms of the number of joules of energy required to push one coulomb of electrons along the circuit. It is measured in joules/coulomb, more commonly referred to as the **volt**. When a coulomb of electrons leaves a battery or generator it has a potential energy but, as it travels round the circuit, this energy is used up. The amount of energy used up by one coulomb in its passage between two points in a circuit is known as the **potential difference**, measured in volts.

Voltage can be thought of as the pressure that pushes the electricity around the system. By applying a voltage to the end of a conductor, we provide an electrical pressure that causes a current to flow. The voltage may be supplied from a battery or a mains supply. A current that is produced by connecting a battery to a circuit is called **direct current** (DC). The electricity that we use in our homes and which is produced by power stations is called **alternating current** (AC).

Resistance

The third item to consider to understand electricity properly is resistance, given the symbol R. To use a comparison with water, think about water being supplied to a bath via a cistern using gravity. The head of water in the cistern will provide the pressure (voltage in electrical terms). With the tap open, the water will flow into the bath at a fixed rate per second (current in electrical terms).

In order to flow from the cistern to the bath the water passes through a pipe of a certain diameter. If we increase the size of the pipe (and tap), the natural resistance to the flow of water will decrease and the bath will fill more quickly. If we decrease the size of the pipe, the resistance to the flow will increase and the bath will take longer to fill. This resistance to current flow in an electrical circuit is given the symbol R and is measured in units called ohms (denoted by the Greek letter omega, Ω). In plumbing terms, the bigger the pipe the lower the resistance to water flow; in electrical terms, the bigger the conductor the lower the resistance to current flow.

It is very important to note that ammeters and voltmeters are connected into live electrical circuits to measure current and voltage, and they pick up the electricity that they need to operate from the live circuit.

Gravity drives the water through this system but for this to happen there must be a difference in level between the supply (the tank) and the load (the bath). This gives the equivalent of potential difference. The tap performs the same function as an electric switch.

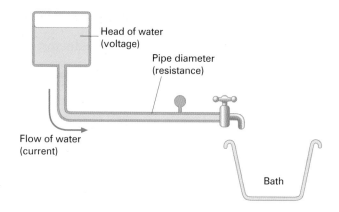

Figure 4.32 Analogy of a water system as an electrical circuit

Ohm's Law

There are three major components that make up an electric circuit: voltage, current and resistance. These are interrelated, and, if we know any two of the quantities, we can calculate the third by using a basic rule known as Ohm's Law.

Ohm's Law describes the relationship between electrical quantities: voltage equals current multiplied by resistance. This can be expressed by the equation:

$$V = I \times R$$

where V = voltage, I = current and R = resistance. It can also be shown in a simple form, often called the Ohm's Law Triangle.

Using the Ohm's Law Triangle

If you know two of the quantities (voltage, current or resistance), you can find the third by covering up the quantity that you are looking for with your finger and calculating the other two.

Suppose you want to determine the voltage (V) of a circuit and you know the current (I) and resistance (R). Covering the V with your finger shows you that you need to multiply the current (I) by the resistance (R) to find voltage (V).

To find the current (I), cover I and you will see that you now need to divide voltage (V) by resistance (R) as V is shown above R.

Example:

For example, a circuit has a resistance of 120 ohms. A current of 2 amps is flowing. What is the voltage applied to the circuit?

$$V = I \times R = 2 \times 120 = 240 \text{ volts}$$

> **Remember**
>
> The resistance of an electric circuit is measured with an instrument known as an ohmmeter. The ohmmeter is connected across the component whose resistance we are measuring.

> **Safety tip**
>
> The ohmmeter has its own internal power supply from a battery, and an ohmmeter must never be connected into a live circuit.

> **Did you know?**
>
> Ohm's Law is named after a German physicist, George Ohm, who defined electrical resistance in 1827.

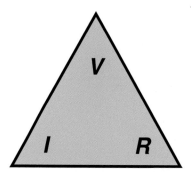

Figure 4.33: Ohm's Law Triangle

How an electric circuit works

If a conductor, such as a copper wire, is bent into a loop and connected to a battery across the two ends, it creates a complete electric circuit and a current will flow. Unfortunately, this would not create a practical circuit that you could use. A single length of copper wire has a very low resistance on its own (depending on the length and cross-sectional area it may have only a few millionths of an ohm of resistance). If you connected this across a normal domestic voltage supply of 230 volts, Ohm's Law shows that $I = \frac{V}{R}$ and R is very low. This means that thousands of amps of current would flow; causing serious damage to the circuit. The flow of electric current causes a corresponding heating effect, and the higher the current, the higher the heating effect; causing a serious fire risk.

Even if you could use the circuit described above, it would still not be practical because there is no load to limit the current, no way to switch off the current and no protection for the circuit from the high current flowing due to the very low cable resistance.

For all practical electric circuits, therefore, we must have:

- conductors through which current can flow (usually copper wire)
- a source of supply (such as a battery) connected across the ends to provide a potential difference (voltage) and to make a complete circuit
- a load such as a lamp or a device that needs electric current to make it work, which will also act to limit the amount of current flowing
- a device such as a fuse or miniature circuit breaker to protect the circuit should too much current flow
- a switch for opening and closing the circuit to turn the current on and off.

Figure 4.34 shows a practical electric circuit containing a battery as a supply source, a lamp as a load, a switch to open and close the circuit and a fuse to protect the circuit from excess current flow.

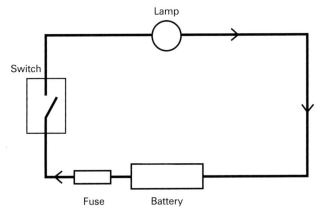

Figure 4.34: A simple electrical circuit

Fuses and circuit breakers

Why are fuses needed?

Fusing is a safety measure preventing a high electrical current from passing through wires that are not designed to carry such large charges. This is important because, when a current flows, it causes a heating effect and a temperature rise in the wire. If a current that is too high for a wire is passed through it, the resulting overheating presents a serious fire risk.

How do fuses help?

There are various different types of fuse. All fuses contain fuse wire, which will melt or 'blow' if an electric current above a specified level is passed through the wiring.

In modern electrical installations you will also work with miniature circuit breakers (mcbs). An mcb is a device that will trip a switch to break the electrical circuit if excessively high current is detected. When electricity flows, it also produces a magnetic field around the conductors as well as the heating effect previously mentioned. The mcb uses a combination of the heating and magnetic effects to break the circuit under fault conditions. Mcbs are more accurate and expensive than fuses but are re-settable, and are now found in all electrical consumer units within newer domestic properties.

You will also come across residual current devices (RCDs). These are very sensitive devices providing a high degree of protection to high-risk parts of electrical systems, such as plug-socket outlets and electric showers. An RCD measures the difference between the current in the electrical conductors in the system (e.g. live and neutral) and measures changes in the electrical current. If a small change occurs, the system is automatically disconnected. A typical operating current for an RCD in the event of a fault would be 30 mA (that is, 30 one thousandth parts of one amp). This may seem very low, but it has been calculated that the average person can 'feel' electricity at current levels of around 1 mA.

Fuse rating

The appropriate size of fuse to be used – the fuse rating – can be worked out using the following simple formula:

$$\text{amps} = \frac{\text{watts}}{\text{volts}}$$

Most items of electrical equipment – lamps, televisions, washing machines, hi-fi systems, vacuum cleaners – are rated in watts, which is the amount of electrical power that the item consumes. Frequently, in domestic environments, fuses are overrated.

Example:

For example, if a lamp contains a 100 watt light bulb, the fuse rating would be calculated as follows:

100 ÷ 230 (the voltage of domestic mains supply) = 0.434 amps

Manufacturers of fuses do not produce fuses rated at 0.434 amps so the next practical size up would be used; in this case a 3 amp fuse. In this example, the bulb may be used in a stand-alone table lamp fitted with a cord and 13 amp plug top, which should be fitted with a 3 amp fuse. Plugs are normally supplied with a 13 amp fuse, but this is unnecessarily high in this example and provides insufficient protection for the user. Technically, this set up would be classed as unsafe.

If the lamp were being used as one of a number connected to a fixed lighting circuit, the entire circuit would usually be fitted with a 6 amp fuse or circuit breaker.

Remember

Fuses come in different sizes to protect against different levels of current.

Figure 4.35: Rewirable fuses

Figure 4.36: Cartridge fuses

Figure 4.37: Miniature circuit breaker (mcb) with a locking-off device fitted for secure isolation

There is potential for confusion here, because 1 amp of current is more than sufficient to kill at 230 volts (the average that a person can 'feel' is 1 mA). A milliamp (mA) is a thousandth of an amp.

Types of electrical circuit

There are two basic types of electrical circuit:

1. Series
2. Parallel.

It is also possible to connect components into a circuit using a combination of these. These are called 'series/parallel' circuits, and they will be considered along with control circuits for central heating and boiler installations.

Series circuits

If you take a number of different resistors (items of electrical equipment such as lamps) and connect them together end to end, and then connect the free ends to a battery, you will find that the current only has one route to take around the circuit. This type of connection is known as a series circuit. Most switches and controls such as thermostats are usually connected into circuits in series, such as in central heating systems.

The main features of a series circuit are as follows:

- The **total resistance** of the circuit is found by adding together all the resistances within the circuit:
  $R = R1 + R2 + R3$
- The **total current flowing (I)** is found, using Ohm's Law, by dividing the supply voltage by the resistance:
  $$I = \frac{V}{R}$$
- The **current flowing** will have the same value at any point in the circuit.
- The **potential difference** or voltage across each resistor is proportional to its resistance. We know from our study of electrons

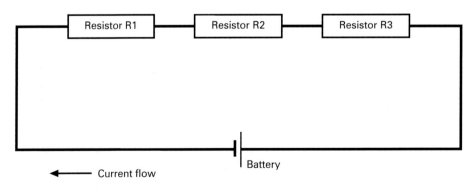

Current flow

Battery

Figure 4.38: The series circuit

that we use voltage to push electrons through a resistor. How much voltage we use depends on the size of the resistor. The bigger the resistor, the more voltage we use for a given current (revise Ohm's Law).

Therefore, the voltage across resistor $R1 = I \times$ value of $R1$ (remember that the current is the same at all points of a series circuit).

Considering the circuit shown in Figure 4.38:

$$V1 = I \times R1 \quad V2 = I \times R2 \quad V3 = I \times R3$$

The supply voltage of the circuit (V) is equal to the amount of the individual voltages across each resistor, as shown by the formula:

$$V = V1 + V2 + V3$$

Calculations with series circuits

Example:

For example, two resistors of 6.2 ohms and 3.8 ohms are connected in series with a 12 volt battery, as shown in Figure 4.39. Calculate:

(a) total resistance of the circuit

(b) total current flowing

(c) the potential difference across each resistor.

(a) Total resistance

For series circuits the total resistance is the sum of the individual resistances:

$$R = R1 + R2 = 6.2 + 3.8 = 10 \ \Omega$$

(b) Total current

Using Ohm's Law:

$$\text{Current} = \frac{\text{Voltage}}{\text{Resistance}} = \frac{12}{10} = 1.2 \text{ amps}$$

(c) Potential difference across each resistor

$$V = I \times R$$

Therefore: across $R1$: $V1 = I \times R1 = 1.2 \times 6.2 = 7.44$ volts

across $R2$: $V2 = I \times R2 = 1.2 \times 3.8 = 4.56$ volts

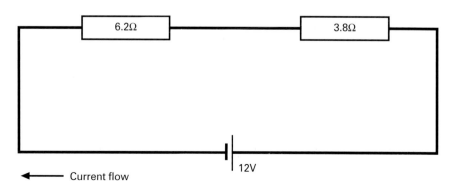

Current flow

Figure 4.39: Total resistance

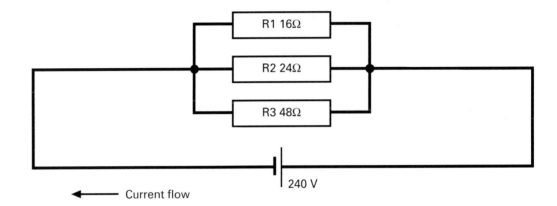

Figure 4.40: The parallel circuit

Parallel circuit

Another way to connect components is via a parallel circuit. These are commonly used for lighting and power circuits, because the same voltage appears across each of the resistances. This is important in lighting circuits where you want all the lamps to burn with the same brilliance, rather than dimly because the available voltage is shared between them.

In a parallel circuit a number of resistors are connected together so that there are two or more routes for the current to flow along, as shown in Figure 4.40.

In a parallel circuit the current 'splits up' and divides itself among the various branches of the circuit. However, while the current divides within a parallel circuit, the voltage across each branch remains the same. This means that if components are removed from the circuit, the remaining ones will still work. In a series circuit, the circuit will be broken if a single component is removed, and will no longer work. It is this feature that makes parallel circuits popular.

The main features of a parallel circuit are as follows:

- The **total current (*I*)** is found by adding together the current flowing through each of the branches:
  $I = I_1 + I_2 + I_3$
- The same **voltage** will occur across each branch of the circuit:
  $V = V_1 + V_2 + V_3$
- The **total resistance** can be found by using the formula:
  $$\frac{1}{R} = \frac{1}{R_1} + \frac{1}{R_2} + \frac{1}{R_3}$$

Some calculations with parallel circuits

Example:

Using the example shown above, calculate the total current flowing through the circuit.

$$\frac{1}{R} = \frac{1}{R_1} + \frac{1}{R_2} + \frac{1}{R_3}$$

Therefore:

$$\frac{1}{R} = \frac{1}{16} + \frac{1}{24} + \frac{1}{48}$$

And therefore:

$$\frac{1}{R} = \frac{1}{16} + \frac{1}{24} + \frac{1}{48} = \frac{3}{48} + \frac{2}{48} + \frac{1}{48} = \frac{3+2+1}{48}$$

Giving us:

$$\frac{1}{R} = \frac{6}{48}$$

Turn this upside down to find R:

$$\frac{R}{1} = \frac{48}{6} \quad R = 8 \text{ ohms}$$

Now, using Ohm's Law:

$$I = \frac{V}{R}$$

We see that:

$$I = \frac{240}{8} = 30 \text{ amps}$$

Direct and alternating current

Figure 4.41 shows the flow of electrons in both AC and DC circuits.

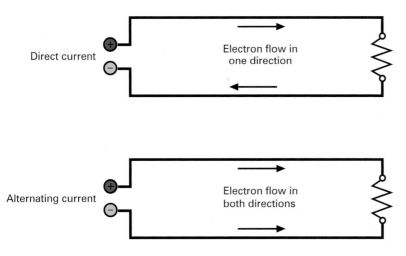

Figure 4.41: Flow of electrons in AC and DC circuits

Direct current (DC)

In a direct current electrical circuit, the electron flow is in the same direction all the time. One example would be from the anode to the cathode of a battery around a simple circuit.

Direct current from batteries is produced by a chemical reaction where plates containing dissimilar metals are placed in a solution known as an 'electrolyte'. When the battery is charged and a load is connected across it, a current flows. Batteries and direct current supplies are generally only used where very small amounts of electricity are required. For example, in torches, light current back-up power supplies and battery-operated power tools.

As a plumber you will probably only use direct current in:

- using small batteries
- heating or boiler control circuits where the direct current has been produced from an alternating current source using a transformer to reduce the voltage and a rectifier (which is a solid state electronic device) to convert the alternating current to direct current.

Alternating current (AC)

Most of the work that plumbing systems are connected into will be supplied with alternating current. The supply in a domestic building will almost always be single phase, and supplied at 230 volts with a frequency of 50 **hertz**.

Electricity is invisible to the naked eye – we cannot see it, or smell it. This said, there is an instrument known as an oscilloscope with a screen like a small television, which allows us to look at electricity.

If you look at direct current, you see a straight line where the electricity flows in one direction from one pole of the battery to the other.

If you look at alternating current, however, you see a shape that is called a 'sine' wave as shown in Figure 4.42.

The sine wave represents one complete 'cycle' of alternating emf, which would be 'induced' into a coil of copper wire when it is rotated within a magnetic field. During the top half of the cycle, the current will flow in one direction, and during the bottom half of the cycle, the current will flow the other way (hence 'alternating'). In the UK, the electricity supply has 50 of these single cycles produced every second. This is known as the frequency of the supply, and the unit used for frequency is the hertz, abbreviated to Hz. The frequency is therefore stated as 50 Hz.

Key term

Hertz – the number of cycles of AC that are produced every second.

Did you know?

Sine is a short name for 'sinusoidal', which you may have come across in school during trigonometry.

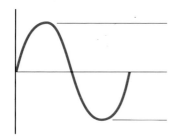

Figure 4.42: The alternating current sine wave

Electromagnetism

While direct current is caused as a result of a chemical reaction in a battery, alternating current is produced as a result of **electromagnetism.**

As well as the heating effect already mentioned, all electrical currents produce a magnetic force. This is the basic principle that underpins the creation of almost all the electricity used today, and is known as electromagnetism. The application of this fact was demonstrated in principle by Michael Faraday in the 1830s. He discovered that electricity could be generated by moving a magnet in and out of, or around, a coil of wire wound on to a soft iron core. The same effect is present if we move a coil of wire on an iron core within a fixed magnetic field.

Earthing

In the case of a short circuit, when the known circuit resistance is by passed or breaks down, very high currents can flow. These can cause serious fire and injury risks, due to the temperature flowing through the current and risk of electric shock. It is this for reason – to protect people from danger – that we earth electrical equipment.

In electrical terms, the earth is a very large conductor at zero potential (voltage). By connecting together all the metalwork of an installation we ensure that the dangerous potential differences cannot exist between the metal parts of the installation and the earth (ground). If they did, people touching the 'live' metalwork could get a very serious electric shock, or even be killed.

To understand earthing in simple terms, think of electricity as being lazy. It is always flowing around a circuit through a load with a known resistance to control the current, and it flows from a high potential to a low potential. As it is lazy, it will always try to find the path of least resistance to get to the zero potential of earth. If it can, it will do that through you (you have a low resistance) rather than fight its way through the circuit.

Earthing provides a very low resistance path to earth through the exposed installation metalwork. In the event of a fault, a very high fault current will flow that will almost instantly operate the circuit protective devices, such as fuses or circuit breakers.

Key term

Electromagnetism – the relationship between electric currents or fields and magnetic fields.

Safety tip

Electrical installations must be earthed to prevent electric shocks.

Remember

Electricity is lazy – it will always find the path of least resistance to get to earth (ground).

Progress check

1. Which metals are said to be the best conductors of electricity (excluding gold)?
2. Which materials are typical insulators of electricity?
3. For practical purposes, which three things need to be present to create an electrical circuit?
4. How is Ohm's Law expressed as an equation?
5. What should never be done with an ohmmeter?
6. Where would an MCB be found in a domestic property?
7. What are the two basic types of electrical circuit?
8. What is meant by the term 'hertz'?
9. Why must electrical installations be earthed?

Check your knowledge

1. How much does water expand upon freezing?
   a   5%
   b   10%
   c   15%
   d   20%

2. With reference to fuel gases, which of the following statements is true?
   a   Propane is heavier than air
   b   Butane is lighter than air
   c   Natural gas is heavier than air
   d   Butane is lighter than propane

3. The Pascal is a unit of intensity of pressure. How can this be calculated?
   a   By multiplying newtons and square metres
   b   By multiplying kilograms and square metres
   c   By multiplying newtons and kilograms
   d   By multiplying newtons and litres

4. Which one of the following materials has the highest coefficient of linear expansion?
   a   Copper
   b   uPVC
   c   Stainless steel
   d   Steel

5. Which one of the following is the unit for power?
   a   Joules
   b   Newtons
   c   Watts (or kilowatts)
   d   Amps

6. Which unit is used to measure electrical current?
   a   Ohm
   b   Watt
   c   Volt
   d   Amp

7. At what temperature does water freeze?
   a   0 °C
   b   2 °C
   c   8 °C
   d   4 °C

8. The process of siphonage uses:
   a   capillary action
   b   specific gravity
   c   atmospheric pressure
   d   vacuum pumps

9. Which plumbing system by-product is directly linked to hard water?
   a   Limescale
   b   Hydrogen
   c   Carbon dioxide
   d   Ferrous oxide

10. The direct result of hard water in domestic properties is most likely to be what?
    a   Water hammer
    b   Dripping tanks
    c   Running overflow
    d   Reduced system efficiency

11. What is relative density?
    a   Density of a substance in comparison to the density of water
    b   Volume of displaced water
    c   Mass of a substance divided by its volume
    d   Force exerted by the mass of a substance

12. Which one of the following metals has the highest rate of heat conductivity?
    a   Steel
    b   Copper
    c   Lead
    d   Iron

13. The sockets on a ring usually have an amp rating of:
    a   9 amps
    b   6 amps
    c   13 amps
    d   15 amps

14. When would a plumber use temporary earth continuity bonds?
    a   Flushing out the hot-water system
    b   Removing a section of pipework
    c   Changing a tap washer
    d   Working on a central heating control panel

15. What is the SI unit of measurement for electrical resistance?
    a   Ohm
    b   Volt
    c   Amp
    d   Watt

Getting ready for assessment

The information contained in this unit of the book, as well as the continued practical assignments that you will carry out in your college or training centre, will help you with preparing for both your end-of-unit test and the diploma multiple-choice test. It will also support you in preparing for the practical assignments you will need to complete to demonstrate your understanding of scientific principles within building services engineering.

There are opportunities throughout the unit to test your progress in and understanding of the required underpinning knowledge, this will enhance your preparation for the forthcoming assessments, so make good use of them.

This unit will be assessed by the following assessment method:

- externally set knowledge assessment.

With regard to the mechanical services industry, you will need to know:

- The standard units of measurement used, including, metre (length), kilogram (mass), second (time), Kelvin (°K), area (m^2), volume (m^3), litres (L), density (kg/m^3), velocity (m/s).

- The properties of materials used, including density air/water, metals – pure, ferrous and alloys including solders, plastics – thermo plastics and thermo-setting, fireclays/ceramics, strength – properties of metals – tensile compressive, hardness, ductility, malleability, conductivity – heat and electricity, atmospheric corrosion/oxidation, UV and heat damage, corrosion – prevention, electrolytic, electromotive series, dissimilar metals in an electrolyte, erosion, liquids – water, refrigerants, anti-freeze/glycol mixes, fuel oils, lubricants/greases boiling/freezing point, volume and pressure, density, steam, capillarity, acidity/alkalinity (pH value), water hardness – soft, temporary hard, permanently hard, gases – air and steam, LPG, natural gas, carbon dioxide, refrigerant gases, pressure – gas, volume occupied by gas, temperature of gases, Charles's law, Boyle's law and heat pump/refrigeration cycle.

- The relationship between energy, heat and power, including, temperature measurement, change of state – melting, freezing, boiling, evaporating, condensing, latent and sensible heat, heat transfer – conduction, convection, radiation, energy – Joules (J), specific heat capacity (kJ/kg/°C, power – Watts (W), calculations – temperature, quantity of heat and the power required to heat a substance.

- The principles of force/pressure and their application, including, acceleration (m/s^2), force – newton (N), pressure (N/m^2), atmospheric pressure, principles of the siphon, flow rate (m^3/s), pressure – bar/millibar, kPa/Psi, metre head, m^3/s, l/s, kg/s, simple force calculations, pressure head, static head, dynamic head.

- Simple mechanical properties and their application, including, mechanical advantage, levers, wheel and axle, pulleys, screws, movement, action/reaction, centre of gravity and equilibrium.

- The principles of electricity, including, electrical flow, conductivity/resistance, DC/AC current, Amps, Volts, Ohms, Watts, Ohm's law, power consumption, protection devices, voltage, current, resistance and earthing requirements.

Good luck!

Understand and carry out site preparation and pipework fabrication techniques for domestic plumbing and heating systems

When working as a plumber there are certain tasks that you will always need to carry out, no matter what the job or location is. These involve preparing the sites where you will be working. As part of this you need to be able to identify the tools and materials you will need to complete tasks. You should also be familiar with some of the basic installation requirements for domestic plumbing and heating pipework.

This unit will cover the following learning outcomes:

- Know the types of hand and power tools used for domestic plumbing and heating work

- Know the types of domestic plumbing and heating pipework and their jointing principles

- Know, and be able to apply, the general site preparation techniques for plumbing and heating work

- Know, and be able to apply, how to use fixings, clips and brackets to support domestic plumbing and heating pipework and components

- Know the installation requirements of domestic plumbing and heating pipework

- Know the inspection and soundness testing requirements of domestic plumbing and heating pipework

Figure 5.1: Large hacksaw

Figure 5.2: Junior hacksaw

Figure 5.3: Plastic pipe cutters

Figure 5.4: Mini tube cutter

1. Know the types of hand and power tools used for domestic plumbing and heating work

As a plumber, you are required to measure, mark out, cut, fabricate, joint and fix a range of materials. In most cases, this involves the use of tools. Good-quality tools of a well-known brand are expensive but are a good investment.

Once you have put together a toolkit, keep the tools clean and well maintained. This should ensure a long life for them so that you do not have to keep buying replacements.

Hand tools

Checklist

Tool safety and maintenance

- Make sure your tools are cleaned regularly.
- Lubricate the working parts of tools.
- Once cleaned, lightly coat the tools with an oil spray to prevent rusting but don't overdo the oiling.
- Always use the right tool for the job. Screwdrivers are not chisels!
- Keep file or rasp teeth clean using a wire brush.
- Never use tools if they have split handles.
- Always replace used, worn or defective hand tools.

Hacksaws

Always:

- Make sure the teeth are pointing away from you and toward the forward cut.
- Check that your hacksaw does not have defective or worn teeth.
- Make sure the blade is tightened correctly when using a large hacksaw.
- Use the right blade type for the job – 32 teeth per inch for light-gauge pipe such as copper and plastics, and 24 teeth per inch for heavier-gauge LCS.

Pipe cutters

Always:

- Replace damaged or blunt cutter wheels.
- Use the correct blade for the material being cut.
- Make sure that the wheel and the rollers are lubricated and move freely.
- Use a pipe cutter or deburring tool to deburr the inside of pipework.

Figure 5.5: Bevelled wood chisel

Figure 5.6: Typical fixed stem tap spanner

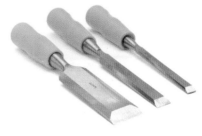

Figure 5.7: Chisels

Wood chisels

Always:

- Check your chisel does not have a split handle.
- Keep chisels sharp using a grinder or whetstone.
- Keep the plastic guard on the chisel.
- Make sure handles aren't loose.

Figure 5.8: Lump hammer

Tap spanner

Tap spanners are:

- designed to allow the tightening of taps and tap connectors, which are hard to reach.
- Available in different types but the most common is either fixed stem or extending stem jaw.

Figure 5.9: Ball pein hammer

Cold chisels

Always:

- Keep the cutting edge sharp using a grinder.
- Keep the striking end of the chisel free from the 'mushrooming effect', again using a grinder.
- Ensure that a chisel has not been ground down so much that it has become too small to handle safely.

Figure 5.10: Cross pein hammer

Hammers

Always:

- Make sure the head is fitted correctly to the shaft.
- Check that a hammer does not have a defective shaft.

Figure 5.11: Claw hammer

Pipe grips and wrenches

Always:

- Keep the teeth free from jointing compounds. If they're clogged up, it could cause the tool to slip.
- Replace the tool once the teeth become worn.
- Check for wear on the ratchet mechanism when using pump pliers. These often slip when under pressure.
- Be careful when loosening a joint or pipe that is difficult to move. It might give suddenly, and you could damage your hands, or even pull a muscle.

Figure 5.12: Adjustable spanner

Figure 5.13: Stubby PZD/SDV, Phillips and normal slotted screwdrivers

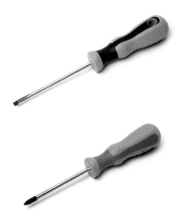

Figure 5.14: Normal slotted and Phillips screwdrivers

Remember

Little maintenance is required for spirit levels but they need to be handled and stored with consideration.

Figure 5.15: Spirit level (boat level)

Figure 5.16: Manual hand threader

Screwdrivers

Ensure you:

- Keep flat-ended screwdrivers for slotted screws to a uniform thickness.
- Never use a screwdriver with a defective handle.

Spirit level

A spirit level consists of a metal bar with three vials all containing an air bubble. The vial that is in the centre of the bar and located on its top edge indicates the horizontal level. The vials towards the end of the bar indicate the vertical level. On some models, one of the vials at the end of the bar can be adjusted to measure specific angles. Spirit levels range in length from 450 mm (1ft 6in) to 1800 mm (6ft). Also available is a pocket version (boat level), which is between 150 mm (6") and 240 mm (10").

Manual pipe threaders

You should:

- Ensure the dies are kept free from swarf.
- Ensure dies are sharp.
- Use approved lubricant when using manual pipe threaders.
- Ensure the handle is not damaged.
- Never run your fingers over the threads.
- If using a ratchet-type threader, ensure that it is set for the correct direction.

Hand-held bending machine

Always read the instructions for the bending machine before using it as each machine is different, although they use the same basic principles. The following is the normal method used by the majority of manufacturers:

- Make sure that the correctly sized former and bending roller are fitted to the machine; one size of former/roller will only bend one size of pipe. Most machines will cater for 15 mm and 22 mm.
- Ensure all moving parts, guides and rollers are kept clean and lightly oiled.
- Do not strain yourself when using the bending machine.
- If using long lengths of pipe, ensure that you do not cause any damage to the surrounding area or others.

The use of bending machines is covered in more detail on pages 224–29.

Additional tools that commonly form a plumber's toolkit

- Blow torch assembly
- Compact tube cutter – for cutting copper tube in tight locations
- Files – smooth and coarse, including a rasp type for chamfering plastic pipework for push-fit connections
- Flooring chisel – for lifting floorboards
- Footprint type wrench – for applications such as pump valves
- High-speed drill bit set – for drilling small holes in timber and metals
- Hole saw set – for preparing holes in storage cisterns
- Immersion heater spanner – for tightening immersion heaters to cylinder bosses
- Insulated screwdrivers – for electrical connections
- Knife – for trimming materials
- Masonry drill set – for drilling holes in masonry wall surfaces
- Pad saw – for cutting small holes, for example through plasterboard
- Pointing trowel – for making good to masonry wall surfaces
- Radiator spanner – for making valves, etc., into radiators
- Spirit level (small and large) – for fixing to correct levels
- Stillson wrench – a heavy duty tool used for tightening steel pipework, etc.
- Tape measure – for measurement of materials
- Water pump pliers – general purpose tool for tightening
- Wood drill bit set – for the drilling of holes in timber to accommodate pipework and components.

Power tools

The main power tools used in plumbing are:

- power drills
- cordless powered screwdrivers
- combined cordless drills and screwdrivers
- power saws.

Power drills

There is a wide range of power drills on the market. Their power varies with the size of the motor, and this will have a bearing on what each drill can do. Some have a 'hammer action' which, when engaged, makes it easier to drill through masonry. As the drill rotates, it also moves fractionally backwards and forwards at high speed, effectively hammering the drill into the material.

Typical power ratings range from 620 W to 1400 W. Most drills are variable speed and some have reversible action.

Figure 5.17: Standard rotary hammer drill (110 V)

Figure 5.18: Battery-operated hammer drill

Figure 5.19: Powered circular saw

Figure 5.20: Jig-saw

Figure 5.21: A frozen pipe

Cordless drills and screwdrivers

Figure 5.19 shows a typical example from a very wide range. Most drills are combined so they can be used as a drill and a screwdriver. They are powered by batteries and are usually supplied with two batteries and a charger, so that one is working and the other is charging. These tools are popular because they are much safer than ones with cords, and you don't need to carry a transformer around. Also, if there is no electricity on site you don't have to revert to hand tools.

Power saws

Most powered circular saws run at 110 V. They are used by plumbers for taking up floorboards or sheets to install pipework under floors. Make sure that nails or screws are avoided when sawing. If possible, take up a board by hand first to check what is beneath the area where you are going to use the saw. Always make the cutting depth the same depth as the floor thickness.

Battery-operated circular saws are also available, usually with an 18 V motor.

Another type of saw that plumbers often use for cutting wood is the jig-saw, as shown in Figure 5.20. Similar care should be taken with this type of saw.

Portable pipe-freezing kits

There are a number of portable pipe-freezing kits available. These are designed to save time when draining down a system and refilling it. Most of these will work providing there is no movement of water within the pipe. They can either be electrically operated or use a throw-away canister of refrigerant gases. The electric pipe-freezing kit is far more eco-friendly although gases used now are also more environmentally friendly.

Electric pipe-freezing kit

The electric freezing kit has two leads, which are placed either side of the section of pipe where work is to be carried out. It will freeze a 22 mm pipe in about 8 minutes. Once the work has been carried out the unit will also de-frost the ice core that it created if necessary.

Canister and sleeve pipe-freezing kit

A sleeve is placed either side of the section of pipe where work is to be carried out, with the ends sealed using cable ties. The contents of the canister are then released into the sleeve via a thin tube. This produces an ice plug in 10 minutes that will last for up to 30 minutes. These can be used horizontally and vertically.

Power tool use

- You should have been properly instructed on the use of the power tool before using it on site.
- Each power tool should be visually inspected for signs of damage prior to its use – if it is damaged it should not be used.
- All electric tools should be double insulated or incorporate an earth cable.
- Battery-powered tools are preferable to mains-operated as they are safer. If you use the mains or a temporary power supply, always use 110 V tools as opposed to 240 V.
- If 240 V power tools are used then the supply to the tool must be protected by the use of a Residual Current Device (RCD).
- Check that electrical cables are not damaged or worn.
- Check that plugs are not damaged.
- Equipment should be PAT tested in accordance with your employer's procedures. PAT tests are maintenance records of all portable electrical equipment to ensure it is in safe working order – the maximum interval between tests for equipment used for construction is usually 3 months.
- Cartridge-operated tools are covered in Unit 1 *Safe working practices*. You must receive full instructions before using them.

You will be assessed in drilling and fixing, chasing, lifting flooring and cutting holes in building fabrics as part of your qualification.

Elena is from Lithuania. She has just started at college studying on the NVQ Level 2 course, having recently got a job with a local plumbing company. Her employer has told her to collect the plumbing tools she would like from a local tool supplier. She has been given a budget of £250 (including VAT). Elena has an idea of what she might need, but is unfamiliar with the manufacturers' names and the quality of the tools in this country. She has asked you to help her select some good quality tools within her budget.

- What tools would Elena need to start off her toolbox? Don't forget any power tools.

- Find out prices for the tools. If the tools you are recommending are over the budget, select the tools that are a necessity and leave those which could wait.
- What could Elena say to her employer if she asks for more money to complete the tool purchase?
- Would it be worth looking at cheaper versions of the tools you recommended to enable Elena to keep within budget? This would also mean she could get more tools for her money.

Storage of tools

You will use a variety of materials, tools and pieces of equipment in your work as a plumber. All of them are potentially dangerous if misused or neglected. Instruction in the proper use of materials, tools and equipment will form part of your training and you should continue to follow safe working methods.

Hand tools and manually operated equipment are often misused. You should always use the right tool for the job and never just make do with whatever tool you may have to hand. For example, never use a hammer on a tool with a wooden handle as you may damage the wooden handle and create flying splinters. Here are some general guidelines:

Safety tip

Cartridge-operated tools can be very dangerous, especially if you have an accidental discharge, which can cause serious injuries from ricochets. Any user must be properly trained to prove competence and over 18 years of age, to meet the requirements of the PUWER Regulations.

Key term

Air embolism – pathological condition caused by gas bubbles in the human body.

Safety tip

Never use a circular saw without a guard.

Figure 5.22: A portable circular saw: one of the different types of portable tools you may encounter on site

Safety tip

Remember to wear safety goggles when using drills and saws. These will protect your eyes from dust and any splinters of material that might fly off while you are working.

- Keep cutting tools such as saws, chisels and drills sharp and in good condition.
- Ensure handles are properly fitted and secure, and free from splinters.
- Check that the plugs and cables of hand-held electrically powered tools are in good condition. Replace frayed cables and broken plugs.
- Electrically powered tools of 110 V or 230 V must be PAT tested in accordance with your employer's procedures.

You may also come into contact with high-pressure airlines. Used carelessly, compressed air can be dangerous and cause explosions or blow aside tools, equipment and debris. Never use an airline to blow dust away, never aim it at any part of your body and never point it at somebody else. If high-pressure air enters the body through a cut or abrasion or through one of the body's orifices it can cause an **air embolism**, which is very painful and can be fatal.

Equally, any bit of equipment, such as your toolbox, portable floodlight or conduit bender, can deteriorate with use. Equipment should always be visually inspected before use, used correctly and stored appropriately when not in use. If equipment is damaged or broken, it should not be used. It could otherwise cause an accident or injury.

When using equipment such as grinders or drills, ensure that:

- any required guards are in place
- the equipment is appropriate, undamaged and fit for purpose (e.g. correct grinding wheel or drill fitted)
- vibration exposure limits are not exceeded
- suitable PPE (e.g. gloves, eye protection, respirators, ear defenders) is worn where required
- the user has been trained to operate the equipment
- suitable ventilation is available
- signs and barriers to prevent unauthorised access are provided where required.

Dangerous occurrences and hazardous malfunctions of equipment

Dangerous occurrences are conditions or actions with the potential to cause hazardous malfunctions. Hazardous malfunctions are failures of objects or assemblies, which may cause injury to operators or bystanders.

Both situations must be reported. If not, even if no one is injured, then no one can take action to prevent their happening again.

Untidy working and storage

Tools, equipment and materials left lying about, trailing cables and air hoses, spilt oil etc., can cause people to slip, trip or fall. Clutter and debris should be cleared away to prevent fire hazards.

Tools and equipment left in this state are targets for thieves and make it difficult for an employer to maintain effective levels of insurance cover. After any work all tools should be cleared away and the workplace left in a safe condition. You should plan how the site will be kept tidy and how housekeeping will be managed.

- Keep walkways and stairways free of tripping hazards such as trailing cables, building materials and waste.
- Keep inside floor areas clean and dry.
- Outdoor footpaths should be level and firm and should not be used for storing materials.

Designate storage areas for plant, materials, waste, flammable substances (e.g. foam plastics, flammable liquids and gases such as propane) and hazardous substances (e.g. adhesives or cutting compounds). Flammable materials will usually need to be stored away from other materials and protected from accidental ignition.

Do not store materials where they obstruct access routes or where they could interfere with emergency escape. For instance, do not store flammable materials under staircases, or near to doors or fire exits.

If materials are stored at height, make sure necessary guard rails are in place. If people could fall when stacking or collecting materials or equipment, ensure that suitable access equipment is used to store or collect them.

Materials stored away correctly in locked and secure storage areas will also be harder to access by thieves. This will protect your tools and materials from being stolen, which could lead to you having to pay for costly replacements.

Keep all storage areas tidy, whether in an agreed storage area or on the site itself. Try to plan deliveries to keep the amount of materials on site to a minimum.

> **Remember**
>
> It is especially important to keep emergency routes clear. Make sure that all flammable waste materials (such as packaging and timber offcuts) are cleared away regularly to reduce fire risks.

Figure 5.23: An untidy work area can present many trip hazards

Progress check

1. What is the main disadvantage of buying cheap tools?
2. When using a large frame hacksaw to cut light-gauge tubes such as copper or plastic, how many teeth per inch are recommended?
3. Why is it necessary to deburr a copper tube after tube cutters have been used?
4. Why is it important to ensure that teeth on pipe grips and wrenches are free from the build-up of jointing compounds?
5. What type of surfaces would a hammer drill be used on?
6. What does the abbreviation PAT stand for, and how often should this usually be carried out on electrical equipment used on building sites?
7. What essential piece of PPE must be worn when using electrical drills and saws?

2. Know the types of domestic plumbing and heating pipework and their jointing principles

Pipes, fittings and jointing materials acceptable for Water Regulation purposes are listed in the Water Fittings and Materials Directory. BS 6700 also states the minimum requirements for pipe joints and fittings. You will cover this in much more detail at Level 3. At Level 2, the following factors must be taken into account when selecting materials for use in plumbing systems:

- effect on water quality
- compatibility of different materials
- vibration, stress or settlement
- ageing, fatigue and durability
- internal water pressure
- mechanical factors
- internal and external corrosion
- **permeation**.

In domestic plumbing installations the main materials that you will work with are copper, low carbon steel and plastic. To a lesser extent, and usually only on maintenance work, you will also come across cast iron and stainless steel.

Copper tube and fittings

Copper tube is available in four grades. Under a new BS Standard, the terminology used for copper grades has changed slightly, but the grades shown below are still widely used.

Grade X (R250 half-hard lengths)

Grade X is widely used for domestic installations. It is classified as half hard, and has pipe diameters ranging from 12 mm to 54 mm, with a wall thickness of 0.7 mm. Its pipe diameter is always specified as the external measurement, and the tubes are normally available in 6 m lengths, although most merchants will also supply in 1 m, 2 m and 3 m lengths. Grade X tube should not be used underground. Grade X is also available in chromium plate, and is used where pipework is exposed to the eye and an attractive finish is required.

Grade Z (R290 hard lengths)

Grade Z is not as popular as Grade X, mainly because it is unbendable. This is because its wall is 0.2 mm thinner than Grade X, so the pipe is hardened during the manufacturing process to make it stronger. Grade Z is available in 1 m, 2 m, 3 m and 6 m lengths, and in pipe diameters of 12 mm to 54 mm. Like Grade X, Grade Z is available in chromium plate.

Key term

Permeation – the possibility of microscopic particles of oxygen entering the water supply through the external wall of the pipe.

Remember

You may find it useful to refer back to the section on plumbing materials in Unit 4, pages 145–66 when reading this.

Find out

Can you think where you might see chrome plated pipework in a domestic dwelling?

Figure 5.24: Compression joints

Figure 5.25: Soldered capillary joints

Figure 5.26: Plastic push-fit joints

Chrome-plated piping is most often found in exposed pipe runs to instantaneous showers in a bath or shower room, particularly where the appliance is a new addition to an existing suite.

Grade Y (R220 soft coils)

Grade Y is used for external underground installations. Its wall thickness is 1 mm, making it the thickest grade. It is classified as fully annealed, which means it is soft. The size range is 15 mm to 22 mm, and it is supplied in 25 m coils. Grade Y is easy to form into wide-radius bends by hand, without affecting the bore of the tube, so the use of bending equipment is not required. Grade Y can be supplied in either a blue (water service) or yellow (gas) plastic coating.

Grade W (R220 soft coils micro-bore)

Grade W is used for micro-bore heating and, like Grade Y, is fully annealed. The sizes used for micro-bore are usually 4 mm, 5 mm, 6 mm, 8 mm, 10 mm and 12 mm. The pipe is supplied in coils ranging from 10 m to 30 m. Like Grade Y, the pipe lengths can be formed into wide-radius bends or offsets without the use of bending equipment. For short-radius bends, a number of purpose-made hand formers are available on the market. Grade W can be supplied with yellow (gas) or white (heating) plastic coating or with air channels in the plastic coating to improve thermal insulation. This is used where pipework is installed internally in solid floors.

Methods of jointing copper tube

There are three main methods of jointing:

- compression joints
- soldered capillary joints
- push-fit joints.

Compression joints

Compression joints are of two types:

- manipulative – Type B
- non-manipulative – Type A.

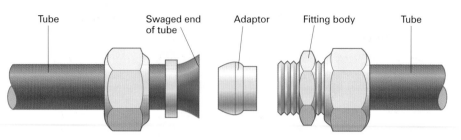

Figure 5.27: Manipulative compression fitting

Manipulative joint

Figure 5.27 shows a typical manipulative fitting detail; these are sometimes referred to as Type B.

The jointing process is as follows:

1. Cut the tube to the required length, using either tube cutters or a hacksaw.
2. Deburr the inside of the pipe (if cut with tube cutters) or the inside and outside of the pipe (if cut with a hacksaw).
3. Place the nut and compensating ring onto the pipe. This acts as a washer when the nut is tightened against the flared tube.
4. Use a **swaging tool** or drift to flare out the end of the tube so that it fits the angle of the adaptor.

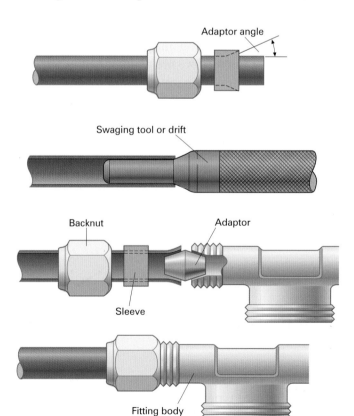

Figure 5.28 Stages of assembly of a manipulative compression fitting

5. Apply this process to each pipe end being used in the fitting (two for a straight coupling or elbow, three for a tee).

6. Assemble the fitting. The angled side of the adaptor fits into the flared end of the tube and the other side fits into the body of the fitting. The joint should be finger tight, with all the fitting components fully engaged. The fitting can then be completely tightened using adjustable grips or spanners.

This type of copper compression fitting is rarely used owing to the increased amount of time required to form the joint. However, compression fittings used on underground copper cold-water supply pipework must always be manipulative joints.

Figure 5.29: Manipulative joint being coupled

Non-manipulative joint

Sometimes referred to as Type A, this is similar to the manipulative fitting. The main difference is that the pipe end is not flared to receive an adaptor. Instead, a **compression ring** or **olive** is used to provide a watertight seal on the pipe end. The cutting and deburring process is the same as for a manipulative fitting. Again, the joint is hand-tightened first, making sure the pipe is fully pushed into the body of the fitting, before tightening is completed with adjustable grips or spanners.

> **Remember**
>
> This type of fitting is suitable for use on grade X, Z and W pipes. Their use is not allowed on underground services.

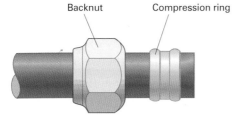

Backnut   Compression ring   Fitting body   Tube

Figure 5.30: Detail of non-manipulative (Type A) joint

Figure 5.31: Non-manipulative joint being coupled

Brass fittings

A range of brass fitting designs is available, which is compatible to pipe sizes. For example:

- straight adaptor
  - with parallel spigot thread
  - with parallel socket thread
  - with taper-made thread
- straight couplers
  - equal compression and reducers
  - imperial to metric
  - with drain valve
  - with air release valve
  - compression to solder-ring capillary

- tap connectors
- stop ends
- tees
  - ○ equal
  - ○ reducing
- elbows
  - ○ equal
  - ○ with drain tap
  - ○ with air release valve
  - ○ compression to parallel socket thread
  - ○ compression to parallel spigot thread
- wall-plate elbows
- bent and straight tank connectors.

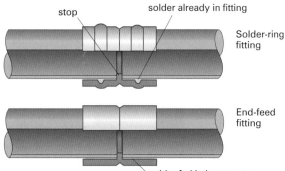

stop   solder already in fitting

Solder-ring fitting

End-feed fitting

solder fed in by operator

Figure 5.32: Soldered capillary joints

Soldered capillary joints

Soldered joints can be classified as soft soldered or hard soldered. Hard-soldered joints, for example using silver and silver alloys including copper, require heating to a much higher temperature than soft-soldered joints. Hard-soldered joints are rarely used on the majority of domestic plumbing installations and will not be discussed here.

Soft-soldered joints are made using two types of fittings:

1. Integral solder ring
2. End feed.

The difference between the two jointing methods is that the solder ring has solder contained in a raised ring within the body of the fitting (integral). When the fitting is heated, the solder melts (between 180 °C and 230 °C) and the solder is drawn into the fitting by capillary action. The same principle is used for the end feed, except that the solder is end-fed separately from a spool of wire solder.

The jointing process is fairly straightforward.

The jointing process

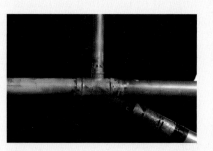

Step 1: The tube is cut square and deburred. The inside of the fitting and the outside of the tube are cleaned to a 'bright, shiny, light orange colour'. This is because the solder will not adhere to the joint correctly if these surfaces are oxidised.

Step 2: Flux is applied to the surface of the tube and fitting using a brush – not your fingers!

Step 3: The joint is assembled and heat is applied to the body of the fitting.

- If using an integral solder ring, wait until you see a complete film of solder around the end of each fitting. If using an end feed, apply solder from a wire spool and feed it in until the joint is complete.

- When the fitting is cool, clean off any excess flux residue with a damp cloth. The system should also be flushed to remove flux residue from inside the pipework, as this could corrode the inside of the pipes.

Table 5.1 summarises what types of fitting can be used on the various grades of copper tube.

Jointing method	Grade X	Grade Y	Grade Z	Grade W
Capillary fitting	✓	✓	✓	✓
Type A compression fitting	✓		✓	✓
Type B compression fitting	✓*	✓		

*Not underground

Table 5.1: Correct fittings for different grades of copper tube

Flux

Flux for use in the plumbing industry is usually in paste form. It may be active (it will clean the pipe when you apply heat) or non-active (it will not clean the pipe when heated). Flux is usually chemical-based, so you must read the COSHH advice on the label. Flux is applied to:

- clean the pipe
- prevent oxidation while heating
- assist with the **whetting** process
- float away impurities
- assist heat transfer through the joint.

Capillary fittings

As with compression fittings, there is a vast range of soldered capillary fittings and the same design patterns apply to solder ring and end feed:

- adaptors
  - capillary to male and female thread
- slip couplings
- straight couplings
- tank connectors
- reducing fittings
- metric to imperial converters
- stop ends

Did you know?

Materials used for pipe fittings include copper and brass. The design will depend on the specific purpose for which the pipe is intended and the ease of manufacture.

Key term

BSPT – stands for British Standard Pipe Thread.

Find out

See if you can find at least three manufacturers of push-fit fittings for copper tube.

- tap connectors
  - straight
  - bent
- return bends
- cylinder unions
- elbows
  - spigot to socket
  - socket to socket
  - capillary to **BSPT**, spigot or socket
- wall connectors (or back-plate elbows)
- tees
  - equal
  - reducing
- offsets.

Push-fit joints

There are a number of types of push-fit joint available for use on hot- and cold-water supplies.

Push-fit joints are made from plastic or metal. In the illustration, a grab ring is used to lock the pipe in place, and a neoprene 'O' ring makes it watertight. You must ensure all pipe burrs are removed as these can cut the seal, causing leaks. These fittings are bulky, so they do not look attractive where they are exposed. Care must be taken when using these fittings as the earthing integrity of the electrical system may be compromised.

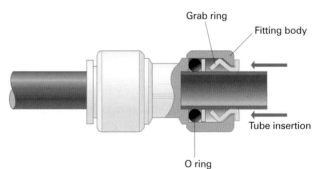

Figure 5.33: Typical push-fit joint

Low carbon steel pipe (LCS) and fittings

Often referred to as mild steel, low carbon steel pipe (LCS) is supplied in three grades:

- Light – colour-coded brown
- Medium – colour-coded blue
- Heavy – colour-coded red.

It is supplied painted in black or with a galvanised coating. Black-painted LCS must only be used on wet heating systems, oil or gas supply pipework, and not on hot and cold supplies.

Generally speaking, light-grade tube is not used for plumbing pipework. You are most likely to work with medium-grade pipes, and occasionally heavy-grade.

Medium and heavy grades are available in approximately 6.5 m lengths, ranging from 6 mm to 150 m in diameter, specified as nominal bore. Nominal bore means that the figure is not the actual bore of the pipe: this will vary depending on the thickness of the pipe wall, which in turn will be determined by the grade.

Methods of jointing LCS pipe

For smaller installations, there are two main jointing methods:

1. Threaded joints
2. Compression joints.

Threaded joints

Jointing LCS pipe can be done by cutting threads into the end of the LCS pipe to give a British Standard Pipe Thread (BSPT), then jointing them together with a range of female threaded fittings made from steel or malleable iron. The threads are cut using **stocks and dies** or a threading machine.

The jointing process is as follows:

- The LCS pipe is cut to length using heavy-duty pipe cutters while the pipe is held securely in a pipe vice. The LCS pipe can also be cut using a large frame hacksaw.
- Any burrs are removed from the inside and outside of the pipe.
- The end of the pipe that is being threaded is **chamfered** to provide a leading edge for the dies to catch the pipe wall.
- Cutting compound is applied to the end of the pipe before the thread-cutting operation is started. A thread is cut that is approximately 1½ to 2 threads longer than the length of the inside of the fitting.
- The excess cutting compound is wiped off.

> **Key terms**
>
> **Stocks and dies** – the stocks are the body and handle of the tool, and the dies are the actual cutter.
>
> **Chamfering** – the process of smoothing the cut edge by slightly bevelling or angling it

Figure 5.34: Pipe vice

Figure 5.35: Pipe cutter

Figure 5.36: Assembling steel pipework using a vice

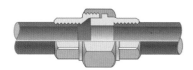

Figure 5.37: LCS union connector

Figure 5.38: Electric pipe-threading machine

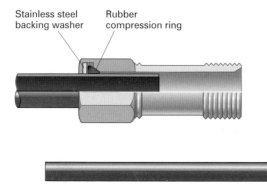

Stainless steel backing washer   Rubber compression ring

Figure 5.39: LCS compression coupling

- Threaded pipe sealant, which includes hemp and paste, PTFE sealing tape or gas thread sealing tape, can then be applied to the thread. The fitting is then screwed in place.
- The pipe and fitting are joined together using adjustable pipe grips or alternatively using a short length of pipe as a lever.

Because of the way in which screwed joints are made and installed (rotated on the pipe), it is sometimes difficult to remove or assemble lengths of pipework. Where this is the case, a union connector, which allows the pipework joint to be 'broken', should be used.

Using a pipe-threading machine

Pipe-threading machines provide a quicker and easier method of forming threads for LCS pipes. 'All in one' machines have a pipe cutter, deburring reamer, and stock head and dies.

You should be fully trained before using these machines.

Threaded pipe fittings for LCS

These fittings can be made of steel or malleable iron. Steel fittings can withstand higher pressure but are more expensive than malleable iron. They are manufactured to BS EN 10241 for steel and BS 1256 for malleable iron.

Malleable cast-iron fittings are adequate for smaller installations and, as with copper tube fittings, there is a wide range available. This includes:

- bends
  - spigot to socket
  - socket to socket
- elbows
  - spigot to socket
  - socket to socket
  - socket to socket union
  - socket to socket reducers
- tees
  - socket to socket to socket equal
  - socket to socket to socket reducing and increasing
  - socket to socket to socket pitcher equal
- crosses
  - socket to socket to socket to socket equal
- sockets
  - socket to socket equal
- unions
  - socket to socket equal
- bushes
- nipples.

Compression joints

There are a number of manufacturers' designs for compression joints. A typical one is shown in Figure 5.39.

The fitting is designed to allow steel pipes to be installed without threading. Made of malleable iron, compression couplings use locking rings and seals, which are tightened onto the pipe. They can be used on gas supplies and water and, although they are more expensive than threaded joints, they do save time on installation.

Plastic tube and fittings

Plastics fall into two main categories:

1. Products of the polymerisation of ethers:
   - polythene
   - polyethylene
   - polypropylene.
2. uPVC and ABS.

Products of polymerisation of ethers are used mainly for service-supply pipework (coded blue for underground services), hot water and hot-water heating.

uPVC and ABS are used mainly for waste distribution pipework and cold-water installations, although the latter will not be discussed here.

Table 5.2 summarises the typical jointing methods for plastic pipes.

Types of plastic	Mechanical joints	Solvent welding	Fusion welding	Push-fit 'O' ring*
Polythene	✓		✓	
Polyethylene	✓			
Polypropylene	✓		✓	✓
uPVC	✓	✓		✓
ABS	✓	✓		✓

*Discharge pipes and overflows only

Table 5.2: Typical jointing methods for plastic pipes

Methods of jointing plastic pipe

Fusion welding: polythene and polypropylene

This process requires specialist equipment and is used mainly for gas and water mains. It will not be discussed here.

> **Safety tip**
>
> Before using any solvents, read the instructions on their use and check the safety warning symbols.

Mechanical jointing

This applies to the jointing of:

- polythene pipework
- uPVC and ABS.

Polythene pipe: compression joint

Used for underground services, polythene pipe is identified by blue colour coding. It is also available, coded black, for internal use on cold-water services, although for domestic situations it is mainly used underground. The joints are made using gunmetal or brass or plastic fittings.

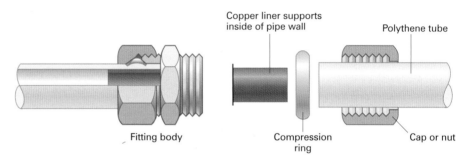

Figure 5.40: Typical polythene pipe fitting

The pipe is cut to length and deburred. The nut and compression ring are slid onto the pipe, and the copper liner inserted into it. The compression ring applies pressure, while the liner stops the plastic pipe from being squashed when the nut is tightened onto the body of the fitting. The pipe is fully inserted in the fitting, and hand tightened, before the tightening process is completed using adjustable grips or spanners.

uPVC and ABS: compression joint

These joints are restricted to use on waste pipes. Traps are a typical example, using a rubber 'O' ring or plastic washer to make the seal.

uPVC and ABS: solvent-welded jointing

This method is used to joint both uPVC and ABS, using solvent cement. The cement temporarily dissolves the surface of the pipe and fitting, causing the two surfaces to fuse together. The joint sets initially within 5–10 minutes but will take 12–24 hours before it is fully set.

This method is used primarily for joints on soil or waste pipes.

uPVC and ABS: push-fit joints

These are used mostly on uPVC or ABS soil and waste applications. The pipe is cut to length, making sure it is square and deburred. The outside edge of the pipe is chamfered to give it a leading edge, which makes it easier to push into the fitting. This also prevents the 'O' ring being dislodged.

Figure 5.41: Applying solvent to joint

A suitable lubricant is applied to the pipe and fitting, and the pipe is pushed home. The pipe must be withdrawn from the fitting to a length of about 10–15 mm to allow for expansion.

Push-fit connectors and pipework

These are also known as **flexible push-fit plumbing systems** for hot and cold water and central heating supplies. A number of manufacturers produce pipe and fittings for these systems, and the fittings can be used on either plastic or copper. These connectors are manufactured from a cross-linked polythene pipe to BS 7291.

Plastic pipe is supplied in diameters ranging from 10 mm to 22 mm and in 3 m and 6 m lengths, or in coils of 25 m, 50 m and 100 m.

Barrier pipes (which prevent **permeation**) must be used for vented and sealed central heating systems.

In most new buildings, construction methods have changed over the years and the materials used in plumbing and heating installations have also changed. Common practices now have the first floor flooring completed prior to any fixing of pipework. The plumber who uses plastic pipe now works like an electrician when it comes to first fixing and feeds the pipework through the joists from below in the same way as the electrician feeds cables. The name given to this practice is the 'cabling technique'.

There is a wide range of fittings available that are similar to their copper equivalents. In a typical system a mix of copper and **polybutylene** will be used: copper for exposed pipework and polybutylene for hidden pipework. The jointing and installation processes are fundamentally different from copper, but generally take less time to complete.

> ### Key terms
>
> **Permeation** – air entering the water supply through the wall of the pipe.
>
> **Polybutylene** – a product of the polymerisation of ethers.

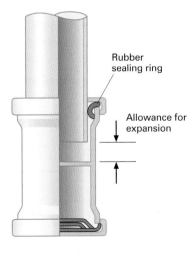

Figure 5.42: Push-fit joint used for low-pressure plastic pipework

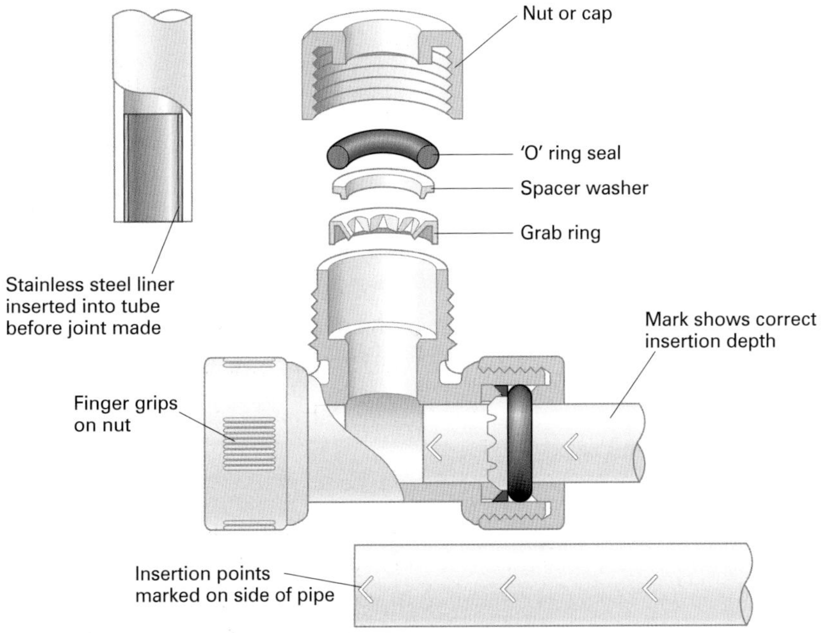

Labels: Rubber sealing ring; Allowance for expansion

Labels (Figure 5.43): Stainless steel liner inserted into tube before joint made; Nut or cap; 'O' ring seal; Spacer washer; Grab ring; Mark shows correct insertion depth; Finger grips on nut; Insertion points marked on side of pipe

Figure 5.43: Typical de-mountable fitting for polybutylene pipe

The cutting and assembly sequence of the HEP$_2$0 flexible plumbing system is as follows.

The cutting and assembly sequence

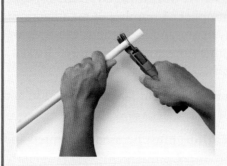

Step 1: Use only HEP$_2$0 specialist pipe cutter to cut the pipe. Never use a hacksaw. Wherever possible, cut the pipe at the 'V' marks provided. To ensure a clean, square cut, rotate the pipe while maintaining pressure on the cutter until the pipe is severed.

Step 2: Ensure the pipe end is clean, and free from burrs and surface damage. Insert a HEP$_2$0 support sleeve into the pipe end.

Step 3: Push the pipe firmly into the pre-lubricated fitting. A secure joint has been made when the end of the retaining cap has reached the next 'V' mark on the pipe. Never knock the fitting onto the pipe, or slacken the retaining cap prior to pipe insertion: these actions will not ease jointing.

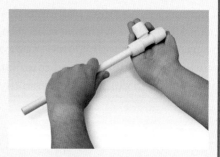

Step 4: Tug back on the pipe to ensure the grab wedge engages correctly. Do not undo the retaining cap after pipe insertion.

Jointing different materials

Fittings are available that are designed to join copper to plastic; usually, these are the plastic push-fit variety. Fittings for jointing copper to LCS are also available.

Figure 5.44: Lead to MDPE

Although lead pipework is no longer allowed for new installations, it may be necessary to join into an existing supply, either to extend a system where a long run of lead pipework cannot be replaced, or where the joint is at the end of an underground service pipe.

There are several types of fitting on the market for jointing plastic to lead for below-ground use, and for jointing lead to copper. Only use WRAS-approved products – joint wiping lead to copper is no longer permitted under the Water Regulations.

Figure 5.45: Lead to copper

Stainless steel

Stainless steel is not used on domestic pipework installations. You may, however, come across it on systems installed during the copper shortage of the late 1960s, when stainless steel was used as a replacement.

Stainless steel was jointed using compression fittings. As the outside diameter was the same as copper pipe, it is logical to connect to copper using a compression fitting if carrying out maintenance work.

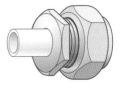

Figure 5.46: Lead to end-feed

Cast iron

You are unlikely to install this on new domestic installations, except on historic or listed buildings, but you could come across it during maintenance work, as it was once used extensively on gutters, rainwater pipes and soil pipes. It is still widely used on industrial/commercial buildings due to its resistance to mechanical damage.

Working life

Ollie has to produce a list of the fittings and pipework required for a bathroom he is going to fit. It's the first time he has had to decide for himself exactly what he will need. The bathroom installation is a complete decommission including the removal of the lead pipework, which feeds both hot and cold water to all the appliances in the room (bath, wash basin, WC). The waste pipework is not under consideration as the work is minimal.

Ollie needs to get this right. Not only is it his first job on his own but it is out of town and getting extra bits will be really awkward. At the same time, he knows that due to the costs and massive range of fittings available, he can't take it all with him.

The bathroom is situated on the ground floor, so has a concrete floor. All the pipework is therefore exposed and will not be boxed in.

- Which pipe material and fittings should Ollie select?
- Under separate headings consider the advantages and disadvantages of both copper and plastic pipework including the fittings applicable to each pipework system.
- Using the size and layout of any bathroom, measure up and decide which fittings will be required to complete the installation. The list should be exhaustive and include any sundries required.
- What are the regulations (including health and safety) for the removal, handling and reconnection to the lead pipework?

Bending pipework

Bending, measuring and marking out are essential basic skills for any plumber. We will concentrate here on bending copper pipe, which can be done by hand or machine, and low carbon steel pipe, which is usually bent by hydraulic machines. The pipe can also be clipped into steel pre-formed 90° brackets for tighter bends.

Bending pipe, rather than using ready-made fittings, has the following advantages:

- It produces larger-radius bends than elbow fittings (larger-radius bends have less frictional resistance).
- Using bends costs less than using fittings.
- Long sections of pipework can be prefabricated before installation, saving time.
- There is less chance of leaks.

Bending methods – copper pipe

The pipe grade suitable for bending is R250 Grade X. R290 Grade Z is tempered and has thin walls, and cannot be bent. R220 Grade Y is softer, with thick walls, and is supplied in coils; it is not usually used internally in dwellings. There is also R220 Grade W, which is used for micro-bore installations.

Copper pipe can be bent by hand or by machine.

Bending by hand

This is a method of bending pipe when carrying out maintenance and repair work. You may be working in a loft space, for example, where it would be much quicker to fabricate the pipe in situ. The best way of doing this is by using a **bending spring**.

Bending springs can be used externally or internally. In either case you pull the bend against your knee to get the desired angle. When using an internal spring there are a few things to remember:

- Don't try to bend a piece of pipe that is too short. You would have to apply a good deal of physical pressure, which could lead to injury, or kink the pipe, or both.

Figure 5.47: Using an external spring

Figure 5.48: Using an internal spring

- Don't make the radius of the bend too tight, or you might find that you can't remove the spring.

- Always pull the bend slightly further than the required angle, and then pull it back to the required angle. This will help to release the spring.

- The spring has a 'hook' at the end. This can be twisted clockwise and will tighten the spring coil inwardly, making it easier to remove the spring.

Spring pipe bending can be set out with the same accuracy as machine bending. All bends appear to gain length when bent. This is because, before the pipe is bent, the length A to B to C is longer than the actual bend A to C (see Figure 5.50). It is possible to work out accurately the length of pipe to be bent in relation to fixed points by finding out the length of the pipe actually occupied by the bend as follows.

Figure 5.49: Pipe bending with internal spring

- Deciding the centre-line radius of the bend. This is usually four times the diameter of the pipe (4D).

- Apply the following formula:

$$\frac{\text{Radius} \times 2 \times 3.14}{4}$$

For example, for a 15 mm diameter pipe:

radius of bend = 4D = 4 × 15 = 60 mm

To find the length of bend, substitute this figure into the formula above:

$$\frac{60 \times 2 \times 3.14}{4}$$

This gives 94.2, rounded to 94 mm. This is the length of pipe required to form the bend.

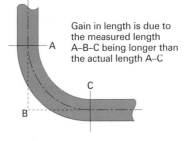

Figure 5.50: Apparent gain in length when tubing is bent

- Now that you have calculated the length of pipe taken up by the bend, add this to the pipe lengths to give an overall pipe length. For example:

500 mm + 400 mm + 94 mm = 994 mm total pipe length.

You would usually be given an end-to-centre measurement to work with, and from which to deduct the gain.

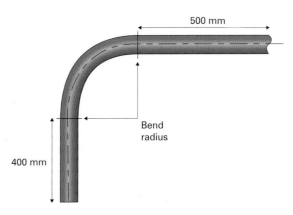

Figure 5.51: Finding the length of pipe required to form a bend

Figure 5.52: Hand bender

Figure 5.53: Small mini-bore (8/10mm) bender

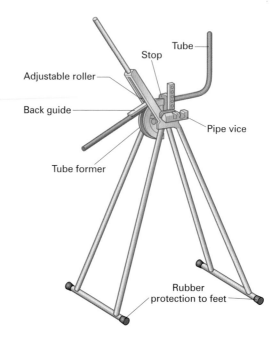

Tube
Stop
Adjustable roller
Back guide
Pipe vice
Tube former
Rubber protection to feet

Figure 5.54: Free standing (stand bender)

Bending by machine

This is the most common method used for bending copper tube. Bending machines can be either hand-held (**hand bender**) or free-standing (**stand bender**), and they work on the principle of leverage.

You will need to prove competence in bending and joining tube for your plumbing qualification.

The small hand-held bender is used for pipe sizes of 15 mm and 22 mm, and is light and portable. The free-standing bender can handle pipes up to 42 mm and uses a range of sizes for the back guide and former. It is important that the machines are set up properly. If the roller is too loose, this will cause rippling on the inside of the radius of the pipe. If it is too tight, it will reduce the pipe diameter at the bend. This is called **throating**.

Types of machine bend

There are three main types of bend:

- 90° or square
- offset
- passover.

90° or square bend

You may be given a measurement from a site drawing, but it is more likely that you will take a site measurement yourself. Measurements should be taken from a fixed point to the back of the bend.

Key term

Throating – when a pipe's diameter has been reduced at the bend by the roller, guide and former being too tight.

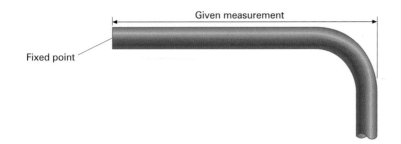

Figure 5.55: Taking the measurement

A pencil mark is made from the fixed point (end of tube) to the required length.

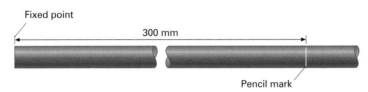

Figure 5.56: Marking and tubing

The tube is set up in the machine with the mark squared off from the outside of the former. Then, the back guide is positioned, the roller correctly adjusted and the bend pulled. Return bends can be made using the same technique, only now the back of the first bend becomes the fixed point.

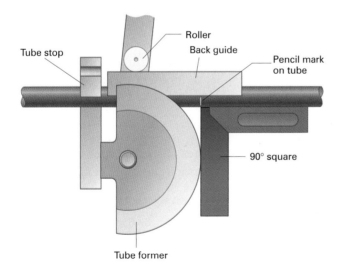

Figure 5.57: Positioning the tube in the bender

Offset

The first set is made to the desired angle. The first angle is not critical, but will usually depend on the profile of the obstacle you want to offset the pipe around.

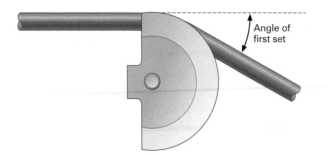

Figure 5.58: The first set

Now that you have made the first set, the tube is reversed and returned to the machine.

A straight edge is placed against the former, parallel to the tube, and the measurement for the offset is taken from the inside of the tube to the inside edge of the straight edge.

Once the tube has been adjusted against the stop of the machine, it is a good idea to mark the edge against the former, in case of any movement.

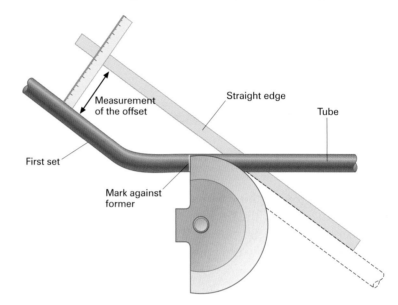

Figure 5.59: The second set

Note the dotted line on Figure 5.59 showing the finished position of the second set. This shows that the measurements are taken from the inside to the back of the tube, so in effect it is a centre-to-centre line measurement.

Passover bend

This would be used to clear other obstacles, such as another pipe, and could be a passover offset or a crank passover bend.

Remember

You can place a piece of pipe alongside the former to help you visualise and measure where the pipe will be when it has been bent to its finished form.

Figure 5.60: Passover offset

Figure 5.61: Crank passover bend

The measurements for a passover bend are taken in the same way as for an ordinary offset. The angle of the first pull will be governed by the size of the obstacle it has to 'pass over'.

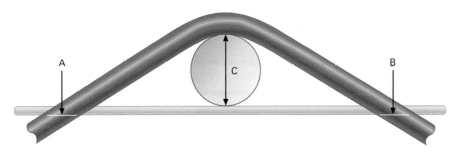

Figure 5.62: Measurement of crank passover

A straight edge is placed across the bend at the distance of the obstacle and the pipe marked. This will be the back of the finished offset.

The pipe is then returned to the machine and, when the first mark lines up with the former, it can be pulled to the position shown. The pipe is then turned around in the machine, the second mark lined up in the former and the pipe pulled to complete the passover.

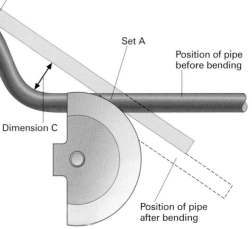

Figure 5.63: The final set in a crank passover bend

Working life

Mark has six 15 mm copper pipes, which have to rise vertically in the corner of a room and then turn through 90° and travel horizontally at ceiling level across the top of the wall.

He has bent three so far when the customer comes along and comments that 'they look awful'. The customer accepts that the pipes will have to be on show but he is unhappy about the distances between the bends because they vary so much.

Mark has been careful in his setting out and has allowed 50 mm centre to centre to each pipe, but the customer is right they do look a mess where all the bends are. They certainly aren't 50 mm apart at that point.

Note: the bends must be 'pulled bends'. It is an unacceptable response to use elbows to resolve the issue; likewise, the pipework may not be boxed in.

- What is the problem here? Why are the pipes at different centres where the bends occur? Suggest a solution that would both satisfy the customer and enable Mark to be once again be proud of his work.
- Indicate the start/finish centre line/end markings, which would be required to make your solution work on six pieces of scrap tube/paper.
- At the end of the scenario it was pointed out that it would not be acceptable to use elbows. Why would elbows be an unacceptable solution to the problem? Do not consider cost in this case.

Bending methods – low carbon steel pipe

LCS pipes are usually bent using a hydraulic pipe bender.

Only medium- and heavy-grade tubes are used in plumbing systems; they are manufactured to BS 1387, and are supplied in black-painted or galvanised coatings. You are more likely to work on medium-grade tube. Galvanised pipe should not be bent, as this will cause it to lose its coating. It is unlikely, however, that you will come across this material outside maintenance work.

Bending with a hydraulic machine

Hydraulic machines are needed to bend low carbon steel tubes, owing to the strength of the material and the thickness of the pipe. For this reason, the pipe does not need to be fully supported with a back guard, unlike copper pipe. Hydraulic bending machines are used to form most bends, including 90° and offsets.

The hydraulic mechanism is usually oil-based. Liquids are incompressible, so they can exert a considerable force on the pipework once under pressure.

Figure 5.64: Hydraulic bending machine

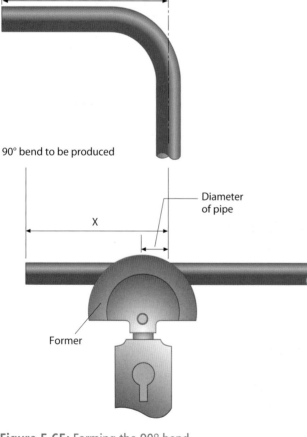

90° bend to be produced

Diameter of pipe

X

Former

Figure 5.65: Forming the 90° bend

Six steps to forming a 90° bend

1. Mark a line on the pipe at a distance from the fixed point, where the centre line of the finished bend is required.
2. From this measurement you deduct the nominal bore of the pipe. This is because there is a gain in length of one pipe diameter when bends are made.
3. Make sure the correct sized former is in the machine.
4. Put the pipe in the bending machine and line up the mark with the centre of the former.
5. The machine can then be worked to apply pressure and bend the pipe to 90°.
6. Due to the elasticity of the metal, you need to take it to approximately another 5° over 90° to allow it to 'spring back'.

Five steps to forming an offset

It is a good idea to make a template from steel wire to help you achieve the required offset profile. The method of marking out the offset is similar to that of copper, as follows:

1. Mark off the required measurement for the first set onto the pipe.
2. Place the pipe in the machine but do not make any deduction. The measurement X mm is from the fixed end of the pipe to the centre of the set (see Figure 5.67).
3. Pull the first set to the required angle.
4. Take the pipe from the machine and place a straight edge against the back of the tube. The measurement of the offset is marked at Point A.
5. Replace the tube in the machine and line up the mark with the centre of the former (see Figure 5.68). Pull the second set and check against the wire template. Again, allow a 5° overpull for the spring back. You will need to level the pipe in the machine to prevent any twist or distortion in the set.

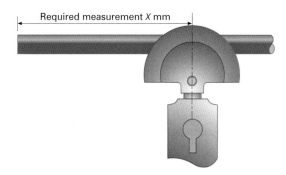

Figure 5.66: Locating the pipe for the first bend

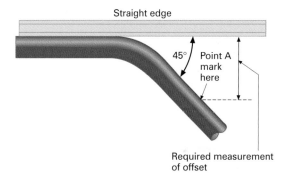

Figure 5.67: Setting out for the second set

Faults and defects

The main problems encountered when bending low carbon steel pipe are usually caused by incorrect marking out, such as not making the deduction for the nominal bore, so the bend does not fit when it is pulled. In the case of offsets, if the pipework is not set up level in the machine, then the resulting offset will be screwed or twisted.

> **Remember**
>
> During the bending operation, the pipe can become wedged into the former. Do not hammer the former or the pipe to remove it. You can place a timber block on the pipe and give it a sharp tap with a hammer. Alternatively, remove the pipe and former. Place a wooden block on the floor and strike the end of the pipe on the block. Try to hold the end of the former so that it doesn't fall to the ground.

Charlie is set the task of forming a total of 20 low carbon steel pipe pre-fabricated bends for a contract his company is working on, a heating system at a new school.

A tube-bending schedule has been made up, which listed the bends (see Figure 5.68).

Charlie used a similar hydraulic centre-bending machine a few years ago when he studied at college and so he assumed that he didn't need any instruction.

Charlie again uses his previous knowledge to do some quick calculations (see Table 5.3) so that he could cut each tube to the required length and save time later by not having to cut the tubes again. His confidence was so high that he not only cut the tubes but threaded them before he bent them.

When the pre-fabricated tubes where delivered to site, they would not fit into the required positions. The whole idea of this exercise had been to save time and money. Unfortunately, by the time the site engineers had sorted through all the issues and had to start again anyway, it cost the company more money in the loss of the materials and many hours of unnecessary work.

- Where did it all go wrong?
- Who is in the wrong here: Charlie, his employer or both of them?

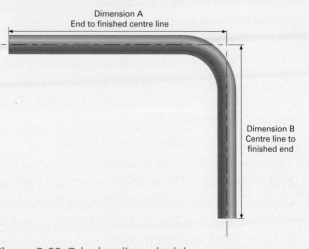

Figure 5.68: Tube-bending schedule

- What did Charlie forget to allow for when cutting, threading and bending the tubes?
- What is the missing dimension, which Charlie had forgotten about, and why would it be so important?
- How much LCS tube would be required to complete the work Charlie was asked to complete? Assume he got it right first time.

Reference number	Quantity	Tube diameter (mm)	Start to centre measurement (mm) Dimension A	Measured end threaded	Threaded both ends	End to centre measurement (mm) Dimension B
1	6	15	270	Yes	Yes	300
2	4	15	350	Yes	Yes	345
3	6	15	400	Yes	Yes	360
4	4	15	550	Yes	Yes	550

Table 5.3: Charlie's calculations

1. Which grade of copper tube is the most commonly used for general purposes? Its temper is classified as half hard.
2. Name the three common methods of jointing copper tube.
3. Name the two types of compression joint and soldered capillary joint.
4. Which type of solder is not allowed on pipework for either hot or cold supplies?
5. How many grades of LCS pipe are there? How are they classified and what colours are used to refer to each one?
6. On smaller installations, what are the two main methods of jointing LCS pipe?

7. What is meant by the term 'cabling technique'?
8. Other than spring bending (internal/external) and machine bending, what is another means of successfully bending copper tube?
9. Why is over bending necessary when using a hydraulic bending machine?
10. After using a hydraulic bending machine, how should the former be removed from the bent piece of LCS tube both correctly and safely?
11. What will be the result if the pipework is not set up level in the machine producing such bends as offsets?

3. Know, and be able to apply, general site preparation techniques for plumbing and heating work

Activities when working on plumbing and heating

This section looks at the activities you need to carry out when you are preparing sites for work.

Designing and selecting materials and equipment

At Level 2 the design and material specification will be created for you by others. However, you must make sure that what you are installing is suitable and fit for purpose, and that the work carried out meets regulations.

Installing systems and components

When installing any system and components you should always follow the manufacturers' instructions and recommendations along with any other regulations such as British Standards, Water Regulations and Building Regulations.

Maintaining and dealing with faults on systems and components

Installations have components that have moving parts or safety devices that may from time to time fail or need to be serviced and/or checked to ensure that they still function. Sometimes there is also failure due to fatigue, which may result in a leak.

Commissioning a system

There are very specific guidelines for commissioning a system.

- The system should be thoroughly flushed on completion of the installation.
- A corrosion inhibitor must be added to the system.
- The system must be commissioned in line with manufacturer requirements.
- The property owner must be fully instructed in the effective operation of the system and all user operating instructions must be left on site.
- A commissioning record must be left on site.

The installation of a boiler, central heating or hot-water system is identified under the Building Regulations as work on a controlled service or fitting. Therefore, such an installation requires one of the following:

- a notice must be given to the local authority before the work begins
- a compliance certificate must be issued on completion of the work by an installer who is a member of a competent persons scheme.

Decommissioning systems and components (temporary and permanent)

When working on a system either for maintenance, repair or replacement of a section, you may need to decommission it. This means that you need to take it out of operation and may involve draining it down or simply isolating it. You also need to take into consideration others who may be affected by your actions. This could be classed as temporary decommissioning.

If you are removing an old system, or part of it that is no longer required, then this could be classed as permanent.

Information passed to the customer

It is good business practice, as well as a requirement of the L1 Building Regulations *Approved Document (Central Heating Systems)*, to provide information to the users on the control and operation of a new system installation.

You will probably find that a high percentage of callouts are from customers who do not know how to work their system controls properly. Here is a list of what should be done for the customer:

- Provide a company folder containing:
  - an emergency contact number for burst pipes, etc.
  - a general advice line number – you might be able to sort a problem out over the phone
  - customer guidance leaflets that most manufacturers produce; also go through these with the customer.
- Label the various fittings or components that the customer may need to use in an emergency, e.g. stop valve, gate valve.
- Show them, or tell them, where the service valves are, and what they do.
- Walk them around the system: show them the various components and tell them what they can touch and what they must not touch.
- Explain what the components do in simple terms, e.g. 'This is the cylinder thermostat, and, when the water in this cylinder gets to that pre-set temperature, it shuts off the heat to the cylinder; you don't have to alter that setting.'
- When you are finished, give your customer the chance to ask questions.

- Leave the manufacturer's installation and user instructions on the job for future reference purposes.
- Ensure that a full commissioning record is completed for the work – it's not only good sense, it's also the law.

Checking the site for damage and protecting the site

When working on site you will use a wide range of tools and equipment. Many of these have been covered earlier in this unit. Unit 1 covers the health and safety issues you need to be aware of when working with and using tools.

Depending on the size of the job, the following checklist will cover most of the preparation items:

- Have you confirmed with the client the start time and date?
- Are all the tools, materials and equipment required to do the job on site?
- Are electrical tools safe to use?
- Is the work area clean and safe to work in?
- If using access equipment (ladders etc.) is it safe to use?
- If you are working in a loft, do you have adequate lighting to see what you are doing (torch, inspection lamp, etc.)?
- If working in an occupied dwelling, have you checked the work area for any pre-work damage? If you have noticed any damage, have you notified the customer prior to starting the work? This could be done by using a pre-site inspection report form or a statement given to the customer.
- Again, in an occupied dwelling, have you made sure you have protected the customer's property before you start?
- Use dust sheets (but not where they could be dangerous, such as on stairs).
- If working in a garden of flower beds, lay down walking boards to avoid damage.
- Request that vulnerable furniture is moved.
- Request that carpets are taken up if necessary.
- Cover up sanitary appliances if working in the bathroom area.

> **Safety tip**
>
> Never leave tools such as hammers or spanners where they could be accidentally knocked, such as into sanitary appliances.

Work methods for preparing buildings for installation work

At the end of this section you should be able to:

- describe the procedures for lifting and replacing floor surfaces
- state the requirements for cutting holes and notching timber joists

Figure 5.69: Crowbar (wrecking bar)   **Figure 5.70:** Wood chisel   **Figure 5.71:** Pad-saw

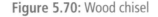

- describe the procedures for cutting holes and chases through a range of building materials
- making good masonry surfaces.

You will probably appreciate by now that a plumber's job is both interesting and varied. As a result, you will need to learn some additional skills to carry out work other than just plumbing. For example, you will have to run pipes under timber floors, which involves notching joists.

In existing properties, gaining access to work under the floor will mean lifting floor surfaces, so you will need to know how to do that. Often, both on new work and maintenance work, you will be required to cut or drill holes in brickwork, blockwork, concrete and timber, so that will also be covered.

Lifting floor surfaces

Lifting floorboards using hand tools

The hand tools shown in Figures 5.69–5.72 are used for lifting floorboards.

Figure 5.72: Bolster chisel

Lifting a length of floorboard to run pipework through joists

This usually involves lifting a single length of floorboard. Lifting a full length is easier, because you won't have to cut across the board.

Using a hammer and sharp bolster, carefully cut the tongue-and-groove joint down either side of the floorboard. Alternatively, you could use a pad-saw.

Good practice requires punching down the nails to enable the board to be removed. Alternatively, a wrecking bar, or draw bar, can be used to prise up the floorboard and nails. Once the board is partially lifted, pushing it down slightly will reveal the nail heads, allowing their removal with a claw hammer.

If you need to lift only part of a board, you will have to make one, possibly two, cuts across a joist, so that when the board goes back it has a firm fixing point. If you cannot locate a joist, you will have to insert timber cleats. You can locate a joist by finding the floorboard nails.

Cross-cuts on a floorboard can be made using an extremely sharp wood chisel or a purpose-made floorboard saw.

Lifting floorboards using power tools

This is done using a circular saw.

Remember the following when using a circular saw:

- Use 110 V supply only.
- The depth of the cut must not exceed the depth of the floorboard.
- If possible, remove a trial board using hand tools to check for electrical cables or hidden pipework.
- A guard must always be fitted to the saw blade.

The saw can be used to cut down the full length of the tongue and groove on each side of the board. A cross-cut is made, again over a joist, making sure that the blade does not hit the nails.

Cutting traps in floorboards

Cutting a trap uses the same method as removing a single floorboard, using either hand or power tools. A trap requires you to cut more boards, but in shorter lengths.

Replacing floorboards and traps

The floorboard length or trap should be screwed back into position to make future inspection or maintenance easier. Use countersunk wood screws. When re-fixed over pipework, the board surface should be marked accordingly, e.g. 'hot and cold pipework'.

If it is not possible to find a joist to refit the board or trap, cleats must be used to support the board end.

Figure 5.75 shows a trap or board replacement over joists and using cleats.

Removing chipboard

This is more difficult than removing floorboards, as chipboard is laid in wider sheets.

The best way to remove a chipboard section is by using a circular saw. If a power supply is not available, a section of board can be cut using a floorboard saw. The section of board to be removed should be marked out across the board, creating guidelines to follow for the cut. If a pad-saw is used to make the cut, it is helpful to drill holes in each corner of the area to be lifted in order to start the cutting process.

If this method is used, a new piece of chipboard will be needed to replace the removed section.

Figure 5.73: Circular saw

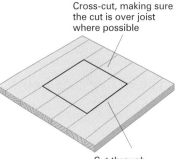

Cross-cut, making sure the cut is over joist where possible

Cut through tongue using hand or power tools

Figure 5.74: Cutting a trap

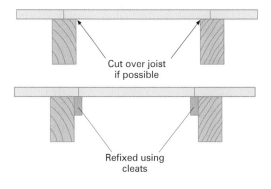

Cut over joist if possible

Refixed using cleats

Figure 5.75: Replacing floorboards

Holes drilled in each corner to start the cut with either pad-saw or jig-saw

Figure 5.76: Preparation for removal of chipboard

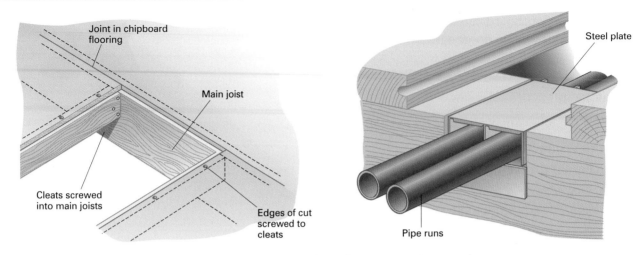

Figure 5.77: Re-fixing chipboard

Figure 5.78: Pipe guards

Replacing chipboard

As with floorboards, chipboard should be screwed back into position.

Pipe guards

Pipe runs under floors should be marked. A more effective way of protecting pipes that pass through joists is to use pipe guards. Figure 5.78 shows a typical example of their use.

Timber joists

In a lot of newly constructed dwellings, you will not be allowed to drill or notch the joists as this would have a serious impact on the structure of the building. In this type of dwelling the joists will have knock outs in them, which have been designed for use by the manufacturer.

It is inevitable that joists will have to be drilled or notched to permit pipe runs under timber floors. The preferred method would be to drill the joist in the centre of its depth, as this is where there is least stress. In practice, unfortunately, apart from when using plastic hot- and cold-water supply pipe, this tends to be impractical.

The main thing to remember, therefore, for either notches or holes, is not to weaken the joists. This also applies to the distance from the wall where the joist is notched or drilled.

The Building Regulations set out requirements for notching or drilling joists, and these must be followed at all times.

A worked example for notching a joist (see Figure 5.79)

Consider a joist 200 mm deep (D) and 2.5 m long (the span, S).

Any notch must have a maximum depth of $\dfrac{D}{8}$ mm. Therefore, for our joist, the depth of the notch is $\dfrac{200}{8} = 25$ mm

The minimum distance from wall is $7 \times \dfrac{S}{100}$ mm.

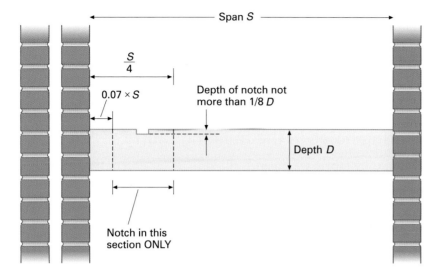

Figure 5.79: Notching joists

For our joist, this is $\dfrac{7 \times 2500}{100} = 175$ mm

The maximum distance from wall is $\dfrac{S}{4}$ mm from its bearing, giving a

maximum span for our joist of $\dfrac{2500}{4} = 625$ mm

Joists are normally cut by hand, using a hand or floorboard saw. It is cut to the required depth and width, and the timber notch removed using a hammer and sharp wood chisel.

The width should be enough to give freedom of movement, in order to allow for expansion and contraction, particularly in the hot-water supply.

Drilling a hole in a timber floor joist

There are similar requirements laid down for drilling holes in joists, although the calculation figures are slightly different:

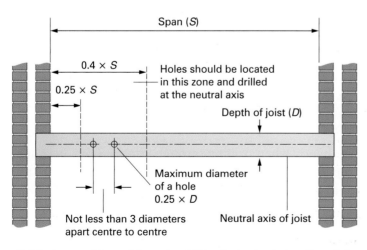

Figure 5.80: Acceptable positions for drilling holes in timber floor joists

Remember

When drilling a joist, make sure that what you are doing will not weaken the joist.

- The maximum hole diameter is one quarter of the depth of the joist.
- Holes should be positioned no closer together than 3 times their diameter.
- The minimum distance from the wall is 0.25 × the span.
- The maximum distance from the wall is 0.4 × the span.

Cutting holes in the building fabric

By building fabric we mean:

- brickwork
- concrete
- blockwork
- timber.

Hand or power tools can be used to cut holes.

Power tools

The selection of drill will depend on the job in hand (see Table 5.4).

Rotary hammer drill	500 watt	850 watt	1400 watt diamond core drill
Drill bit required for brick/block	30 mm	42 mm	152 mm
Drill bit required for concrete	24 mm	30 mm	120 mm

Table 5.4: Choosing the appropriate drill bit

Drill bits are designed for specific tasks and are purpose made for brick, block, concrete, steel or wood. A diamond-core drill uses either diamond- or tungsten-tipped bits.

Core drills are excellent for drilling through brickwork, blockwork or concrete where you need to pass large-diameter waste or soil pipes, or flue pipework. They provide a much neater finish to the job, and leave less making good than hand tools.

Figure 5.81: Selection of wood bits

Wood bits or wood-boring bits

Wood bits are useful for drilling holes in joists or other timber constructions for the passage of pipes.

Hole saws

Hole saws are handy when drilling through kitchen units in order to pass waste pipes through the side or back of cabinets. They are also used to drill holes in plastic cold-water storage cisterns.

Hand tools

Hand tools for brick, block and concrete include cold chisels, brick bolsters (preferably with guard), plugging chisels and club hammers.

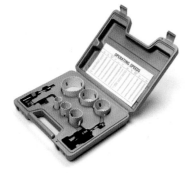

Figure 5.82: Hole-saw set

Tools for timber include claw hammers and wood chisels, and brace and bits for cutting or boring holes in timber. However, power tools have generally replaced the brace and bit.

Hand tools such as club hammers, cold chisels and bolsters are used to cut holes and chases in brickwork, blockwork or concrete. Care must be taken to cut out only the minimum of material required for the passage of the pipe, in order to keep making good to a minimum. Plugging chisels are used to chase out mortar between brickwork, and the plumber does this to let in sheet lead flashings or timber-fixing plugs.

A brace and bit, or cordless drill, is used to bore holes through timber in the same way that power tools are used. However, the power drills can use much larger-diameter bits, and do not require as much hard work.

Making good

For most jobs, a mortar mixture of 4 parts sand to 1 part cement will be adequate for pointing any brickwork joints disturbed while doing the job, and for **making good** the gap around the pipe penetration. Mastic sealant, either clear or close to the colour of the pipework, can be used as an alternative.

> **Key term**
>
> **Making good** – repairing and finishing off the brickwork, blockwork and concrete.

Working life

Jamie is a Level 2 plumbing apprentice, working on a new-build site. His supervisor sends him to 'first fix' Plot 13. Jamie has carried out first fixing on quite a few sites, so he feels confident. Once inside the building, he notices the floor joists on the first floor are slimmer than usual and look different from the others he has worked on. He believes this was done to save money, but the joists were manufactured from compressed chipboard, which are incredibly rigid for their weight.

Jamie carries out some calculations with regard to the positioning and depths of his holes and their required diameters. He decides on the preferred routes around the building within the constraints of the calculations.

By the end of the day he has completed all the drilling and has most of the first fix completed as well. When his supervisor arrives to see how Jamie has got on, he is furious. The joists were of a new design where drilling or notching was not allowed, and the strength of the joist has been compromised.

- Who was to blame: Jamie, his supervisor or both of them? What should Jamie have done when he noticed something was different?
- How is this type of joist allowed to have cables/pipework passing through it? If you can't drill them, how is this done? Would the joists have to be repaired or replaced? What could happen to the structure of the building if this is not corrected?
- Which member of the construction team would take a decision regarding what happens next?

Progress check

1. When handing over an installation to a customer, what information should you pass on, by hand and verbally?
2. Produce a checklist of things you need to do and be aware of when you are preparing to start work.
3. What is the purpose of a pipe guard?
4. The depth of a joist is 225 mm. What is the maximum depth of the notch allowed?
5. The span (length) of a joist is 3.1 metres. What is the maximum distance of a notch away from the wall?
6. What are the minimum and maximum distances from the wall at which a hole may be drilled into a timber joist, when multiplied by the span?
7. Define the term 'making good'.

4. Know, and be able to apply, how to use fixings, clips and brackets to support domestic plumbing and heating pipework and components

This section covers the various types of fixing devices used to secure pipework so that it looks neat and is kept in its proper position. Fixings should provide sufficient support to withstand possible accidental damage from people treading on pipework, children pulling at it and so on.

As a plumber you will be required to fix pipework, sanitary ware and appliances to various surfaces. You will also need to know how to re-fit boards and access traps in timber floors.

Pages 243–48 cover the measuring and marking-out of fixings for pipework, and the clip and bracket types used for particular pipework. Refer to this section to see when some of these fixings are used.

Fixing devices

These include:

- screws
- brass wood screws
- self-tapping screws
- turn-threaded wood screws
- steel countersunk screws
- chipboard screws
- mirror screws
- plastic plugs
- plasterboard fixings
- cavity fixings
- nails.

Screws

Screws are specified by their length in inches, and gauge. The usual lengths and gauges used in plumbing range from $\frac{5}{8} \times 8$ for fixing 15 mm saddle clips to $2\frac{1}{2} \times 12$ for fixing radiator brackets.

This is just a rule of thumb, and as a plumber you will often have to make a decision about the length and gauge that is right for a particular situation. For example, if the wall you are fixing to is not in good condition, you might have to use a longer screw and thicker gauge. Trial and error and experience are factors here.

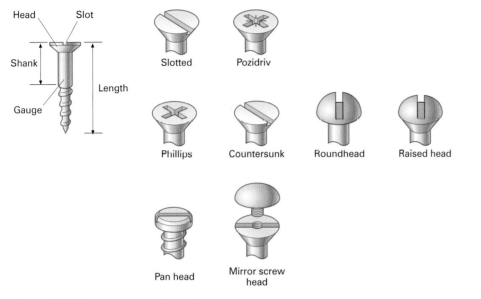

Figure 5.83: Some commonly used screws

Brass or alloy screws are used internally where they could be affected by moisture; this would include use in sanitary appliances, for example. They are also used externally (mostly in alloy form due to costs) for soil and rainwater fixings and wall-plate elbows for hose union bib taps.

Self-tapping screws are used when fixing into metal sheet. This is not typical in domestic installations but could apply if you needed to clip to a metal stud partition. This involves drilling a pilot hole in a smaller gauge than the screw, using a steel drill bit, and then screwing into it.

Steel countersunk screws are used for general purposes, such as fixing clips and radiator brackets.

Chipboard screws, as the name suggests, are used for chipboard fastenings. For example, if you had to fix an access trap in a chipboard floor following some installation work.

Mirror screws are used where appearance is important, and are often used for fixing timber or plastic bath panels.

Plastic plugs

Plugs come in a range of gauges that are appropriate to the gauge of the screw (see Table 5.5). They are colour coded for ease of selection.

Screw size (gauge)	Drill size (mm)	Plug colour code
6–8	5	Yellow
8–10	6	Red
10–14	7–8	Brown
14–18	10	Blue

Table 5.5: Size matches for screws, drill bits and plugs

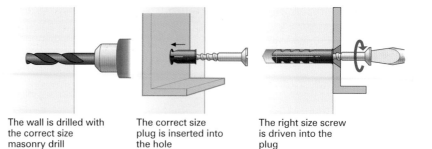

The wall is drilled with the correct size masonry drill

The correct size plug is inserted into the hole

The right size screw is driven into the plug

Figure 5.84: Fixing operation using plug and screw

Plasterboard fixings

Plasterboard fixings are used when fixing to plasterboard stud partitions, where there is nothing solid to fix into at the back of the plasterboard. There are several types available. Spring toggle, rubber-nut and cavity fixings can be used for many fixing jobs.

When using a spring toggle, drill a hole in the plasterboard big enough to take the toggle when folded. This is inserted through the hole, and when it is in the space behind the board, the toggle is pushed open by the spring.

Rubber-nut fixings work by drawing the nut mounted in the rubber towards the screw head. As it tightens, the rubber is squashed to form a flange at the back of the board.

Cavity fixings work on the same principle as rubber-nut fixings, but this time an aluminium body is 'squashed' to form the flange. When using this type of fixing, a small pilot hole is drilled into the plasterboard and the complete fixing is screwed into the board. The screw is then removed, leaving a fixing point similar to a plastic plug.

Other fixings

The fixings described so far should be adequate for day-to-day work. There will be occasions, however, when you need larger fixings, perhaps for a heavy appliance, or where the fixing surface is in poor condition.

Coach screws are normally supplied with their own purpose-made plastic plug, and are fixed like a normal plug and screw. They are tightened using adjustable grips or a ring spanner.

Did you know?

Although plasterboard fixings are very useful, a stronger fixing can be obtained by ensuring that a 'backboard' of plywood is fixed in position before the plasterboarding takes place.

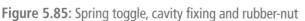

Before

After

Figure 5.85: Spring toggle, cavity fixing and rubber-nut

A wall bolt is an extremely strong masonry fixing. A hole is drilled in the masonry just large enough for the gauge of the bolt. The bolt is inserted and the nut tightened. This pulls the tapered end forward, expanding the segments against the masonry (similar to a cavity fixing). The nut is then removed, the item put in position, and the nut and washer reattached.

Figure 5.86: Wall bolt

Nails

Plumbers use a range of nails, particularly on maintenance, repair and refurbishment work. This includes procedures such as re-fixing floorboards (although you should use screws where access to pipework is required), pipe boxing, skirtings and boards for clipping pipe runs. If you are working for a small business, you may have to build your own support platforms for cold-water storage cisterns.

You will probably carry the following in your tool bag:

- panel pins
- masonry nails
- oval brad/lost head
- round head, plain and galvanised.

All these nails are available in a range of lengths and diameters.

Clips and brackets

It is likely that the majority of your work will take place in domestic dwellings where the use of copper or plastic clips is adequate for supporting copper and plastic pipework. Once again, there is a range of sizes of clips available.

However, there may be times when you will work on other buildings: schools, hospitals or small industrial units, for example, where the clips or brackets need to be strong and robust.

On large jobs, clips or brackets will be specified, but frequently you will have to decide what type of fixing to use.

> **Remember**
>
> You should always aim to fix pipework to meet the spacing requirements summarised in Table 5.6.

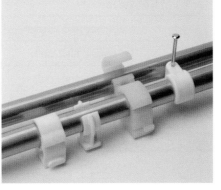

Figure 5.87: Selection of metal and plastic pipe fixings

Plumbers also use low carbon steel pipe (LCS) fixings, which are similar to the brass fixings shown for copper.

Table 5.6 lists the recommended spacings to use between internal pipework fixings.

Pipe size		Copper		LCS		Plastic pipe	
mm	in	Horizontal (m)	Vertical (m)	Horizontal (m)	Vertical (m)	Horizontal (m)	Vertical (m)
15	½	1.2	1.8	1.8	2.4	0.6	1.2
22	¾	1.8	2.4	2.4	3.0	0.7	1.4
28	1	1.8	2.4	2.4	3.0	0.8	1.5
35	1¼	2.4	3.0	2.7	3.0	0.8	1.7
42	1½	2.4	3.0	3.0	3.6	0.9	1.8
54	2	2.7	3.0	3.0	3.6	1.0	--

Table 5.6: Recommended spacing between pipework fixings

Working life

Ken has gained a contract for the installation of a central heating system in a large Victorian house. The installation will consist of a condensing wall mounted boiler, located in the kitchen and radiators in each room. See the schedule provided in Table 5.7.

Assuming that there are two fixing holes per radiator bracket and two brackets per radiator:

- Which type and quantity of fixings (including any screws, lengths and gauges) will Ken require to install the radiators?

- Which type of power tool and voltage will Ken require?

- Which type of bits will he require and what would their diameters need to be (depending on the type of securing device)?

The boiler has been supplied with fixings, but when Ken starts to drill into the wall there are signs of dampness. He is concerned that the manufacturer's fixing will be inadequate to hang the boiler here.

- What alternative means are there to support this fairly heavy unit on a wall?

Quantity	Description	Location to be fixed	Wall surface being fixed to
1	Wall mounted condensing boiler	Kitchen	Brick with plaster
4	Radiators	Bedrooms	Lath and plaster
3	Radiators	Hallway and kitchen	Glazed brick wall
1	Radiator	Ceramic tiled wall	Brickwork/plaster/ceramic wall tile
3	Radiators	Living/dining rooms	Plastered brick walls
1	Radiator	Landing	Plastered brick walls

Table 5.7: Schedule of works for Ken

1. Make a list of different fixing devices used to secure clips, brackets and sanitary ware to the building fabric.
2. How are screws specified?
3. In which plumbing situations would you use a brass screw?
4. Where would a mirror screw be used on sanitary appliances?
5. Which masonry drill bit size would be required to prepare a hole for a brown plastic plug?
6. What is the disadvantage of using a spring or gravity toggle type plasterboard fixing?

5. Know the installation requirements of domestic plumbing and heating pipework

Methods of installing domestic plumbing

There are several common methods of installing domestic plumbing and heating pipework. The key methods are introduced below.

Prefabrication of pipework

This is where pipework is assembled or bent to shape prior to being installed. In larger installations this could involve pipework being fabricated in a workshop before delivery and installation on site. For a domestic property it is normally fabricated on site before being installed permanently.

Installing pipework in situ

This is pipework installed on site at the time of the installation work. In most cases, pipework will be installed **in situ**.

Use of sleeves

Similar to most items, pipes expand and contract due to temperature changes. Where a pipe passes through a wall, the pipe will move within a sleeve placed around it, avoiding causing damage to the plaster. Pipework without sleeves is more likely to cause the plasterwork to crack under movement, as the movement of the pipe would be pressing directly onto the plasterwork rather than within the sleeve.

Sleeves also protect pipes from corrosion. Where copper pipes come into contact with cement, the cement will in time cause the copper to corrode and then leak.

Key term

In situ – in the workplace. *In situ* is a Latin phrase meaning in its original position or place.

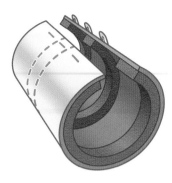

Figure 5.88: Fire sleeve for pipes

If a gas pipe is passing through a cavity wall, the sleeve is used to bridge the cavity so gas cannot leak out and enter the cavity.

Fire stopping for pipework

Pipes passing from one floor to another in a block of flats need to be sealed to prevent the spread of fire from one dwelling to another. This is normally done by placing a fire sleeve around the pipe (see Figure 5.88). The outer part of the sleeve is sealed in place during construction and the pipe is installed through the inner part. If a fire occurs, the material immediately around the pipe expands and keeps the seal between the pipe closed to prevent fire spreading from one compartment to another.

Selecting pipework materials

Identification of pipework

British Standards are produced to standardise the colour coding of pipework. This is particularly relevant in larger installations where a number of services are used. In domestic installations, pipework colour coding is not essential in most situations, but it is used on service pipes below ground, as follows:

- cold-water service pipes = blue
- domestic natural gas = yellow
- central heating pipework for use in solid floors is covered in white plastic.

Plumbing symbols

In order to provide some standardisation of symbols used on drawings, the BSI has produced a specification for pipework symbols; the main symbols are shown in Figure 5.89.

Measuring jobs and estimating materials

This topic includes:

- estimating from a drawing
- estimating from site.

Estimating from a drawing

It is unlikely that you will be required to carry out an estimate from a complex services drawing, but you may be required to estimate from a line drawing, such as the one shown in Figure 5.90.

On the drawing in Figure 5.90 the dimensions are given. The detail could have been given drawn to scale, in which case you would need to use a scale rule to measure the dimensions.

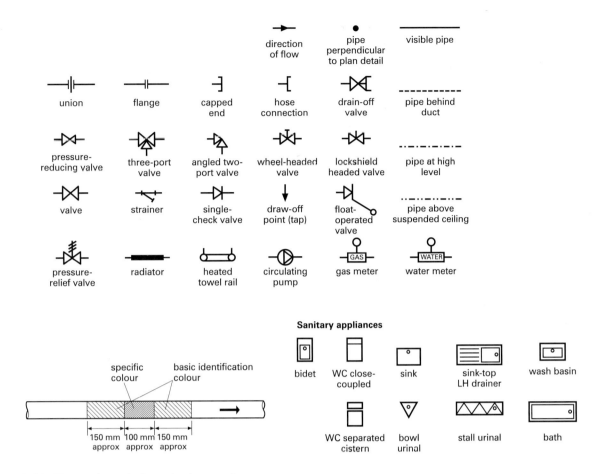

Figure 5.89: Pipework symbols and colour coding

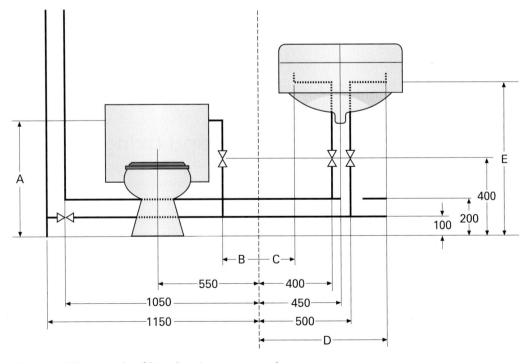

Figure 5.90: Example of line drawing, not to scale

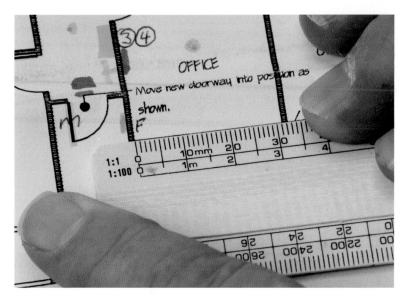

Figure 5.91: A scale rule

The photograph in Figure 5.91 shows an example of a scale rule, used to measure dimensions on a drawing.

Estimating from site

Some of your estimating will take place on site, without drawings or specifications. On one-off new installations, it will be carried out in two stages:

- first fix
- second fix.

The **first fix** involves installing all the pipework before the floor covering goes down and plasterboard is finished (sometimes referred to as carcassing). The **second fix** is the completion of the installation – fitting the bathroom suite, other sanitary appliances or radiators.

You will need to have a thorough understanding of the system requirements and where the intended pipe runs are. The next stage is to take measurements, using as a minimum a 3 m, and preferably a 5 m, tape and recording your measurements in a notebook. An estimate of fittings can also be done at the same time.

Once completed, the dimensions can be totalled to give an estimate of the pipe requirements.

On larger contracts, a large housing project for example, the dimensions, pipe and fitting requirements will probably already have been worked out. These are often provided to the plumber in the form of a specific materials pack, booked out from the stores on a house-by-house basis using a requisition order, so that the stock can be monitored.

Find out

Estimate the material requirement for a typical installation – maybe the installation in your own home.

Installing, positioning and fixing pipework

One of the main tasks when installing domestic plumbing is the ability to work with pipework. We covered the types of pipework, and the tasks involved in using them, earlier in this unit on pages 208–30. The main tasks you will need to carry out with pipework during installation are as follows:

- measure, mark and cut pipework
- bend domestic pipework to clear obstacles
- joint domestic pipework to specifications.

Progress check

1. Define the term 'prefabrication'.
2. Define the term 'in situ'.
3. List three reasons why pipes are 'sleeved'.
4. What are the British Standard colour codings for pipes?
5. Explain the term 'first fix'.
6. Explain the term 'second fix'.

6. Know the inspection and soundness testing requirements of domestic plumbing and heating pipework

Soundness testing on completed hot- and cold-water systems is essential to ensure that there are no leaks. Even the most competent plumber can make the occasional mistake, so testing is always required.

BS 6700 provides the standard for soundness testing on hot- and cold-water systems.

Soundness testing

Soundness testing of hot- and cold-water systems includes:

- visual inspections
- testing for leaks
- pressure testing
- final checks.

Visual inspection

This includes making sure that all pipework and fittings are thoroughly inspected to ensure that:

- they are fully supported, including cisterns and hot-water cylinders
- they are free from jointing compound and flux
- all connections are tight
- terminal valves (sink taps etc.) are closed
- in-line valves are closed to allow stage filling
- the storage cistern is clean and free from any unwanted material (e.g. the remains of plastic from when the holes were cut).

It is useful at this stage to advise the customer and/or other site workers that soundness testing is about to commence.

Safety tip

Testing pipework with air first can prevent water damage if a leak is present.

Remember

Any particles left in the CWSC will get in to the pipework and could cause a blockage.

Testing for leaks

When testing for leaks you should follow the checklist below.

> **Checklist**
>
> **Testing for leaks**
>
> 1. Slowly turn on the stop tap to the rising main.
> 2. Slowly fill, in stages, to the various service valves, and inspect for leaks on each section of pipework, including fittings.
> 3. Open service valves to appliances, fill the appliance and visually test for leaks again.
> 4. Make sure the cistern water levels are correct.
> 5. Make sure the system is vented to remove any air pockets prior to pressure testing.

Figure 5.92: Hydraulic pressure tester

Figure 5.93: Using a hydraulic pressure tester

Pressure testing

Pressure testing of installations within buildings is done using hydraulic pressure-testing equipment. BS 6700 has separate procedures for testing rigid pipes and plastic pipes.

The procedure for testing rigid pipes (e.g. copper):

- Make sure any open-ended pipes are sealed, e.g. vent pipes.
- Once the system has been filled, leave it to stand for 30 minutes to allow the water temperature to stabilise.
- Pressurise the system using the hydraulic testing equipment to a pressure of 1½ times the system maximum operating pressure.
- Leave it to stand for 1 hour.
- Check for visible leakage and loss of pressure. If sound, the test has been satisfactory.
- If not sound, repeat the test after locating and repairing any leaks.
- Complete a test certificate.

The procedure for testing plastic pipes

BS 6700 has two test procedures for plastic pipes – Procedures A and B. (See Water Regulations for more details.)

Test A procedure:

- Apply test pressure (1½ times maximum working pressure), which is maintained by pumping for a period of 30 minutes. Visually inspect for leakage.
- Reduce pressure by bleeding water from the system to 0.33 times maximum working pressure. Close the bleed valve.

- Visually check and monitor for 90 minutes. If the pressure remains at or above 0.33 times working pressure, the system can be regarded as satisfactory.

Test B procedure:

- Apply test pressure and maintain by pumping for a period of 30 minutes.
- Note the pressure and inspect visually for leakage.
- Note the pressure after a further 30 minutes. If the pressure drop is less than 60 kPa (0.6 bar), the system can be considered to have no obvious leakage.
- Check and monitor for 120 minutes. If the pressure drop is less than 20 kPa (0.2 bar), the system can be regarded as satisfactory.

Final system checks

After the system tests have been completed, the system should be thoroughly flushed out to remove any debris or swarf before carrying out a final visual check for leaks. Advise the customer and/or other site workers that testing is complete. At this point in the test procedure a performance test of the system is conducted, plus a handover; both of these will be covered at Level 3.

Working life

Nick is a qualified plumber who gained his qualification several years ago. He has just completed the installation of a hot- and cold-water supply and hot-water heating system in a new-build property. The installation has been completed very quickly using plastic tubes and fittings.

Nick tests the system using a hydraulic pressure-testing kit, but because his machine has a leak on it he is unable to gain a pressure at 1.5 times the system pressure. Not only that but he is in a hurry so, after checking round the building for any leaks, which there was no evidence of, he completed the test certificate and left.

Several months later another plumber was asked to commission the system. He was aware that a test certificate had been issued and so carried out the commissioning, checked round for visual leaks and handed back the keys. The same afternoon the owners moved in. During the night, a fitting on the cold-water supply under the floor in the bathroom 'blew off', damaging the carpeting, kitchen ceiling and causing water damage to the brand new fitted kitchen units.

- Who was at fault? What had gone wrong? Why did the fitting not 'blow off' during the day after commissioning, and when Nick initially carried out his hydraulic test?
- Who would be responsible for the insurance claim and would the insurance company pay, bearing in mind a certificate had been issued? What would be the probable cause/s of the plastic fitting being able to 'blow off'?

Progress check

1. Which British Standard provides the standard for soundness testing on hot- and cold-water systems?
2. What means is there of soundness testing, other than the use of water?
3. What should soundness testing of a hot- or cold-water system include?
4. How many test procedures are there for plastic installations of hot- and cold-water supplies, as stated in BS 6700 and the Water Regulations?
5. What should be done after the system soundness checks have been completed with regard to hot- and cold-water supply systems?

Check your knowledge

1. Which of the following statements means a fixed price that a plumber's merchant will supply material for?
   a Quotation
   b Estimate
   c Specification
   d Invoice

2. Where would find details of the quantity of materials available for use on site be found?
   a Stock control
   b Security
   c Bill of quantities
   d Site meeting

3. When notching joists for pipework, what depth can you remove, according to the Building Regulations?
   a $\frac{1}{4}$ of the depth of the floor joist
   b $\frac{1}{5}$ of the depth of the floor joist
   c $\frac{1}{6}$ of the depth of the floor joist
   d $\frac{1}{8}$ of the depth of the floor joist

4. Which type of fixing should be used to fix a wash hand basin to a plaster partition wall?
   a Plastic plug
   b Coach bolt
   c Masonry nails
   d Toggle type fixing

5. Hydraulic levels must be checked regularly. What item is used to check these levels?
   a Threading machine
   b Pillar drill
   c Electrical saw
   d Steel pipe bender

6. Which of the following is **not** a type of machine bend?
   a Offset
   b Passover
   c Square
   d Overtake

7. When should a joist not be drilled or notched?
   a When you have a headache
   b In a modern constructed dwelling
   c If the joints are too thick
   d If the work will disturb others

8. On a scaled drawing, what should be used to calculate the actual size of things?
   a A ruler
   b A tape measure
   c A scale ruler
   d A piece of string

9. Which of the following documents are used in the plumbing industry?
   a Water Regulations
   b Building Regulations
   c British Standards
   d All of the above

10. What grade of copper pipe is best suited to being bent using an internal bending spring?
    a X
    b W
    c Z
    d Y

11. What is the recommended clipping distance for a vertical 15 mm copper pipe?
    a 2 m
    b 1.8 m
    c 1.6 m
    d 1.2 m

12. What is the name of the tool used to hand thread low carbon steel pipe?
    a Stock and dies
    b Stocks
    c Dies
    d A pipe vice

13. What is the recommended size drill bit for a screw with a gauge of 8–10?
    a 5 mm
    b 6 mm
    c 7.8 mm
    d 10 mm

Getting ready for assessment

The information contained in this unit of the book, as well as the continued practical assignments that you will carry out in your college or training centre, will help you with preparing for both your end-of-unit test and the diploma multiple-choice test. It will also support you in preparing for the practical assignments you will need to complete to demonstrate your understanding of and enabling you to carry out site preparation and pipework fabrication techniques for domestic plumbing and heating systems.

There are opportunities throughout the unit for you to test your progress in and understanding of the required underpinning knowledge; this will enhance your preparation for the forthcoming assessments, so make good use of them.

This unit will be assessed by the following assessment methods:

- externally set knowledge assessment
- externally set assignments.

With regard to the mechanical services industry, you will need to know:

- The types of hand and power tools, including, their purpose, identification, safe use and maintenance.
- The types of pipework, their jointing principles, sizes and their installation, including, couplers, elbows, bends, equal tees, reducing tees, reducers, tap connectors, flexible connectors, manifolds, specialist fittings. The types of copper tube – soft, half hard, hard, jointing - solder ring, end feed, compression (type A and B), push and press fit, bending – 90° bends, sets and offsets, passovers, formed by

bending springs and bending machines. Low carbon steel (LCS) – medium, bending by hydraulic machine, jointing – threaded, compression. Plastic pipework – polyethylene (MDPE), polybutylene, PVC-u, polypropylene, MUPVC, ABS, jointing – push-fit, compression, proprietary fittings, ring seal, solvent weld, and compression, cabling techniques.

- Site preparation techniques, including, safe access/ exit, tripping hazards, risk assessment, PPE, selection and safe use of hand and power tools, correct calibration of tools and equipment.

- How to measure/mark out fixings to pipework components, including, fixing devices – nails, screws, heavy duty fixings, fixing devices, clips and brackets.

- How to apply fixings and brackets, including, measurement/marking out, recommended spacing for brackets and clips when using copper, LCS and plastic pipework.

- The installation requirements of domestic plumbing and heating systems, including, prefabrication, in-situ pipework, sleeving, firestopping to pipework, selection of pipework materials and fittings from instructions/plans/drawings.

- Inspection and soundness testing requirements, including, visual inspection, procedures and requirements with regard to soundness testing metallic and plastic pipework, filling pipework with water at normal operating pressure and checking for leaks.

Good luck!

UNIT 6

Understand and apply domestic cold-water system installation and maintenance techniques

Supplying fresh wholesome cold water to people's homes fulfils a basic human need. As a plumber, it is your job to get the water to the taps so that it is clean and fit for human consumption. Otherwise, severe illness can occur. Most people take for granted the supply of cold fresh water to their homes; few will appreciate what goes into providing this service.

It is important that you have a full understanding of water supply legislation as well as the installation requirements of cold-water systems. You should gain a general understanding of all the areas relating to your work at this stage.

This unit will cover the following learning outcomes:

- Know the cold-water supply route to dwellings

- Know the types of cold-water system and their layout requirements

- Know, and be able to apply, the site preparation techniques for cold-water systems and their components

- Know the installation requirements of cold-water systems and components and be able to install them

- Know, and be able to apply, the service and maintenance requirements of cold-water systems and components

- Know, and be able to apply, the decommissioning requirements of cold-water systems and components

- Know, and be able to apply, the inspection and soundness testing requirements of cold-water systems and components

1. Know the cold-water supply route to dwellings

Water is clear, tasteless and colourless. It is defined in chemical terms as a compound of the two gases, hydrogen and oxygen, in the ratio of two parts hydrogen to one part oxygen (H_2O).

The classifications of water

The most important property of water is probably its ability to dissolve gases and solids to form solutions; this is referred to as its **solvent power** and has a bearing on how soft or hard the water eventually becomes. Water is classified according to its hardness. Hard water can result in limescale build-up, as seen in Figure 6.1. This can be tackled by using water treatment methods, which are covered on pages 277–79.

The hardness of water is measured as parts per million or milligrams per litre. Table 6.1 summarises how hardness is classified.

> **Did you know?**
>
> To be an **approved contractor** a company needs to meet a number of key requirements, such as employing competent staff, completing proper records and allowing work to be checked on site.

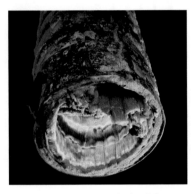

Figure 6.1: Limescale build-up in pipework

Water type	Dissolved gases (mg/litre)
Soft	0 to 50
Moderately soft	50 to 100
Slightly hard	100 to 150
Moderately hard	150 to 200
Hard	200 to 300
Very hard	Over 300

Table 6.1: Water hardness

> **Did you know?**
>
> **Solvent power** can be a problem for the plumber to deal with. It causes the corrosion and 'blocking up' of systems and components.

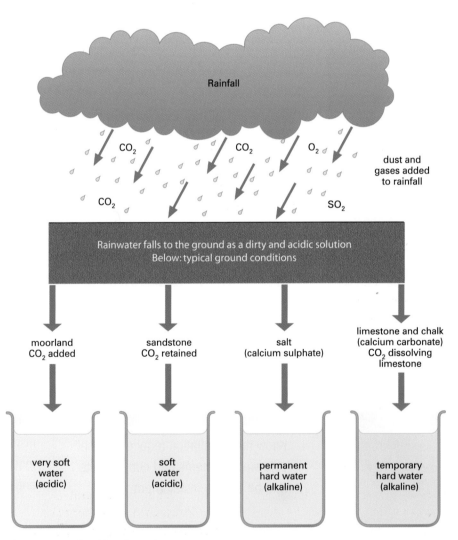

Figure 6.2: Classification of water type

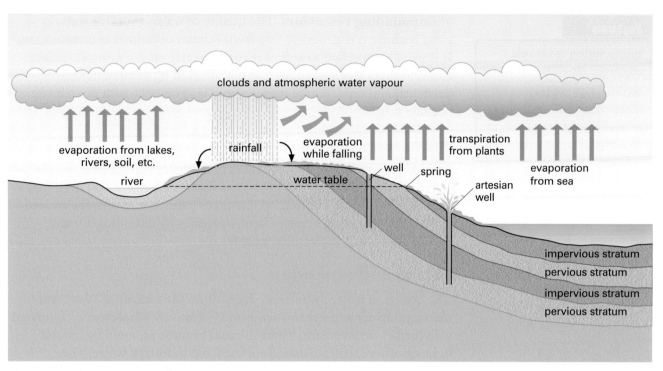

Figure 6.3: The rainwater cycle

Where does water come from?

Water evaporates from the sea, rivers, lakes and the soil. It forms clouds containing water vapour, which eventually condenses and falls as rain. When it hits the ground, some of the rainwater runs into streams, rivers and lakes; some soaks into the ground, where it may collect temporarily and evaporate, or soak away and form natural springs or pockets of water to be accessed by wells (see Figure 6.3).

Water companies obtain their water for public consumption from two main sources:

1. Surface sources:
   - upland surface water
   - rivers and streams
2. Underground sources:
   - wells
   - artesian wells
   - springs.

> **Remember**
>
> All surface and underground water sources are dependent upon rainfall.

Surface sources

Upland surface water

This category covers impounding reservoirs, lakes and natural reservoirs. This source of water is mostly found in the northern part of the UK. Here, the landscape is hilly or mountainous, allowing lakes to form naturally, and the damming of streams allows the formation of **impounding reservoirs**. The quality of upland surface water is good because it is generally free from human or animal contamination. It is usually classified as soft as it runs directly off the ground surface and into the water source, so it is not affected by passing through a particular soil type. Where water comes into contact with peat, it can become acidic.

Artificial reservoirs

Artificial reservoirs are built to meet the ever-increasing demand for water in both domestic and industrial sectors. They are also used in areas where insufficient natural resources exist. Reservoirs are created by flooding low-lying areas of land; normally by damming a water course. The water is classified as soft.

Rivers and streams

The quality of water from rivers depends on their location. Water from moorland rivers and streams tends to be relatively **wholesome** compared with further downstream, where it could become polluted by natural drainage from farmyards, road surfaces and industrial waste. The hardness of river water varies in hardness depending on the nature of the ground where it originated. Water from upland river sources is generally soft compared with that of the lower reaches, which is usually hard.

Underground sources

Wells

In the past, before the existence of Water Authorities (water companies), wells were used to provide the water supply to dwellings or small communities. There are two types of well: shallow and deep. Surprisingly, this classification does not necessarily refer to the actual depth of the well, but whether it penetrates the first impervious stratum of the earth.

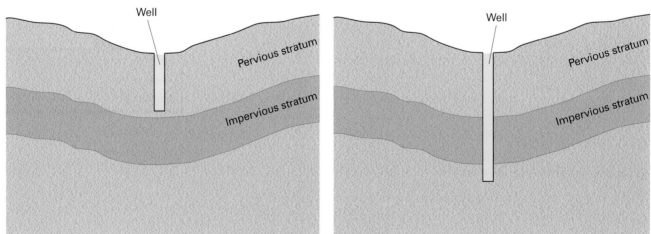

Figure 6.4: A shallow well does not penetrate impervious stratum. A deep well penetrates through impervious stratum

Some water companies still retain deep wells as a back-up to their supplies from other sources, or as a standby in case of drought. A shallow well has a greater risk of contamination but water from a deep well should be pure and wholesome.

Artesian wells

An artesian well penetrates the impervious stratum and enters a lower porous zone containing water. The outlet of the well is situated below the water table, so the water is forced out by gravity through the mouth of the well.

Springs

The quality of spring water varies depending upon the route that the water has taken to the surface. If it has travelled for a long distance through rock formations, it will probably be free from contamination but is likely to be hard.

Table 6.2 summarises the various water sources and their level of contamination before treatment.

Key term

Water table – the natural level of water under the earth.

Classification	Source	Contamination level
Wholesome	Spring water Deep well water Upland surface water	Very palatable
Suspicious	Stored rainwater Surface water from cultivated land	Moderately palatable
Dangerous	River water Shallow well water	Non-palatable

Table 6.2: Summary of water source classification

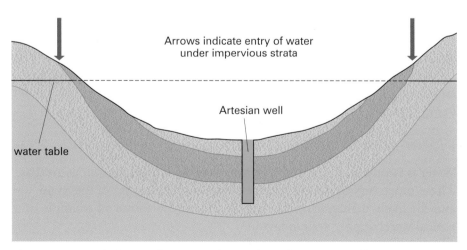

Arrows indicate entry of water under impervious strata

Artesian well

water table

Figure 6.5: Head of water forces supply out through artesian well

Water is becoming more costly and the shortage during the summer is increasing. This places more importance on different ways of saving and re-using water. There are two main methods used:

1. Collecting rainwater: known as rainwater harvesting
2. Using waste washing water: known as grey-water harvesting.

Both rainwater and grey water are not considered wholesome. However, they can be used for things like flushing toilets, washing machines, car washing and garden watering.

Water storage and supply

Water is stored by water companies either in its untreated state in impounding reservoirs or lakes, or as wholesome water in service reservoirs. It is worth noting that water companies usually aim to store enough drinking water in their service reservoirs to maintain supply for about 24 hours in an emergency. This safeguards against failure of pumps or mains and allows time to repair any faults before supplies run out.

Water treatment

It is the responsibility of the water companies to ensure water is wholesome (fit to drink). All water must be treated before it is put into the supply system. How it is treated will depend on its source and the impurities it contains. Some impurities are actually essential to our health and will be retained; others are harmful and must be removed during the treatment process.

Deep wells and bore holes

The quality of water from deep wells and bore holes is already quite good, due to the natural filtering process as the water passes through the rock strata, so the only treatment needed is **sterilisation**. Sterilisation serves to keep both the water and supply pipework free from bacteria as the water is piped into our homes.

Rivers, lakes and impounding reservoirs

Rivers, lakes and impounding reservoirs are our main sources of supply. However, this water is usually dirty and polluted, so treatment needs to be extremely thorough. The treatment process usually involves the water being strained, filtered, chemically treated and sterilised.

Water supply and distribution

Water is supplied to our homes via a network of pipes known as **mains**. These pipes vary in diameter depending on the purpose of the main and the likely demands on the supply. Figure 6.6 shows an example of a typical water supply system layout.

The local mains shown will provide the 'final leg' of the supply of water to your home.

Key terms

Sterilisation – purification by boiling, or, within the industry, dosing the supply with chlorine or chlorine ammonia mix.

Water mains – the network of pipes that supply wholesome water to domestic and commercial properties.

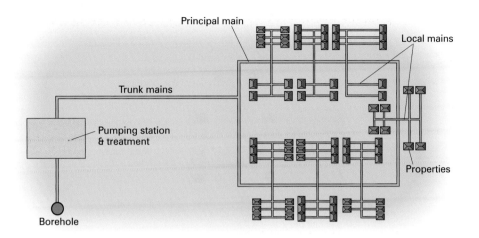

Figure 6.6: A typical mains water distribution system

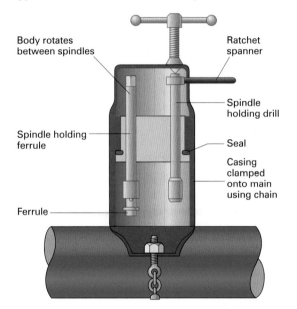

Figure 6.7: Section through a mains tapping machine

Mains connection

Water mains are constructed of asbestos cement, steel, PVC or cast iron. PVC is now used extensively on new installations and mains replacements. The connection to the mains is the responsibility of a water company, and new connections to existing mains, operating under live pressure, require the use of specialist equipment. Figure 6.7 shows an example of a connection to a cast iron main using a mains tapping machine, where the new connection can be made without the need to isolate the live main.

The connection to the main is made using a **ferrule**. The ferrule is in fact an isolating valve. Once the threaded section has been inserted into the mains, the valve can be isolated, allowing the mains tapping machine to be removed and the service pipework to be connected to the ferrule outlet.

Key term

Ferrule – a metal fitment used to connect to a main allowing isolation.

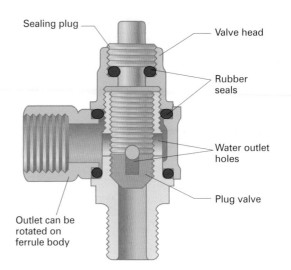

Figure 6.8: Section through a brass ferrule

Labels: Sealing plug, Valve head, Rubber seals, Water outlet holes, Plug valve, Outlet can be rotated on ferrule body

Cold-water supply and treatment

Water regulations

The purpose of the Water Supply (Water Fittings) Regulations 1999 is to prevent:

- the contamination of a water supply
- the waste of water
- the misuse of a water supply
- the undue consumption of water
- erroneous measurement (fiddling the meter).

The regulations are also designed to permit the introduction of new products and ideas, as well as supporting environmental awareness.

Enforcing the regulations

The water supplier is responsible for enforcing the regulations. Under the Water Regulations, water companies are encouraged to set up approved contractor schemes. These require that approved contractors certify to the water company that water fittings installed are in compliance with the regulations.

Although there is no legal requirement for a person working on water services to be qualified, anyone who carries out such work can be prosecuted for an offence against the Water Regulations and, if convicted of an offence, can be fined.

Building Regulations

The Building Regulations 2000 Part G also cover cold-water systems, including the requirements of the supply, and water efficiency of fittings and appliances. These regulations also cover the consumption of wholesome water, stating that the consumption should not be more

Remember

Water supply installations are covered by the Water Supply (Water Fittings) Regulations 1999, which require water companies to supply 'wholesome' water.

than 125 litres per person per day. This is set out in the Water Efficiency Calculator for New Dwellings.

Building Regulations are enforced by the local authority.

2. Know the types of cold-water system and their layout requirements

The supply from the mains to the building

The next stage is for the water to be delivered into the building from the mains. Figure 6.9 shows a full installation from the mains to the stop-and-drain valve wherever it enters the building. A plumber's work usually starts from the external stop valve.

There are some key points to remember about service pipes:

- In order to protect against frost damage, mains and service pipes should be at least 750 mm beneath the surface of the ground.
- The maximum depth of cover should not exceed 1.35 m, as this would prevent ease of access.

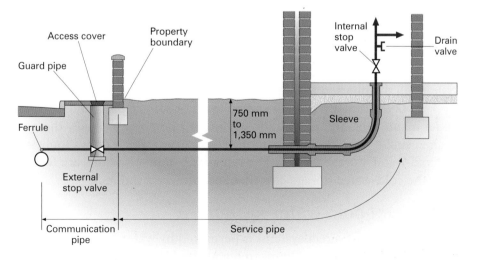

Figure 6.9: Supply pipework to the building

- Blue-coloured medium-density polyethylene pipe (MDPE) is normally used for underground service pipework into a property. Plastics are now widely used to make service pipe connections as they can be jointed easily and are not subject to corrosion. Care needs to be taken, however, when installing standard plastic pipes in ground containing contaminates, such as petrol or oil, as these pipes are susceptible to **permeation**, which may cause the water to be contaminated. When installing service pipes across contaminated land, a special barrier plastic pipe, or metallic pipe such as copper, should be used.

- Metal pipes should be protected against possible corrosion from the soil, particularly from acidic soils. This can be done by:
  - using plastic-sheathed pipe
  - wrapping the pipe with anti-corrosive tape
  - installing the pipe inside a duct.

- The minimum size allowed for a cold-water service pipe to a dwelling is 15 mm. Service pipes to new dwellings are now usually provided with a 25 mm internal diameter in order to meet the flow rate demands required by higher pressure mains plumbing systems.

Installation of external stop valves

As Figure 6.9 shows, the service pipe to the building can be isolated from the mains by using the external stop valve. Ease of access to the stop valve is very important, so it should be located in a stop valve chamber, constructed from 150 mm PVC or earthenware pipe, sited on a firm base, and finished off at the top with a stop valve cover. The stop valve cover is usually made from steel plate, or a combination of steel plate and plastic.

Figure 6.9 shows a PVC stop valve chamber; these are now widely used on new installations. The guard pipe is cut to the desired length to suit the installation as part of the fitting process. The pipework from the mains up to and including the external stop valve is the responsibility of the **water undertaker**.

Types of cold-water system

There are two types of cold-water system: direct and indirect. In a direct system, all the pipes to the draw-off points (sink, bath, hand basin, WC, etc.) are taken directly from the rising main or service pipe, and operate under mains pressure. In an indirect system, one point – usually the kitchen sink – is fed directly from the rising main, which then supplies the cold-water storage cistern. The remaining draw-off points are fed from the cold-water storage cistern; hence the term indirect.

The direct cold-water system

The installation of direct systems is permitted by water companies in domestic properties in medium- to high-pressure areas where the supply can provide adequate quantities of water at sufficient pressure to meet the building's needs.

A direct system has the following advantages:

- It is cheaper to install because:
  - less pipework is required
  - the storage cistern is smaller (100 litre minimum); BS 6700 recommends that the storage cistern to a direct system should be at least the same size as the hot-water storage cylinder.
- Drinking water is available from all draw-off points.
- There is less risk of frost damage due to a smaller amount of pipework.
- Less structural support is required for a smaller cistern.

However, a direct system also has the following disadvantages:

- Higher pressure may make the system noisy.
- There is no reserve of cold water if the mains or service supply is shut off.
- There is more wear and tear on taps and valves due to high pressure.
- There is higher demand on the mains at peak periods.

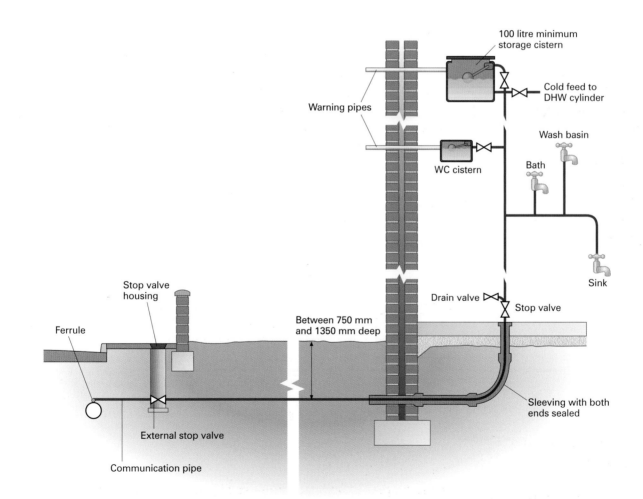

Figure 6.10: Pipework layout for a direct cold-water system

The direct system is the most commonly installed cold-water system in domestic properties because its installation is cost-effective and the pressure of supply available is usually relatively high. It is a form of direct system that normally provides the cold-water supply to an instantaneous water heater or a combination boiler.

The indirect cold-water system

The draw-off points in an indirect system are fed indirectly from the cold-water storage cistern; only one outlet is fed directly from the supply pipe. The system is designed to be used in low-pressure water areas where the mains supply pipework is not capable of supplying the full requirement of the system. This type of system also has a reserve of stored water in the event of mains failure.

The advantages of indirect systems are that there is:

- a reserve of water should the mains supply be turned off
- reduced risk of system noise due to lower pressures
- reduced risk of wear and tear on taps and valves, again due to lower pressure
- lower demand on the main at peak periods.

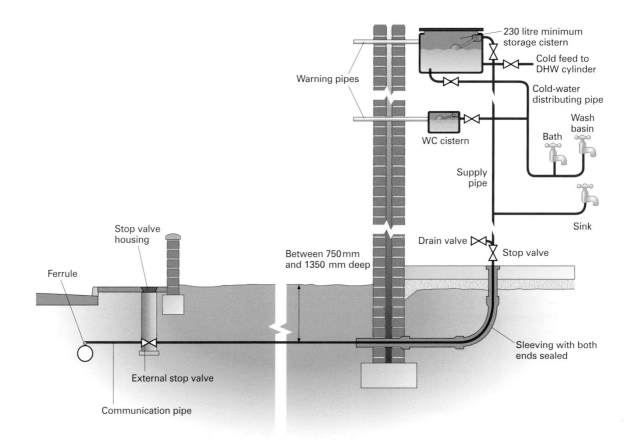

Figure 6.11: Pipework layout for an indirect cold-water system

The disadvantages of indirect systems are:

- an increased risk of frost damage
- the space occupied by the larger storage cistern (230 litre minimum – the minimum size for supplying both hot and cold outlets in dwellings as laid down by BS 6700)
- the additional cost of the storage cistern and pipework
- water might not have been potable before the 1999 Water Regulations.

The basics of backflow prevention

When installing a hot- and cold-plumbing system, you must ensure that the system remains hygienic at all times and is not subject to contamination. Never forget people will drink from these systems, so contaminants inside the system could be dangerous to their health.

There are two major ways in which water in a system may become contaminated:

1. The use of a non-approved material such as lead in the supply system.
2. Backflow in the system, which can cause contaminated water to be drawn off inside the system.

The first point is dealt with by always installing approved pipework, components and fittings in hot- and cold-water systems.

The second point is dealt with by installing **point-of-use backflow prevention devices** at all water outlets and fitting connections.

Backflow in the system is quite complicated and will be covered in detail at Level 3. Here we will cover the basics, and how they relate to typical domestic plumbing systems.

There are two ways in which backflow can occur in a plumbing system:

1. Back-pressure – reversal of water flow in a pipe caused by a pressure decrease in the system.
2. Back-siphonage – siphonage from a cistern, appliance or water storage vessel back into the pipe feeding it.

There is an extensive range of backflow prevention devices on the market, which provide backflow protection at the point of use, such as a tap outlet or a plumbing component. The selection of the correct type of device is based on an assessment of the risk presented by the fluid that the plumbing fitting may come into contact with.

Fluid risk categories

Before selecting a suitable backflow prevention device, you must assess the fluid risk category that the water fitting may come into contact with. In the UK a five-level risk system is used to undertake the assessment. This is summarised in Table 6.3.

> **Did you know?**
>
> Backflow is water flowing in a reverse direction from its intended direction of flow.

Fluid category	Description	Examples in domestic properties
1	Wholesome water provided by the water undertaker	Water used for drinking taps
2	Water that has had its aesthetic quality changed, such as by a change in temperature, taste or odour	Mixed hot and cold water discharged from mixer taps Water discharged from a base exchange water softener
3	Fluid that represents a slight health hazard	Water in primary central heating circuits The water in the bowl of a bath, basin or shower tray Hand-held garden hose used with a hose union tap The contents of a washing or dishwashing machine
4	Fluid that represents a significant health hazard	Water used in a mini-irrigation system in a garden, such as a porous hose or a sprinkler system with sprinkler heads less than 150 mm above ground level
5	Fluid that represents a serious health hazard	The contents of a sink bowl, WC pan or bidet

Table 6.3: Fluid risk categories

Find out

How can the temperature surrounding a cold-water pipe affect the quality of drinking water?

Examples of typical backflow prevention devices used in domestic properties

Taps and shower roses for baths, basins and shower trays

The contents of the bowl (a bath, basin or shower tray) are assessed as a Fluid Category 3 risk. To prevent the contents of the bowl being drawn back into the hot- or cold-water system, a Type AUK2 backflow prevention device is usually installed. This uses a simple air gap between the outlet of the water fitting and the spill-over level of the bowl of the appliance; this fixed air gap prevents the contaminated water from being drawn back into the system. This is known as a non-mechanical backflow device.

The air gap dimension must be a minimum of:

- 20 mm for an outlet with a diameter of 15 mm or less
- 25 mm for an outlet with a diameter between 15 mm and 25 mm
- 70 mm for an outlet with a diameter over 25 mm.

In the event that the air gap cannot be provided to the dimensions listed above, then a double-check valve (Fluid Category 3 protection) will need to be installed to both the hot and cold pipes feeding the individual appliance. This is known as a mechanical backflow device.

Taps and shower roses for domestic sinks

In domestic sinks, the contents of the sink bowl are regarded as a Fluid Category 5 risk – the most serious – owing to the bowl's potential contents. For instance, the bowl may be used for the washing of animal blood products or muddy vegetables. Protection against the Fluid Category 5 risk is provided by a Type AUK3 air gap. The minimum air gap dimensions are 20 mm or twice the diameter of the inlet pipe

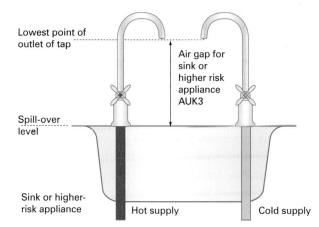

Figure 6.12: Air gap above water fitting supplying domestic sink

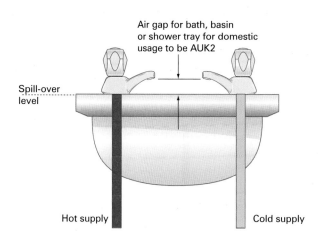

Figure 6.13: Air gap above water fitting supplying domestic bath, basin or shower tray

supplying water to the fitting. For a 15 mm pipe, the air gap should be 30 mm.

In this situation, you must ensure that the air gap is created as otherwise you cannot fit another type of device. For example, a double-check valve connected to the inlet pipes only provides up to Fluid Category 3 protection.

> **Remember**
>
> A double-check valve only provides protection up to Fluid Category 3.

Other examples of simple backflow protection in domestic properties

- A **close-coupled WC suite** has a Type AUK1 air gap built into the design of the appliance. This ensures that the contents of the WC pan cannot be drawn back into the pipe supplying cold water to the pan. This air gap is built into the design of the WC suite so there's nothing more for you to do.

- A **new outside hose union bib tap** requires Fluid Category 3 backflow protection to protect against the end of the hose being left in potentially contaminated water. Protection is provided here in the form of a double-check valve fitted to the tap inlet (preferably sited inside the building) providing Fluid Category 3 protection. This is shown earlier in this unit on page 268.

- The water supplied to a **pressurised (sealed) central heating system**, such as a combination boiler system, is fed through a double-check valve built into the filling loop. This provides Fluid Category 3 protection against contamination from the contents of the central heating system.

- **Mixed hot and cold water in a mixer tap** is identified as a Fluid Category 2 risk. The hot and cold supplies to the mixer tap must therefore be fitted with single-check valves unless the supplies to the appliance are from dedicated distributing pipes from a storage cistern.

- The **cold-water supply to a water conditioning device**, such as a base-exchange water softener, is protected by installing a single-check

Figure 6.14: Backflow prevention applied to mixer taps with mixing of water in the valve body

valve to provide Fluid Category 2 protection against mild contamination of the water supply through backflow from the salt additives in the device.

- A **washing machine or dishwasher** will normally have a built-in device (double-check valve) to stop any contamination. For domestic purposes, this will provide Category 3 protection. In a commercial/industrial location, the category risk will be higher so the protection will have to be higher, depending on the use.

- **Bidets** that are not of the over-the-rim type require Fluid Category 5 protection. These should not be supplied from a common hot or cold supply. This will be covered in greater depth at Level 3.

Working principles of cold-water system components

Taps and valves

Taps and valves are used to isolate supplies, reduce the flow rate through pipework, permit drainage from systems and provide an outlet to an appliance. These are used in appliances such as sinks, baths and washing machines.

Water Regulations require that all fittings must be:

- suitable for their purpose
- made of corrosion-resistant materials
- sufficiently strong to resist normal and surge pressure
- capable of working at appropriate temperatures
- easily accessible to allow the renewal of seals and washers.

Taps and fittings for both hot and cold water should conform to BS 1010 Parts 1 and 2, BS 1552 and BS 5433.

Taps and valves are usually made of brass pressings or castings, and are chrome plated when appearance counts. Chrome plating also makes them easier to clean.

Stop valve

The stop valve pictured in Figure 6.15 is a typical screw-down valve. The washer and jumper fit into the threaded part of the spindle, which is raised or lowered by turning the crutch head. When lowered to its maximum extent onto the seating of the valve, the incoming supply is isolated.

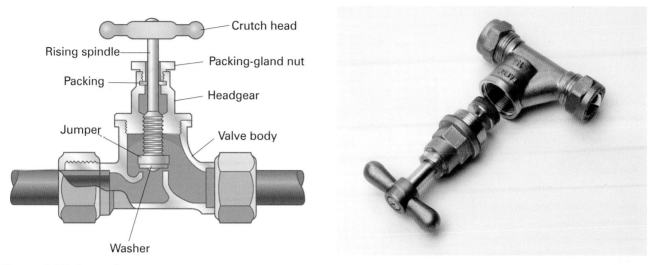

Figure 6.15: Stop valve

The stop valve is usually located on high-pressure pipelines such as between the incoming mains and rising main. It is therefore used as a main supply isolation valve.

Stop valves used below ground must be made of a corrosion-resistant material such as gunmetal or bronze to avoid **dezincification**, and marked CR in the UK or DRA in the rest of Europe.

Gate valve

Gate valves are sometimes referred to as full-way gate valves, because when they are open there is no restriction through the valve. The gate valve is fitted with a wheel head attached to the spindle. When the head is turned anticlockwise, the threaded part of the spindle is screwed into the wedge-shaped gate, raising it towards the head.

Gate valves are usually located in low-pressure pipelines, such as the cold feed from the cold-water storage cistern (CWSC) to the hot-water storage cylinder. You may also find them used on supplies to shower valves where the shower is fed from low pressure via a CWSC.

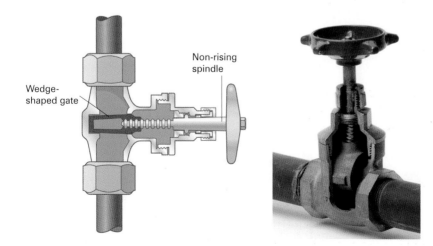

Figure 6.16: Gate valve

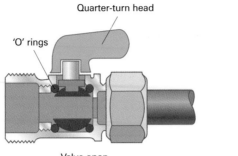

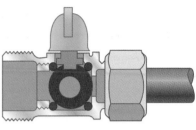

Valve open   Valve closed

Figure 6.17: Lever-operated spherical plug value

Find out

Why should a quarter-turn valve be closed slowly on high-pressure pipework?

Quarter-turn valves (spherical plug valves)

The 'ball valve'-type quarter-turn valve (spherical plug valve) can be used as a service valve to a cistern. Here, it will have a compression nut and olive on each side of the valve. It can also be used on the hot and cold supplies to a washing machine or dishwasher, when it will have a flexible hose connection. These valves are used to isolate the supply for the service, repair or maintenance of components and appliances within the hot and cold system.

Figure 6.18: Ball-o-fix showing a reduced bore type valve

The valves work by the quarter-turn operation of a square head (or slotted head) located at the top of the valve, which aligns the hole in the valve with the hole in the pipe. When the head is in line with the direction of the pipe, the valve is open. The quarter-turn head could be made from plastic and operated by hand – as used on washing-machine supplies – or it could have a slotted head, operated by a screwdriver, as fitted on supplies to cisterns. It can be used on low- or relatively high-pressure installations for servicing purposes. Care should be taken with the use of spherical plug valves on low-pressure installations; for this type of installation, the valve should be of the full bore type rather than the reduced bore type shown in Figure 6.18. A reduced bore in the valve control mechanism will greatly increase the water pressure loss through the valve body, which will have an impact on the water flow rate available.

Stop-and-drain valve

This section deals with the supply from the stop-and-drain valve inside the building.

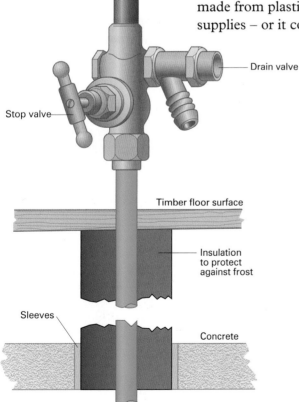

Drain valve

Stop valve

Timber floor surface

Insulation to protect against frost

Sleeves

Concrete

Figure 6.19: Combined stop-and-drain valve

Remember

The supply pipe must contain an isolation valve as close as possible to the point where it enters the property, together with a drain-off facility at the lowest point. These arrangements can be made up using separate fittings as an alternative to the combined valve shown.

The purpose of the stop-and-drain valve is to:

- turn on/off the water supply to pipework fittings and components in order to enable system maintenance
- drain down all the system pipework fittings and components to enable system pipework repair/replacement.

Did you know?

Other servicing valves are included in the system to isolate components and appliances.

Draw-off taps

Draw-off taps, such as bib taps and pillar taps, work in a similar way to screw-down valves. The more modern type do not work as screw-down valves, but allow two polished ceramic discs to turn and align with two port holes through which the water can pass.

The bib tap

The tap shown in Figure 6.20 includes a hose union attachment used for garden hosepipes.

You are unlikely to find the bib tap shown in Figure 6.21 ('supatap') in most modern domestic dwellings but you may come across it when carrying out maintenance work on other properties.

The pillar tap

The pillar tap has a long vertical thread, which passes through pre-drilled holes in the appliance. It is held in position with a backnut and the supply pipe is attached to the thread via a tap connector. Its method of operation is very similar to the stop valve covered earlier, on page 271. The tap shown in Figure 6.22 is a pillar tap with non-rising spindle. Here the tap spindle is fixed in position by means of a circlip and does not rise when the tap is opened.

The tap shown in Figure 6.23 uses quarter-turn ceramic discs. This type of tap is in common use. However, they often have maximum water pressures within which they operate. If you disregard the manufacturer's

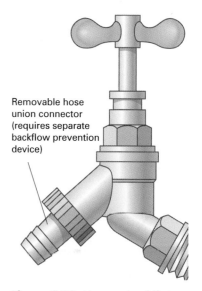

Removable hose union connector (requires separate backflow prevention device)

Figure 6.20: Hose union bib tap

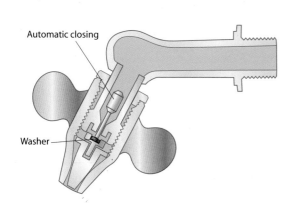

Automatic closing

Washer

Figure 6.21: Bib tap: 'supatap'

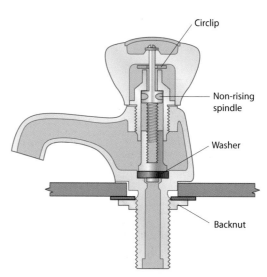

Circlip

Non-rising spindle

Washer

Backnut

Figure 6.22: Pillar tap

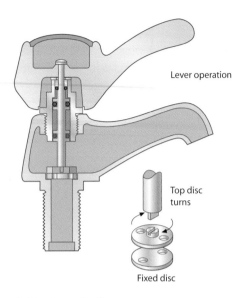

Lever operation

Top disc turns

Fixed disc

Figure 6.23: Ceramic disc tap

Figure 6.24: Ceramic disc

requirements during installation, the discs can shatter under operating conditions.

Globe taps

You may come across globe taps when carrying out maintenance work on an old bath installation. Instead of having a spigot thread connection to the supply, like a bib tap, they use a socket thread. The spigot threaded outlet from the water supply is connected through prefabricated holes in the bath, and the globe tap is then screwed on to the fitting. These can pose a real threat of water-supply contamination, as they discharge below the spill-over level of the bath. Their method of operation is similar to the stop valve.

Find out

Obtain a varied selection of taps and valves and undo all the components so that they are loosely assembled. Dismantle the various taps and valves and, using the information in this section, cross-reference the various parts to the diagrams in order to see what they look like and how they work.

Do the same for float-operated valves.

Mixer taps

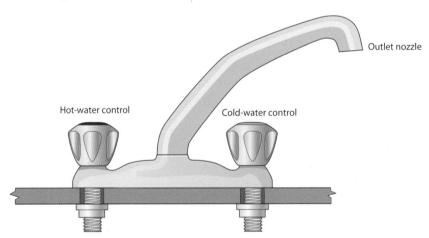

Outlet nozzle

Hot-water control

Cold-water control

Figure 6.25: Mixer tap

There are a number of design variations in mixer taps, but they all work on the principle of allowing the water from the hot and cold supply to flow from one outlet.

There are two types of mixer tap:

1. Single-flow outlet
2. Twin-flow outlet.

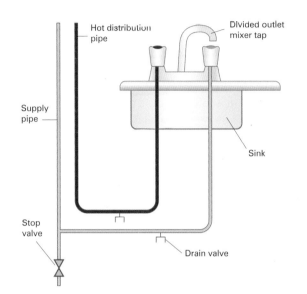

Figure 6.26: Mixer tap pipework

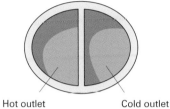

Hot outlet   Cold outlet

Figure 6.27: End view of outlet (twin-flow)

In the case of the twin-flow outlet nozzle, as shown in Figure 6.27, the water is not allowed to mix in the tap swivel outlet. In a single-outlet unit, mixing can take place and cross-contamination of water can occur between the hot and cold supplies.

The drain-off tap

Conforming to BS 2879, the drain-off tap is located at the lowest point of any system and has a 'ribbed' outlet to enable a good grip for a hosepipe connection. These can be supplied so that they can be soldered into a fitting or threaded so that they can be screwed into a fitting. You can also get a type fitted with a 'lockshield head', which means it can only be operated by a purpose-made key. Combined stop valves with drain-off taps are also available (see Figure 6.29).

Figure 6.28: Drain-off tap

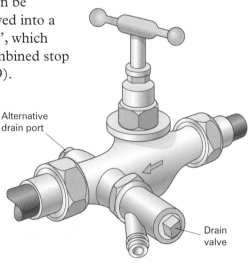

Figure 6.29: Combined stop valve with drain-off tap

Working life

Marcus was asked to re-washer the appliance taps on a hot- and cold-water supply system. The property was about 30 years old and, by the look of them, the appliances and taps had been installed when the property was new.

Marcus found the internal stop valve. Although very stiff, he managed to turn it but, even when fully closed, the valve was still letting water pass. This valve would now require his attention as well, creating more work.

Outside the property, on the pavement, he found that the cast-iron lid of the external stop tap had been lifted. To his dismay the chamber had a lot of silt in it, which would need cleaning out before he could gain access to the valve head.

Later, when he had cleared most of the debris from the chamber, he probed about with his stop tap key until he eventually found the head of the valve. When he tried to turn the key, the valve was too tight and would not move.

Someone at the house shouted that there was a cup of tea ready for him. Having had so much trouble getting the stop tap key on the valve, he decided to leave it in position on the pavement while he decided what to do next.

Consider the following points:

- What should Marcus do next?
- Should he contact anyone? If so, who?
- Was leaving the stop tap key on the valve the right thing to do?

Marcus eventually managed to get the cold-water supply to the property turned off. He renewed the internal stop valve after informing the customer that it was beyond repair.

Next he found a gate valve on the cold feed to the domestic hot-water storage cylinder in the airing cupboard and turned it off to isolate the hot-water supply.

Firstly, Marcus re-washered the hot and cold taps on the kitchen sink. He noticed that the seating of the cold tap was badly damaged and repaired it.

In the bathroom he started to work on the bath taps. There was no problem with the hot tap. When he opened the cold tap, water was delivered. He thought this was just in the pipework but it kept on running. It didn't stop until the cold-water storage cistern was empty.

The cold tap on the basin was very difficult to unfasten; Marcus protected the tap and used a pair of grips to hold the tap body while he tried to break the tap headgear from the body. It was so tight that he cracked the basin due to the tap body moving in the appliance.

Now consider the following points:

- What was wrong with the seating of the cold tap on the kitchen sink? What did Marcus do to repair the problem?
- Why did the cold-water taps in the bathroom continue to deliver water until the cold-water storage cistern had emptied? The mains supply was already turned off.
- What should Marcus have done before attempting to remove the cold tap on the washbasin?
- What are the reasons why the washbasin broke while Marcus tried to unfasten the headgear?
- What should Marcus do next?

Water filters

A water filter removes impurities from water using a fine physical barrier, a chemical process or a biological process. Filters cleanse water to various extents depending on the type of filter. A simple filter such as a line filter, which has a fine wire gauze inside it, will reduce the amount of solid impurities like sand or grit. This can normally be removed, washed out and replaced as required. This type of filter can be found on the cold supply to an un-vented hot-water cylinder.

Figure 6.30: A simple line filter

Another filter used is the cartridge type, which helps to reduce particles and substances in the water, such as chlorine, to give it a more pleasant taste.

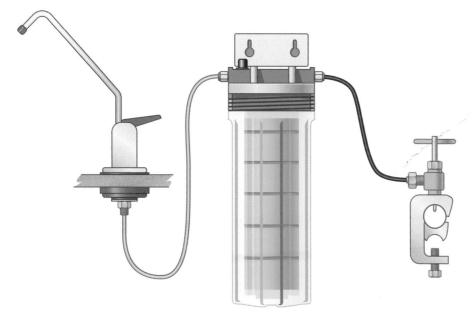

Figure 6.31: A typical cartridge water filter

Use of water conditioners

One way to prevent hard-water damage to water systems is to install a water conditioner in the cold-water supply. This softens the water prior to it being heated, which reduces scale formation.

There are three main methods of treating water hardness:

1. Base exchange softeners
2. Scale reducers
3. Magnetic water conditioners.

Base exchange softeners

These work by passing the hard water through a tank containing resin particles. The resin attracts and absorbs the hardness salts – mainly calcium and magnesium – from the water. At the same time it replaces them with sodium from the resins. After a while, the resin becomes saturated with hardness salts and needs to be regenerated, using salt solution to put sodium back into the resin. The hardness salts are released from the resin and washed down the drain.

The unit requires regular maintenance and checking, and an annual service. A check on water hardness also needs to be carried out by a service engineer. The salt, however, can usually be topped up by the end user. BS 6700 makes recommendations for the installation of base-exchange water softeners. Its main consideration is preventing backflow and contamination of the water supply. As there is a backflow risk associated with the use of a base-exchange water softener, where salt additives are put into the cold-water system, the installation must be protected by using a single-check valve. For further details on the basics of backflow prevention see pages 267–70. BS 6700 is the key reference source, together with the manufacturer's instructions, when installing the water softener. Backflow and its prevention are also covered in the Water Regulations.

When installing a base-exchange water softener, at least one tap (normally the kitchen sink) is not connected via the softener.

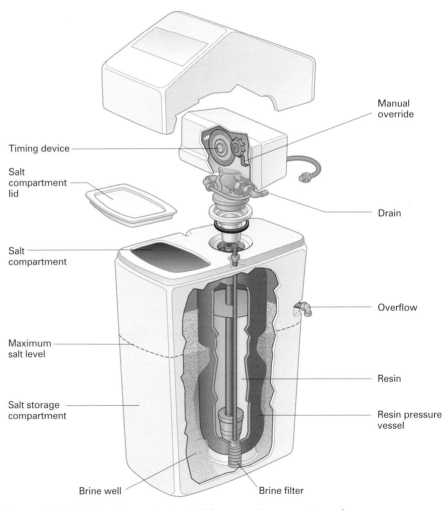

Figure 6.32: Section through a typical base exchange water softener

Scale reducers

Scale reducers are chemical conditioners designed to reduce limescale in water; they dispense chemical additives into the system to treat the water. The chemicals alter the molecular structure of the calcium salts in the water, preventing them from bonding together. The scale reducer will need periodic replenishing with chemical additives. It is commonly used to provide protection to individual system components such as multipoint water heaters or combination boilers.

Magnetic conditioners

The molecular structure of the salts in the water is changed by a small electrical field set up by a magnetic device, which is clamped to the pipework. There are many variations on magnetic conditioners, some simply comprising a strong magnet, others connected to the mains via a transformer to generate the electric current.

Figure 6.33: Scale reducer

Installation of water meters

The installation of water meters for domestic premises is becoming more widespread, particularly on new housing developments. They measure the amount of water used by the householder, who then pays only for the water used. Water meters can be installed internally or externally and are the responsibility of the water undertaker.

External water meters (an example is shown in Figure 6.36) usually incorporate a service stop valve and are found on new domestic

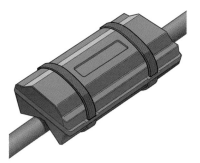

Figure 6.34: Magnetic conditioner

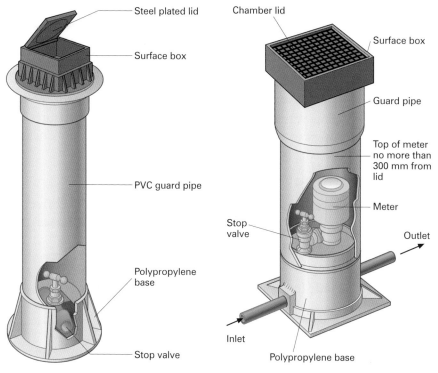

Figure 6.35: PVC stop valve chamber without meter

Figure 6.36: External water meter installation

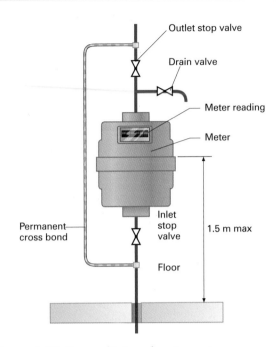

Figure 6.37: Exposed internal water meter

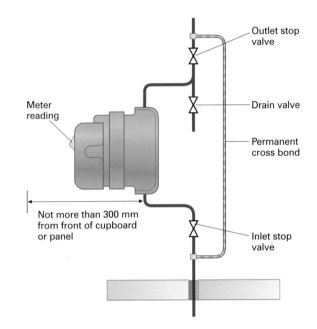

Figure 6.38: Concealed internal water meter

installations. The parts come already assembled, with the guard pipe being cut to length during the fitting process carried out by the water undertaker.

Figures 6.37 and 6.38 show the installation of internal water meters. These are normally seen in commercial or larger domestic properties but may be found elsewhere.

Cold-water storage cisterns

Under the Water Regulations, a storage cistern supplying cold-water outlets, or feeding a hot-water storage system, should be capable of supplying wholesome water. Therefore, various protection measures are included in the design of the cistern to ensure the water supply does not become contaminated.

Cisterns must therefore be:

- fitted with an effective inlet control device to maintain the correct water level
- fitted with service valves on inlet and outlet pipes
- fitted with screened warning/overflow pipes to signal overflow
- covered to exclude light or insects
- insulated to prevent heat loss and undue warming
- installed so that the risk of contamination is minimised
- arranged so that water can circulate preventing stagnation
- supported to avoid distortion or damage leading to leaks
- readily accessible for maintenance and cleaning.

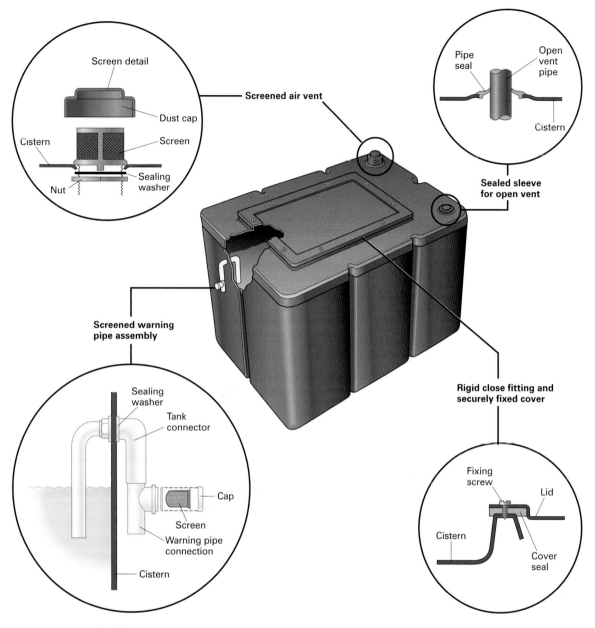

Screen detail

Dust cap

Cistern

Screen

Nut

Sealing washer

Screened air vent

Pipe seal

Open vent pipe

Cistern

Sealed sleeve for open vent

Screened warning pipe assembly

Sealing washer

Tank connector

Cap

Screen

Warning pipe connection

Cistern

Rigid close fitting and securely fixed cover

Fixing screw

Lid

Cistern

Cover seal

Figure 6.39: Protected cold-water storage cistern

Materials for cisterns

In the past, galvanised low carbon steel was the main material used to make cold-water storage cisterns. You may still come across this on maintenance jobs but most new installations use cisterns made from plastic such as polyethylene, polypropylene and polyvinyl chloride.

The majority of cisterns are polypropylene because this material allows them to be:

- light
- strong
- hygienic

- resistant to corrosion
- flexible enough to be manoeuvred through small openings.

Cisterns are available in square, rectangular or circular shapes, and are produced in black to prevent the growth of algae. However, because they are flexible, the base of the cistern **must** be fully supported throughout its entire length and width, as covered in the Building Regulations Part G. Holes for pipe connections should be cut out using a hole saw. The joint between the cistern wall and fitting should be made using plastic or rubber washers.

The hole in the cistern must not be made by heating a section of pipe and using it to make a hole in the cistern – this degrades the plastic and will result in cistern failure.

Connections to cisterns and control valves

Inlet controls

Water Regulations require that a pipe supplying water to a storage cistern be fitted with an effective adjustable shut-off device, which will close when the water reaches its required level. For most domestic applications, a float-operated valve is used. This must comply with BS 1212 (Parts 2–4), and the following types are available:

- Portsmouth type (Part 1 valve)
- diaphragm valve made of brass
- diaphragm valve made of plastic.

The Portsmouth valve, which complies with BS 1212 Part 1, is not widely used on new installations because it does not provide an effective air gap between the water level and the point at which the valve discharges. However, you may see this valve on existing installations. Brass or plastic diaphragm float valves, which comply with BS 1212 Parts 2 or 3, are commonly used in new storage cisterns. These valves provide an effective air gap between the water outlet from the valve and the water level in the storage cistern.

BS 1212 Part 4 refers specifically to diaphragm equilibrium float valves (not illustrated), which are designed primarily for use in WC flushing

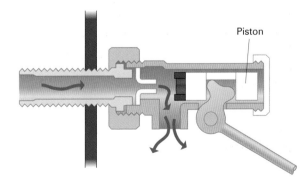

Piston

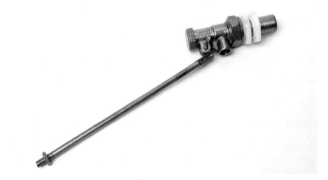

Figure 6.40: Portsmouth float valve to BS 1212 Part 1

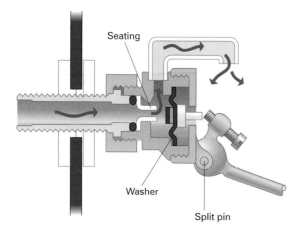

Figure 6.41: Diaphragm float valve to BS 1212 Part 2 or 3

cisterns. The Portsmouth and diaphragm valves work by the principle of leverage, while the equilibrium valve uses the water pressure. However, equilibrium valves are available with a similar external appearance to the Portsmouth and diaphragm valves.

Inlet pipes to cisterns must have a servicing valve fitted immediately before connection to the cistern, to enable maintenance to be carried out on the cistern without turning off the whole water supply. This also applies to WC cisterns. The valve usually installed is a spherical-type plug valve, suitable for use with higher-pressure cold-water supplies.

Outlet pipes

The outlet pipes should be connected as low in the cistern as possible and there is a move in the new Water Regulations to preferred connections being in the bottom of the cistern rather than the side. This prevents the build-up of sludge in the bottom of the cistern. Outlet pipes, such as cold feed and distribution pipes, should be fitted with servicing valves, and these should be located as near to the point of connection to the cistern as possible, while still being accessible. The valve type used in this position is usually a wheelhead gate valve, suitable for isolating lower-pressure cold-water supplies.

Overflow and warning pipes

When water in a cistern rises above a pre-set level, usually due to a faulty float-operated valve, the water is allowed to flow through a pipe away from the cistern. An **overflow pipe** is used to discharge water where it will not cause damage to the building. A **warning pipe** is a pipe used to give warning to the occupiers of a building that a cistern is overflowing and needs attention.

Small cisterns of up to 1000 litres (i.e. domestic cisterns) must be fitted with a warning pipe and no other overflow pipe. Larger cisterns (between 1000 and 5000 litres) are fitted with both.

Location of a warning pipe in a small cistern

The warning pipe must be located in the cistern so that a minimum air gap is maintained between its point of discharge and the normal water level in the cistern. The position of the float-operated valve is also crucial, to ensure that a minimum air gap is maintained between its outlet and the spill-over level of the cistern at the warning pipe. You should remember the following points:

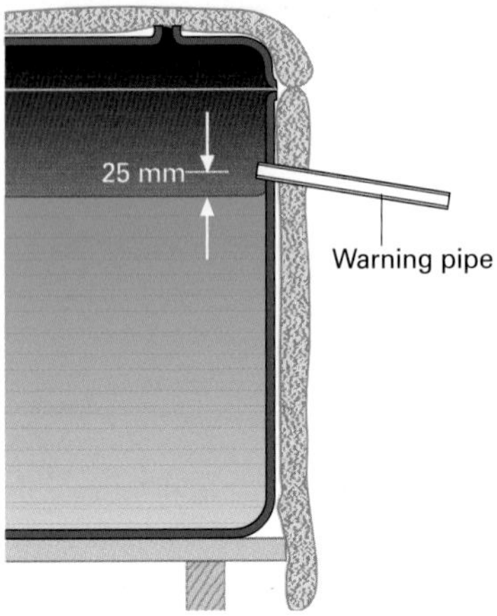

- If the float-operated valve becomes defective, the warning pipe should be capable of removing excess water without becoming submerged.
- The minimum pipe size is of 19 mm internal diameter, although the pipe may need to be larger than this with higher inlet flow rates to the cistern.
- The warning pipe should fall continuously from the cistern to the point of discharge.
- Warning pipes should discharge where the water will be noticed, usually outside the building.
- Warning pipes should be fitted with a screen or filter to exclude insects.

Figure 6.42: Warning pipe positioning

Connecting two or more cisterns together

It is crucial to avoid contamination and stagnation in cisterns, which may occur if water cannot fully circulate throughout the cistern. This problem is more likely to occur when two or more cisterns are joined together – perhaps because the access hatch was not big enough. When jointing cisterns we must ensure proper water movement throughout the cisterns to avoid stagnation, as illustrated in Figure 6.43.

When linking larger cisterns in domestic properties, attention must be given to:

- the provision of inlets to both cisterns fed via a float-operated valve
- the method of connecting the distribution pipes via a manifold arrangement taking equal draw-off from both cisterns simultaneously
- siting outlet connections at the opposite end to inlet connections to allow effective water distribution across the cisterns
- the method of linking the cisterns using lateral connections between them.

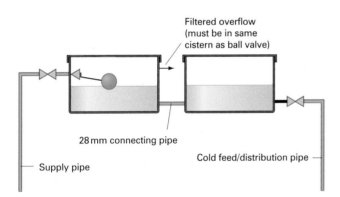

Figure 6.43: An example of linking two or more storage cisterns together

Lucy has recently qualified as a Level 3 plumber. Part of the requirement at this level was that she had to gain a Water Regulations certificate. Lucy is called out to a property, which was only completed six months earlier. A warning pipe at the rear of the property was leaking water and damaging the newly laid flowerbeds. The problem was obviously in the roof space as the warning pipe emerged through the fascia board.

On gaining access to the roof space, Lucy was amazed to find that the 100-litre cold-water storage cistern had been fitted with total disregard for the requirements of the Water Regulations.

The installation consisted of:

- an open-topped cold-water cistern, pushed tight under the roof trusses
- a BS 1212 : Part 1 float-operated valve, which had been modified by fitting a silencer tube from its outlet into the water to deaden the noise
- the warning pipe, which was fitted at the same height as the float operated valve

- the cold feed and the cold distribution pipes, which were likewise fitted at the same height as one another
- the cistern, which was simply sitting across the ceiling joists.

Lucy went downstairs to find the mains stop valve to isolate the supply. She reported her findings to the customer at the same time.

- What should the installation have consisted of so that it would conform to the current Water Regulations? Produce a list.
- What should Lucy do next? Who, if anyone, should she report this situation to?
- Could there be other similar installations on the same site?
- What issues would there be with the current positioning of the float-operated valve and the warning pipe (at the same height in the cistern)? Where should they be sited?
- With regard to the cold feed and the cold distribution pipe, again fitted at the same height as one another, what could be an issue? Which one should be fitted higher than the other and why?

1. How far below ground should mains and service pipes be placed in order to protect against frost damage?
2. There are two types of cold-water system. What are they? What are the advantages and disadvantages of each one?
3. What can water of Fluid Risk Categories 1 and 5 be used for?
4. Where is the stop tap usually located?
5. How do draw-off taps work?
6. Where is the drain-off tap usually located?
7. What does a water filter do?
8. A number of protection measures must be built into the design of the cold-water storage cistern. What are they?

3. Know, and be able to apply, the site preparation techniques for cold-water systems and components

Preparing a site for plumbing work, and the principles surrounding this, is covered in Unit 5 on pages 231–39. You will use many of the same practices when preparing to work with cold-water systems and components.

You will need to know about the following areas of site preparation:

- Selecting clips and brackets (pages 240–245)
- Regulations, standards and technical instructions (pages 245–48)
- Preparatory work on the building fabric (pages 231–39)
- Protection measures for the building fabric and customer property (page 233)
- Pipework materials and fittings required for cold-water systems and components (pages 208–21)
- Hand and power tools needed to work on cold-water systems and components (pages 200–205).

4. Know the installation requirements of cold-water systems and components and be able to install them

Many of the key points surrounding the installation of cold-water systems and components are covered earlier in this unit. You can find information about each area on the following pages:

- using MDPE, copper and lead pipework to connect water supply pipework to incoming service pipework (page 264)
- positioning requirements of components in cold water systems (pages 287–90)
- measuring, marking out and drilling plastic storage cisterns and pipework connections to cisterns (pages 280–85)
- suitable methods for connecting new pipework connections, using copper, plastic, lead and galvanised steel (pages 263–66).

Additionally, the positioning, fixing and connecting of new cold-water outlets, including bath taps, wash hand basins, sink taps and WC flushing cisterns will be covered later in Unit 10 on pages 389–436.

System pressure and flow rate

The incoming water pressure and flow rate supplied via the mains have a key bearing on the size of pipework and fittings used in the system. Pipe sizes are covered in more detail at Level 3. However, you may be required to assist with taking readings of incoming water pressure and flow rate as part of the installation process.

Figure 6.44: Water pressure reading being taken

Water pressure measurement

This is carried out using a pressure gauge as shown in Figure 6.44. The gauge usually reads in bar pressure.

Water flow rate measurement

This is carried out using a water flow measuring device as shown in Figure 6.45. The reading is usually quoted in litres per second (l/s).

A common fault with domestic systems installation is that water pressure readings are not taken at the right stage in the job – that is, at the design or pricing stage. Insufficient supply pressure and flow rate can result in big problems. For example, if the wrong system components (e.g. combination boilers) are installed, they will not work correctly. If the pressure to an existing dwelling is poor, check first that there is no burst on the service from the external stop tap. You can usually tell by putting your ear against the stop tap key while it is on the stop tap and listening for a hissing noise.

Positioning and fixing requirements for cold-water systems

This section looks at some of the basic installation aspects of system pipework and the reasons why you need to think carefully about how you route pipework.

Accessibility

Water Regulations require that water pipes and fittings are easily accessible so that they can be inspected and repaired. However, customers and installers are not keen to see exposed pipework, so the tendency is to hide it under floors or behind wall surfaces.

BS 6700 gives recommendations about the accessibility of pipes and water fittings.

Water fittings that are difficult to access are commonly described as concealed fittings. Examples include:

- a fitting installed below ground
- a fitting passing through or under any wall, footing or foundation
- a fitting enclosed in any chase or duct
- a fitting in any position that is inaccessible or renders access difficult.

Figure 6.45: Water flow rate reading being taken with a weir cup

Remember

Keep the weir cup **level** when taking a water flow rate reading.

Remember

Water companies have a responsibility to supply water at a minimum supply pressure and flow rate but this figure is quite low, and not suitable for all systems.

Find out

Find out the Water Regulations and British Standards requirements for pipework installed in non-accessible locations.

Figures 6.46 and 6.47 show acceptable methods of placing pipework in concealed locations.

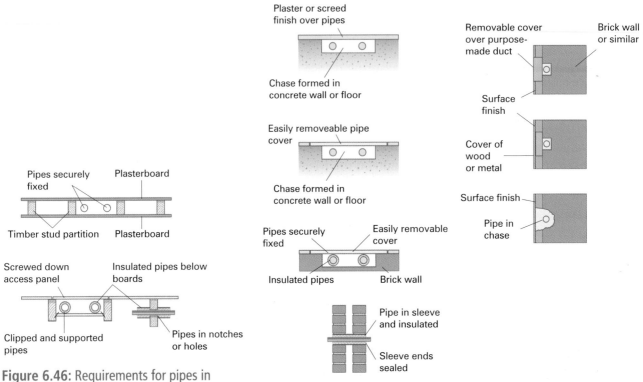

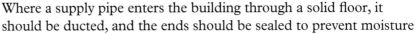

Figure 6.46: Requirements for pipes in solid walls and floors

Figure 6.47: Requirements for pipes in timber walls and floors

Solid and timber floor installations

Where a supply pipe enters the building through a solid floor, it should be ducted, and the ends should be sealed to prevent moisture

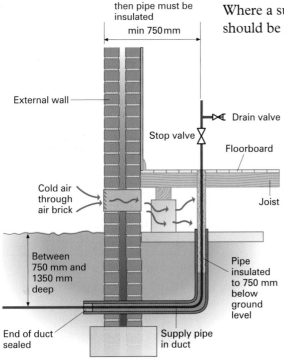

Figure 6.48: Suspended timber floor supply pipe entry

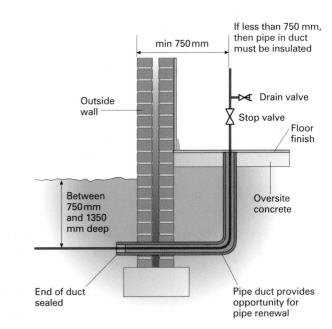

Figure 6.49: Solid floor supply pipe entry

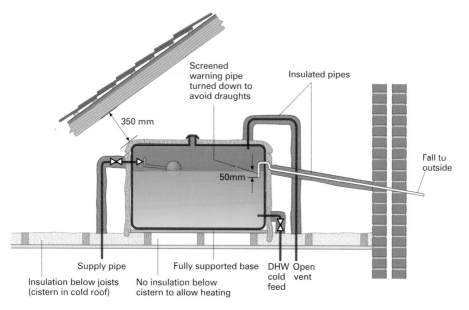

Figure 6.50: Cold-water storage cistern installation in roof space

entering the pipe. Depending how close the pipe is to the outside wall as it rises to the property, it may need to be insulated as well. If it enters the building through a suspended timber floor, it should be ducted, and the ends sealed, and it should be insulated as well. This is because the space between the suspended timber floor and the oversite concrete is exposed.

Pipes and cisterns in roof spaces

Figure 6.50 shows the correct installation of a cold-water storage cistern and associated pipework in a roof space. You should remember the following points:

- a 350 mm gap to the roof surface is provided to allow access for maintenance and frost protection
- both the pipework and cistern are fully insulated
- the space under the cistern is left un-insulated to allow heat from the property to warm the cistern.

Water regulations also require that cold-water systems should not exceed 25 °C. The use of insulation on pipework and components acts in hot weather to prevent stored water from becoming overheated.

Other outside applications

Outside taps or standpipes for garden hoses are common. If the tap has a dedicated underground supply, it requires protection.

Figure 6.51 shows a typical garden tap installation. Note that a drain valve is located below the pipework. This enables the pipework to be drained once the stop tap is turned off. A double-check valve is also fitted to protect against backflow.

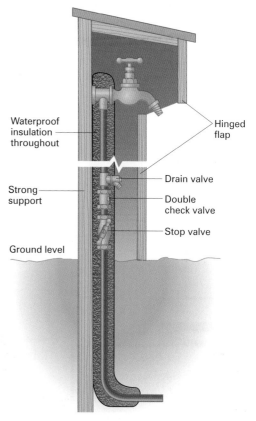

Figure 6.51: Installation requirements of externally sited and fed hose union tap

Location of pipework

When you are planning pipe runs, you should try to avoid areas that are hard to keep warm, such as:

- outside locations or outside walls
- roof spaces, cellars and under-floor spaces
- garages and other outbuildings
- areas close to windows, air bricks or ventilators.

Obviously it is not always possible to avoid these locations, so on these occasions you will need to provide some protection to the pipework.

Frost protection

The major problem associated with freezing temperatures is the dramatic effect they can have on plumbing systems. Be aware that when water freezes it expands by about 10%, causing damage to unprotected pipes and fittings.

When water freezes within the confines of, for example, a copper pipe, it will expand. The water either side of the 'ice plug' cannot be compressed, so the expanding ice will cause the wall of the pipe to split. Once the ice thaws, you have a burst pipe. Water damage in domestic properties can prove to be very expensive. It also causes an undue waste of water, which goes against the requirements of the Water Regulations.

The Water Regulations and BS 6700 both deal with frost protection.

Protection of pipes and fittings

Where pipework is at risk of freezing, it should be protected using pipe insulation material. The thickness of the insulation will depend on the type of insulation used and its insulation properties.

Table 6.4 gives the minimum recommended thickness of insulation (in millimetres) for 15 mm, 22 mm and 28 mm water pipes, to comply with BS 5422 and Water Regulations Schedule 2, Paragraph 4.

External diameter of pipe (mm)	Thermal conductivity of insulation materials (W/mK)				
	0.02	0.025	0.03	0.35	0.04
15	20	30	25	25	32
22	15	15	19	19	25
28	15	15	13	19	22

Table 6.4: Minimum recommended insulation thickness (mm)

Table 6.4 shows how thick the insulating material must be for products with varying insulation properties. Note that all the entries in the table require an insulating material that is thicker than the typical 10 mm wall thickness you buy off the shelf at the merchant – so be careful.

Remember

When water freezes, it expands by about 10%.

Did you know?

The minimum depth for pipes to be laid underground is 750 mm. This is to protect them from frost damage.

Did you know?

The wrong thickness of insulating material provided is a common fault identified by many water company inspectors when checking systems.

Remember

Instead of using a secondary circuit, trace heating can also be used to overcome the problem of long dead leg runs in hot-water systems.

Generally, the insulating material must be resistant to or protected from:

- mechanical damage
- rain
- moisture
- subsoil water
- vermin.

Trace heating

Trace heating involves attaching a low-temperature heating element to the outside of the pipe, controlled by a thermostat which activates the heater when the temperature is low, thereby preventing the pipe from freezing. It is more common on industrial installations, but there are domestic products available.

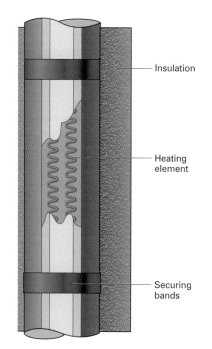

Figure 6.52: Trace heating element protecting pipework against frost

5. Know, and be able to apply, the service and maintenance requirements of cold-water systems and components

Maintenance

Hot- and cold-water system maintenance at Level 2 requires you to understand the principles of maintaining some of the key components. Maintenance is important as it ensures that the system components operate as they were designed to. It also reduces the chances of wasting water and energy, by preventing dripping taps, float-operated valves and showers.

What is maintenance?

Maintenance work usually includes turning off hot- and cold-water supplies and then turning them on again. You need to ensure that the customer is notified when you are doing this, so that they can fill kettles and turn off washers and dishwashers before the maintenance work begins. Don't forget that co-workers on site may also need to store water for the period that the supply is going to be turned off. If a cold-water system needs to be isolated for a prolonged period of time, it might be necessary to establish a temporary water supply.

The principles of maintenance/servicing are quite simple:

- Find out what the fault is with the component to be serviced and ensure you have any parts that may be required. This may be done through a visual inspection (see Unit 5, page 249)
- Isolate the supply.
- Strip the components.

Remember

An important aspect of maintenance, which applies to all the areas considered in the section, is liaison with the customer.

Check that draw-off taps are turned off when turning the supply back on.

- Repair or replace the defective part.
- Reassemble the components.
- Turn on the supply and test the component for correct operation.
- Report back to the person who asked for the work to be carried out, informing them that the fault has been repaired.

Manufacturer's instructions and job maintenance schedules

Manufacturer's instructions should always be followed when installing items as they detail the approved method of installation. It should be noted that manufacturer's instructions will override all other documents you may use concerning that particular item. The instructions will be based on compliance with all Regulations currently in force. If manufacturer's instructions are not available then British Standards, Water Regulations and Building Regulation information should be followed in their place. If their goods have not been installed correctly then manufacturers will not be liable to provide insurance or warranty cover for their goods.

Some items will require routine maintenance to keep them in correct working order and, again, the manufacturer will give this information in their instructions.

Routine checks on components

Visual inspection of pipework for leakage

As stated in the Water Regulations, a key aim is to reduce the amount of water wasted. Water is wasted by allowing leaks to continue without being rectified. Where pipes can be accessed this should not be much of a problem as you will be able to see any signs of leakage. One common sign that a joint is leaking on a copper pipe/joints is that the water turns the fitting or pipe green and leads to a build-up of limescale. Before carrying out any work it is good practice to do a visual inspection of the system and report your findings to the customer. Where pipes and fittings are hidden, the task is not so easy and usually people are unaware of the problem until some damage is caused.

Hot- and cold-water tap – kitchen sink

The servicing requirements of hot and cold taps are generally the same, but the isolation of the supply, and turning it on, differs.

Common faults

Depending on the type of taps fitted, faults tend to be:

- dripping taps
- leaking taps (glands)
- tap heads that are difficult to turn
- noise when the cold tap is turned on.

Did you know?

You may come across sink taps using ceramic discs rather than the more traditional washer and seating methods of control. The ceramic discs do wear and allow leakage. These discs can be replaced, but you may have difficulty locating the parts. It may be more cost-effective to replace the headgear completely.

On some properties, servicing valves may have been fitted to the hot and cold supply under the sink. This is good practice but is not a requirement of the Water Regulations. If servicing valves are fitted, isolation is achieved by turning them off.

The checklist below assumes servicing valves are not fitted.

Remember

If the cover of the tap head is plastic, don't over-tighten the fixing screw or you could split the plastic.

Checklist:

1. Isolate the supply
   - On a direct cold-water system, turn off the cold water supply to the system and drain through the sink tap.
   - Turn off the gate valve to the domestic hot-water cylinder for hot-tap repairs.
   - On mains-fed hot-water systems close the service valve on or near the unit.
   - Drain down the system pipework through the sink tap.
2. Repair or replace the defective part
   - This procedure will depend on the type of taps fitted. Most taps fitted to kitchen sinks are chrome-plated brass but some may be plastic.
   - Remove the cover from the tap head assembly; this will usually be screwed through the top of the tap cover and concealed behind a chrome or plastic cap. Lift the cap with a small flat-headed screwdriver; the cover screw (which is usually cross-headed) can then be removed.
   - You are then left with a headgear similar to the stop tap (see page 271).
   - The process for re-washering the tap is the same as for the stop tap (see page 270).
3. Reassemble the component
   - This is the same procedure as for the stop tap (see page 270), but you will also need to replace the cover.
4. Turn on and test
   - Make sure the hot and cold taps to the sink are turned off.
   - Turn on the stop tap if it is a direct system.
   - Turn on the service valve on the cold-water distribution pipe if it is an indirect system.
   - Turn on the gate valve to the hot-water storage cylinder.
   - Return to the sink taps and make sure the supply pressure to the cold-water tap is satisfactory (direct) and that the cold-water (indirect) and hot-water supplies are flowing smoothly.
   - Test the operation of the taps.
   - Report back to the person who asked for the work to be carried out, informing them that the fault has been repaired.

Float-operated valve

There are three types of float-operated valve:

- Portsmouth
- diaphragm
- equilibrium.

The trend for new installations is to use plastic diaphragm float-operated valves both on CWSC and WC cisterns. In maintenance work, however, you could come across a range of different types.

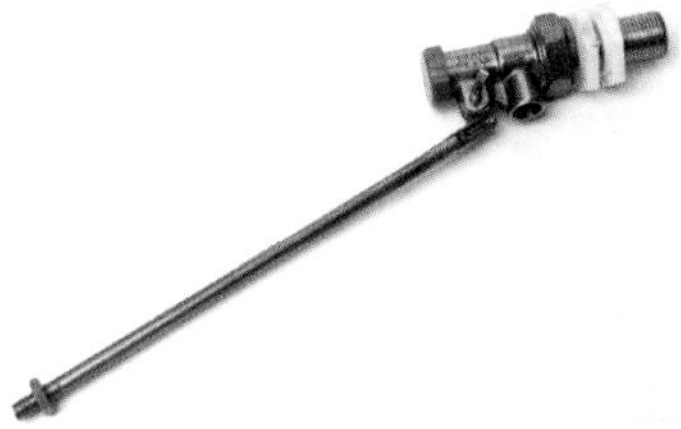

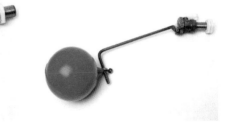

Figure 6.53: Portsmouth float-operated valve

Figure 6.54: Diaphragm float-operated valve

It is a Water Regulation requirement that servicing valves are fitted as close as possible to the cistern in order to isolate the float-operated valve for maintenance work.

Common faults:

- Probably the most common fault is a running warning or overflow pipe. This is caused by the float-operated valve not closing off the supply properly.
- There is no water in the cistern, the toilet will not flush or there is no hot water. These faults are usually because the hole in the seating is blocked with tiny bits of debris that got into the system when it was installed.
- There is excessive noise in the system, particularly when the WC cistern is flushed or hot/cold water is drained off from the cold-water supply cistern.

Checklist:

- Isolate the supply.
- If a service valve is fitted then you can simply turn this off.
- If a service valve is not fitted, you will have to turn off the supply at the stop tap.
- If you are working on a CWSC, run the level down to enable you to work easily on the float-operated valve. If it is a WC cistern, pull the flush to allow the float to drop. You might think this wastes water, but you need to check the correct operation of the float valve by watching it working as it fills up again.
- Repair or replace the defective part (when dealing with overflows).
- If the float valve is plastic, then it is either the washer in a Portsmouth valve, or the diaphragm in a diaphragm ball valve, that is defective.

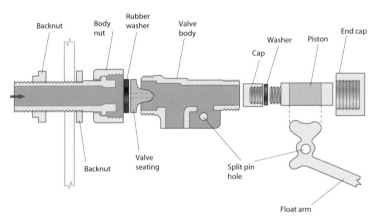
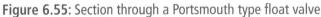

Figure 6.55: Section through a Portsmouth type float valve

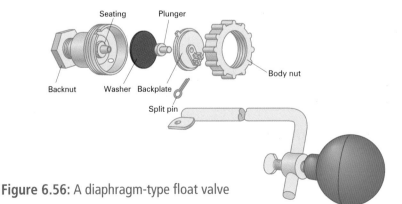

Figure 6.56: A diaphragm-type float valve

Brass Portsmouth ball valve

You may come across this type of ball valve on maintenance work.

Common faults

Most problems are caused by defective washers. Others are caused by oxidisation or a build-up of scale between the lever arm and piston.

Checklist

Once you've split the body from the body nut, follow this sequence:

1. Remove the end cap.
2. Remove the split pin from the lever arm. This frees the piston.
3. Remove the piston cap, take out the defective washer and replace it.
4. Reassemble the components as follows:
   - Refit the cap.
   - Make sure the piston, and the end of the lever arm that locates the piston, are cleaned thoroughly.
   - Replace the piston, relocate the arm and split pin.

- Replace the end cap.
- If a service valve is fitted, turn it on; if not, turn on at the stop tap.
- For both WC cisterns and CWSCs, allow the cistern to fill naturally. Make any final adjustments using the float adjustment.
- Service valves should be fitted in installations where these do not exist.
- Report back to the person who asked for the work to be carried out, informing them that the fault has been repaired.

Internal stop tap

Common faults

- The stop tap won't turn off the supply properly. This is usually due to a worn-out washer or a defective seating.
- The stop tap is stiff and hard to turn. This is due to infrequent use and lack of regular maintenance. The spindle and packing-gland nut oxidise over time, which 'bonds' the two together.
- The stop tap is leaking. If the stop tap is a compression type, the nut and 'O' ring joint could be leaking. Similarly, with a capillary fitting, the solder joint could be leaking. It is more likely, however, that the packing-gland nut is leaking. Another possibility is that condensation has been mistaken for a leak. The incoming main is very cold, and the stop tap is usually located in a kitchen, where moisture levels are high. If the pipe is not insulated, the moisture will condense on the cold surface of the pipe and trickle down it to the stop tap, giving the impression of a leak.

Figure 6.57: Tap reseating kit

Checklist

1. Isolate the supply:

   - Make sure you have a stop-tap key in readiness.

   - Locate the external stop tap. It should be at the boundary of the property (footpath at the front of the house). On other property types, it may not be easy to locate. In some cases you may need assistance from the water undertaker to isolate the supply.

   - Once you have located the external stop tap, lift the cover and locate the external stop-tap head.

   - Be prepared for the stop-tap box to be filled with silt or other rubbish. You may have to clear that first to find the head.

   - Once you have located the head, use the stop-tap key to turn off the supply (clockwise). If it is stiff, **do not force it** – you may snap the head. Try twisting it both ways to get some movement. Again, if it is impossible to move it, you may have to call the water undertaker.

   - Return to the property and drain down the cold water from the lowest point (usually the kitchen sink). Hopefully, there will be a drain-off tap close to the stop tap. On older properties, drain taps are not always fitted.

2. Strip the component:

   - You will need to remove the headgear from the body. This is not always easy, as it may not have been stripped for years. It often helps to apply heat to the joint between the headgear and body. Remove the headgear using adjustable grips/spanners. If a drain-off tap was not fitted, get ready for a rush of water from the pipework above the stop tap.

3. Repair the component:

   - Strip the headgear by removing the **crutch head** and **packing-gland nut**.

   - Clean the rising spindle and the inside of the packing-gland nut using fine emery cloth (or a file in some cases).

   - Replace the defective washer.

   - Check the seating for defects. If the seating is severely pitted, it might be worth replacing the stop tap. Alternatively, you could re-seat the stop tap using a re-seating tool that 'grinds' the seating flush again.

4. Reassemble the component:

   - Fit all the component parts together, making sure that the spindle and the inside of the packing-gland nut are lubricated.

   - Repack the gland nut using PTFE tape.

   - Refit the crutch head.

   - Before refitting the headgear, make sure the washer between the headgear and body is in place and intact.

   - Refit the headgear and tighten it.

   - If not already fitted, install a drain-off tap to bring the installation up to requirements.

5. Turn on and test:

   - Make sure the internal stop tap is turned off, as well as the drain tap.

   - Turn on the external stop tap.

   - Return to the building, and make sure the cold tap on the sink is open.

   - Slowly open the stop tap and allow water to flow from the sink tap. This will help to remove any bits of debris that may have entered the pipework.

   - Open the stop tap fully and assess the pressure at the sink tap.

   - Turn off the sink tap and allow the supply to charge the system.

   - Check that all the appliances fed by the supply are operating properly; check the cistern water levels.

   - Report back to the person who asked for the work to be carried out, informing them that the fault has been repaired.

The wrong taps

Ben and Zoë are both apprentices working for a large firm on a maintenance contract for a housing estate. They are carrying out routine servicing of sink mixer taps that are fed directly from the incoming cold-water supply. Zoë discovers that the type of sink mixer tap is not permitted by Water Regulations and suggests to Ben that they should inform the supervisor. Ben says 'Don't bother, it will only cause a problem as the firm is already behind schedule, and anyway, it is nothing to do with us.'

- Why does the mixer tap not conform to current regulations?
- Who is right: Ben or Zoë?
- What should they do regarding this issue?
- Who should they report the issue to?
- Should they travel to the nearest plumbing merchants and purchase the correct mixer tap?

Procedures for dealing with defects

As a plumber, you will need to be familiar with several procedures for dealing with common faults connected to cold-water systems.

Cistern failure

WCs that are not flushing or hot tap outlets with no hot water can often be caused by a float-operated valve problem. Such problems are usually caused by small deposits, such as solder or grit, which have got into the system and eventually become lodged in the throat of the seating, preventing the water from getting through. This is rectified by cleaning out the seating.

Incorrect support to cold-water pipework and storage cisterns

If pipework and cold-water storage cisterns are not supported correctly, this can have a number of effects ranging from very minor to very costly. Guidance for pipe clipping distances can be found on pages 243–44. You will also find this information in BS 6700.

The key guidance requirements for cold-water storage cisterns are that the base of the cistern should be fully supported and this support should extend beyond the cistern by 150 mm in all directions. The support should be capable of withstanding the weight of both the cistern and its contents.

Airlock in hot- and cold-water pipework

Airlocks are one of the most common causes of problems in low pressure hot- and cold-water pipework systems. Airlocks are usually caused when pipework from cold-water storage cisterns, hot-water storage vessels and boilers is not installed correctly. Figure 6.58 shows two installation details that highlight the problem.

Key terms

Crutch head – the handle of a stop tap.

Packing-gland nut – nut used to compress packing to make valve spindles watertight.

Safety tip

Remember that the water draining off may be hot if you used heat to remove the headgear.

Find out

What are the main causes of airlocks in a hot-water system? How can these be avoided?

Pipework running horizontally should be level or at an appropriate fall allowing air to escape from the system. Horizontal runs of primary gravity circulation pipework should also be installed at the appropriate fall.

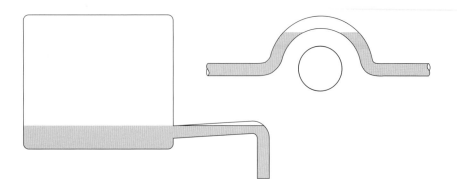

Figure 6.58: Airlocks

Safety tip

On the cold-feed pipework from cisterns, always use purpose-made bends of slightly less than 90° to help avoid airlocks.

Noise in systems

Noise in systems is usually caused by vibration. Vibration is a source of annoyance to the occupier of the building and, in severe cases, it can also damage pipework and fittings, eventually causing leaks.

Noise in systems is categorised as:

- water hammer
- flow noise
- expansion noise.

Remember

Secondary circulation can also be installed on systems using gravity circulation.

Water hammer

This is probably the most common cause of complaint from customers. When a valve is closed suddenly, shock waves are transmitted along the pipework, making a loud hammering noise. The problem is made worse where cold-water supply pipework is not adequately clipped.

Defective float valves and tap washers on cold taps can cause water hammer, so it can be cured through regular maintenance of the valve or taps. The velocity of the cold-water supply will further affect the problem; reducing the flow rate by the installation of a **water governor** (pressure-reducing valve) can help to reduce the possibility of water hammer occurring.

Key term

Water governor – pressure-reducing valve.

Flow noise

Pipework noise becomes significant at velocities over 3 m/s, so the system should be designed to operate below 3 m/s even if this means increasing the diameter of the pipe or installing a water governor.

Flow noise is also sometimes heard from cisterns; this may be splashing noises caused by the incoming water hitting the water surface as the cistern fills.

Silencer tubes on float valves were once used to cure this, but are no longer allowed, except for the collapsible type.

Expansion noise

Expansion noise usually occurs in hot-water pipework. As the system expands and contracts, it causes creaks and cracking sounds. The use of relevant pipe clips, brackets or pads between pipes, fittings and pipework surfaces should help to deal with expansion and contraction.

Leakage of internal and below-ground cold-water pipework

Any leakage of water should be rectified as soon as possible. Not only could it cause damage but it is a waste of water and contravenes the Water Regulations. If the leak is internal:

- Isolate the supply. If there are no service valves, the main stop valve at the point of the incoming cold water will need to be shut off. This is normally located under the kitchen sink.
- Before the leak can be repaired, drain off the system to remove any water from inside the pipe.
- Carry out the repair before turning the main stop tap back on.

If the leak is on underground pipework try to locate it as follows:

- Shut off the stop tap at the property boundary and dig up the ground where the water is showing. In some cases this may not be at the point of the leak. It would be advisable to close the main stop tap in the property before turning the boundary stop valve on. This will allow you to control the water entering the property following the repair.
- Open the kitchen tap and then slowly open the stop tap so that any dirt that may have entered the pipe is flushed out of the tap rather than getting into the system and damaging float-operated valves.
- After repairing the leak leave any excavations open until tested.

Information for a maintenance record

A maintenance record can be as simple or as complex as you want to make it. The reason for having it is to ensure that the items listed are maintained on a regular basis. A record for the plumbing installation in a domestic property will be quite small and simple compared to one for the machinery in a large factory. However, the main details required are the name of the item, its location and the frequency of maintenance. You also need to be able to identify that it has been maintained and on what date.

Ashley is a self-employed plumber who has built up a good business through his reputation, dependability and the variety of skills he is able to offer. One of his clients is a factory. Last winter, due to extreme weather conditions, the building suffered considerable damage following several burst pipes. Ashley carried out the repairs.

The factory's insurance company was very displeased with the cost of the claims made and has now insisted that an annual maintenance schedule is drawn up by a competent person. Ashley has been given the task.

The cold-water system is extensive with three male washroom facilities (including flushing urinals), two female washrooms, a kitchen and a canteen. There are several outside taps fitted around the exterior of the building for washing vehicles and

so on. The hot-water facilities are all localised to each area and may be ignored for the purpose of this exercise.

- What should Ashley's maintenance schedule consist of? Produce a detailed list to identify what he should be checking and servicing on an annual basis.

- What recommendations should he make to improve the cold-water provision in particular? Consider that most of the factory is very cold and it is not feasible to heat the main working areas. These are too large and there is a constant movement of vehicles through the factory. Also consider that much of the pipework is exposed inside the building with little protection.

- How should Ashley present his maintenance schedule and proposals to the factory managers?

6. Know, and be able to apply, the decommissioning requirements of cold-water systems and components

Decommissioning domestic systems means turning off the supply to the system, removing system pipework and components and making sure that the hot or cold supplies are sealed or left so that they cannot be turned back on. The correct term for this is 'permanent decommissioning'.

This might be necessary if:

- an old system is to be completely stripped out of a domestic property and replaced with a new or alternative system (for example, a direct hot-water system being replaced with an indirect hot-water system)

- a system is to be stripped out permanently (for example, prior to the demolition of a building).

Temporary decommissioning is the process of taking a system out of action temporarily so that work can be undertaken on the system, such as the maintenance activities discussed earlier in this unit.

Before commencing work you will need to ensure that people who may be affected by your actions are aware of what is happening. You will also need to make sure that items such as washing machines and dishwashers are not operating.

When decommissioning part of a system, you should consider the use of other parts of the installation and the amount of time during which disruption will occur. The work may have to be planned to coincide with others or you may be able to cut into the pipe and fit valves so you can work without affecting others for long periods of time. You will need to make sure that all open ends of pipes are sealed off correctly either permanently or temporarily. Any dead legs should be removed from the system. If required, you may need to put out warning signs advising people not to use items. If a valve has been turned off you may be able to remove the handle to ensure that someone does not turn it back on.

7. Know, and be able to apply, the inspection and soundness testing requirements of cold-water systems and components

Soundness testing

Soundness testing is covered in depth in Unit 5 pages 249–51. Refer to this section for more information.

Filling cold-water pipework with water at normal operating pressure and checking for leakage

When repairs or extensions to existing systems have been carried out, it may only be possible to test the system by turning the cold supply back on. This could be because the remainder of the system is old and you cannot isolate it from the new works. You will need to turn the existing supply back on and allow the system to re-fill. Testing will only be done at normal working pressure. Once the system is full and subject to the pressure, a visual check of all joints and connections should be made.

The flushing procedures for cold-water systems and components

Following any work on a system it should be flushed to ensure there are no particles or foreign matter within the system that might cause damage or harm. This can be done by opening a tap prior to turning the supply back on; this tap should be located as near as possible to where the work was carried out. The supply should be turned back on slowly until fully open, allowing the water to flow from the tap for a short period of time. The tap can be closed once the water is running clear. This should prevent damage to tap seatings and float-operated valves.

Actions taken when inspection and testing reveal defects in cold-water systems

If, when you carry out a test or visual inspection, you find a leak then the system should be isolated to allow a repair or replacement to be carried out. Once isolated, further investigations can be carried out to see what action needs to be taken. If the water supply is not isolated before you start moving items around the leak, you may find that the water starts coming out much faster and causes more damage. Once the repair has been made then the flushing procedure above should be followed.

Sometimes you may find that noise in the system is caused by a lack of pipe support in the form of brackets or clips. If you need to add additional brackets, this can be done in most cases, without the need to isolate the system.

Check your knowledge

1. What is the recommended minimum diameter of an incoming cold-water service pipe?
   a 12 mm
   b 15 mm
   c 22 mm
   d 28 mm

2. How should incoming water pressure be checked?
   a U gauge
   b Flow pressure meter
   c Pressure gauge
   d Air pressure testing kit

3. What is the minimum distance that a cold-water storage cistern should be sited away from the underside of the roof?
   a 350 mm
   b 400 mm
   c 450 mm
   d 500 mm

4. What is the recommended depth a cold-water service pipe should be sited below ground?
   a 550–1000 mm
   b 650–1350 mm
   c 750–1350 mm
   d 750–2000 mm

5. What is the first valve after the mains connection on a cold-water supply to a domestic property?
   a External stop valve
   b External drain valve
   c Air release valve
   d Internal stop valve

6. With reference to cold-water storage cisterns, a common warning pipe is only used for cisterns under how many litres?
   a 500 litres
   b 1000 litres
   c 1500 litres
   d 5000 litres

7. Which of the following should be installed above the internal stop valve?
   a Service valve
   b Air vent
   c Float-operated valve
   d Drain valve

8. What should be used to check the flow rate of water?
   a Weir cup
   b Flowing cup
   c Flow mug
   d Pressure gauge

9. Which one of the following pipes should be connected to a cold-water storage cistern by means of a filtered connection?
   a Cold-feed pipe
   b Overflow/warning pipe
   c Distribution pipe
   d Supply pipe

10. How many types of cold-water system are there?
   a 1
   b 2
   c 3
   d 4

Getting ready for assessment

The information contained in this unit of the book, as well as the continued practical assignments that you will carry out in your college or training centre, will help you with preparing for both your end-of-unit test and the diploma multiple-choice test. It will also support you in preparing for the practical assignments you will need to complete to demonstrate your understanding in and enabling you to carry out and apply domestic cold-water system installation and maintenance techniques.

There are opportunities throughout the unit to test your progress in and understanding of the required underpinning knowledge; this will enhance your preparation for the forthcoming assessments, so make good use of them.

This unit will be assessed by the following assessment methods:

- externally set knowledge assessment
- externally set assignments.

With regard to the mechanical services industry, you will need to know:

- The cold water supply route to dwellings, the types of cold water system and their layout requirements.
- The site preparation techniques for cold water systems and their components and how to apply them.
- The installation requirements of cold water systems and their components and how to install them.
- The service and maintenance requirements of cold water systems and their components and how to carry them out.
- The decommissioning requirements of cold water systems and their components and how to carry them out.

- The inspection and soundness testing requirements of cold water systems and their components and how to carry them out.

Check it out – using your knowledge from this unit practise and test yourself on the following key terms:

Access/exit, backflow and its prevention, Building regulations, clips/brackets, Codes of practice, cold water storage cisterns (CWSC) - sizes and requirements, commissioning and decommissioning, communication pipe, connection to mains supply, depth of service pipes, direct and indirect cold water supply, distribution, external stop valve, fixings, float-operated valves, flushing procedures, frost protection, hazards, industry standards, installation, insulation, linking of CWSC, main sources of supply, maintenance, maintenance/records, manufacturers technical instructions, measurement and marking out, metering, pipe sizes, pipework selection/specification, pipework/fittings, point of entry, positioning components, PPE, preparation of plastic storage cisterns, preparatory work, pressure/flow rate, protection measures, re-cycled water, reading pressure and flow, remedial work, risk assessment, safe access/exit, selection and safe use of hand and power tools required, service pipe, servicing, site preparation, soundness testing, statutory regulations, stop/service valves, surface/underground sources, system components and appliances, system layouts, the rainwater cycle, tools/equipment, types of cold water system, visual inspection, water treatment and wholesome water.

Good luck!

Understand and apply domestic hot-water system installation and maintenance techniques

It is hard to imagine life without hot water: no hot showers, baths or water for cleaning. It is your job to ensure that hot water is available as required. The design of a hot-water system will depend on the type of building. For example, when installing a toilet in a factory canteen, an instantaneous water heater is a more economical option than a storage system. In a domestic situation, there are instantaneous methods of supplying hot water as well as systems fed from storage.

This unit will cover the following learning outcomes:

- Know the types of hot-water system and their layout requirements

- Know and be able to follow the installation requirements of hot-water systems and components

- Know and be able to apply the site preparation techniques for hot-water system and their requirements

- Know and be able to apply the service and maintenance requirements of hot-water systems and components

- Know and be able to apply the decommissioning requirements of hot-water systems and components

- Know and be able to apply the inspection and soundness testing requirements of hot-water systems and components.

Much of this material is covered in Unit 6. You will need to refer to this Unit, as many of the component, installation, servicing, decommissioning and inspection requirements will be the same.

1. Know types of hot-water system and their layout requirements

A supply of hot water to domestic dwellings only became commonplace after the Second World War, and system design has been improving since then. Hot-water system design is also covered by the Water Regulations and BS 6700. The Building Regulations Parts G & L1 lay down legal requirements for the energy efficient installation of hot-water systems while Part G deals with installation requirements and water efficiency.

These details are covered at some length in the *Domestic Heating Compliance Guide*, and The New Water Efficiency Calculator publication provided by the government department Communities and Local Government. You can access this document by going to www.pearsonhotlinks.co.uk, searching for this title and clicking on this unit.

A wide range of hot-water systems are available for domestic properties. The following factors should be considered in their selection and design:

- the quantity of hot water required
- the temperature during storage and at outlets
- the cost of installation and maintenance
- fuel energy requirements and running costs
- any wastage of water and energy
- safety for the user.

Methods of heating hot-water systems

The following energy sources or fuels are used for heating hot water:

- Electricity:
  - immersion heater
  - instantaneous heater
  - storage heater.
- Gas:
  - boiler
  - water circulator
  - instantaneous heater
  - storage heater.
- Solid fuel:
  - boiler
  - combined cooker and boiler.
- Oil:
  - boiler
  - combined cooker and boiler.
- Renewable energy sources.

Choosing a system

Systems can range from a simple, single-point arrangement supplying one outlet to a more complex centralised boiler system supplying hot water to a number of outlets. BS 6700 sets out a number of ways of supplying hot water, as detailed in Figure 7.1.

The chart in Figure 7.1 is divided into **centralised** and **localised** systems. A centralised system is one where water is heated and can be stored centrally within a building. The heating of the water can be controlled by a thermostat. A system of pipework supplies the heated water to various draw-off points. In a localised system, water is heated where it is needed, such as a single-point water heater sited over a sink. Localised systems are often used in situations where long distribution pipe runs would involve a waste of water and energy.

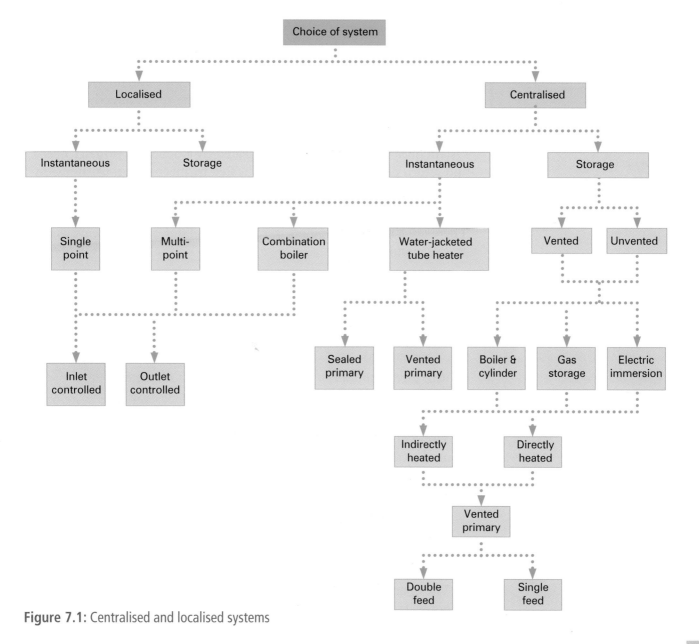

Figure 7.1: Centralised and localised systems

Key terms

Gravity circulation – cold water is heavier than hot water and gravity therefore exerts a stronger pull on it, drawing it down and allowing the hot water to rise through the system.

Immersion heater – an electric element fitted inside the hot-water storage vessel. It can be controlled by a switch and thermostat.

Types of connection

The connection between main and service pipework will depend on the type of hot-water system that is to be installed, but generally will be categorised as direct or indirect.

Direct hot-water system (vented)

Appliances such as instantaneous water heaters, when supplied directly, should be fitted with a servicing valve (ball type) as close to the appliance as possible.

Direct hot-water storage systems are fed via the cold-water storage cistern (CWSC).

You are only likely to work on the type of system shown in Figure 7.2 when carrying out maintenance or repairs: it is no longer widely used due to recent energy efficiency requirements. This direct system works by heating water in a boiler (by gas, oil or solid fuel). The water then rises due to the principle of convection – otherwise known as **gravity circulation**. It rises through the primary flow pipe and into the hot-water storage vessel, heating the contents of the vessel directly. The hot water from the boiler is replaced by the cooler and heavier water moving in the primary return from the lower area of the storage vessel. The system does not always have to be heated by a boiler via primary flow and return pipes. It can be directly heated by means of an **immersion heater**; gas circulators are also used on direct systems, connected directly to the storage vessel.

There are a number of points to remember about direct systems using gravity circulation:

- There are minimum pipe sizes for primary circuits to hot-water storage vessels:
  - 22 mm for short pipe runs
  - 28 mm for longer pipe runs (or from continuous burning appliances).
- Vent pipes should not be less than 22 mm in diameter.
- All pipes must be laid to falls to prevent airlocks and help systems drain down.
- The vent route from the boiler, primary flow and open vent should not be valved.
- The cold feed pipe (normally sized at 22 mm in a small domestic property) should be sized in accordance with BS 6700. The cold feed is the key route by which expanding water is taken up from the cylinder when it is heated, i.e. the heated water from the cylinder moves through the cold feed pipe and the water level rises in the storage cistern.
- The open-vent pipe cannot be taken directly from the top of the hot-water storage vessel. The hot draw-off pipe should incorporate a 450 mm offset between the storage vessel and its point of connection

to the open-vent pipe to prevent one-pipe circulation.

- Corrosion inhibiters should not be used, as the water in the boiler is fed directly to the appliances.
- No other supplies or draw-offs should be connected to the cold feed.

Figure 7.3 shows a direct system heated by an immersion heater. The immersion heater should be controlled by a thermostat.

New immersion heaters are now supplied with a secondary thermostat or overheat thermostat to guard against overheating in the event that the normal control thermostat fails. Following a number of recent high-profile scalding incidents with immersion heaters, it is now advisable to replace any immersion heaters with a single thermostat with the safer dual thermostat variety.

Hot-water cylinders solely heated by electricity are often designed to take advantage of low tariff electricity supplied in the evenings. This means a much larger cylinder than normal is installed, together with two immersion heaters. The bottom heater works on the low electricity tariff and is used to heat the entire contents of the cylinder during the evening. The top heater uses normal tariff electricity and provides daytime top-up if there is a shortage of hot water.

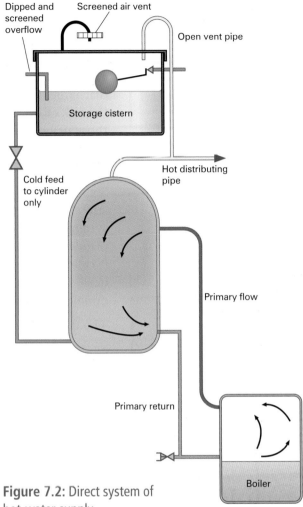

Figure 7.2: Direct system of hot-water supply

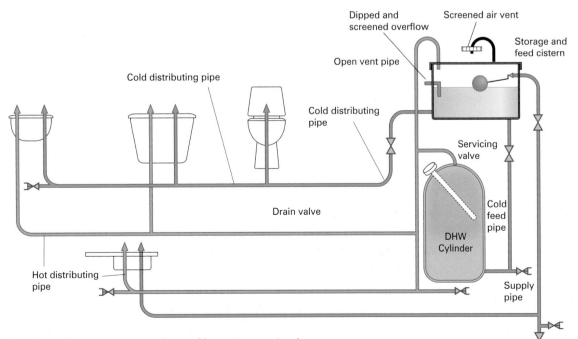

Figure 7.3: Direct hot-water system heated by an immersion heater

The purpose of the vent pipe

The vent pipe maintains atmospheric conditions in the pipework. It allows any air entering the system to escape and, should the water in the system become overheated, it allows it to expand up the vent pipe and discharge into the cistern. The use of better system controls has reduced the risk of overheating but it can occur on direct systems, particularly those using solid-fuel back boilers.

Indirect hot-water system (vented)

This is the most common form of vented domestic hot-water system and allows the boiler to be used for the central heating circuit. The system permits the use of a variety of different metals because the primary circuit is totally separate from the secondary circuit. The system is called indirect because the water contained in the storage vessel is heated indirectly through a heat exchanger. Figure 7.4 shows a vented double-feed indirect system.

The double-feed type of cylinder has the following key points:

- The open vent and cold feed pipes may be connected to the primary flow and return pipes as shown, or fed separately into the boiler.
- Where the vent pipe is not connected to the highest point in a primary circuit, an air release valve should be fitted.

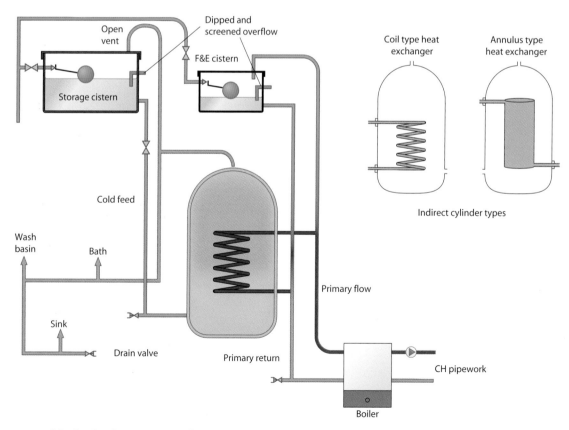

Figure 7.4: Double-feed indirect system of hot-water supply

- A separate feed and expansion cistern needs to be provided to feed the primary circuit. This ensures that where a double-feed cylinder is used, the primary water is kept totally separate from the secondary hot water.

Indirect single-feed system (also known as primatic cylinders)

This system uses a self-venting cylinder, and does not require a separate feed and expansion cistern. The water in the primary and secondary circuits is separated by means of an air bubble. The cylinder must be carefully installed, in accordance with the manufacturer's requirements, to ensure that the air bubble is not dislodged in the cylinder permitting the two waters to mix. It is for this reason that this type of cylinder is not widely used in modern vented types of hot-water system.

Key points about single-feed type of cylinder are as follows:

- Water enters the primary circuit via a number of holes at the top of the vertical pipe immediately under the upper dome.
- The system is self-venting through the air vent pipe while the primary circuit is filling.
- Once the primary circuit is filled, the filling of the secondary supply (the one which feeds the appliances) continues.

> **Remember**
>
> The manufacturer will specify a maximum head of water above the base of the cylinder.

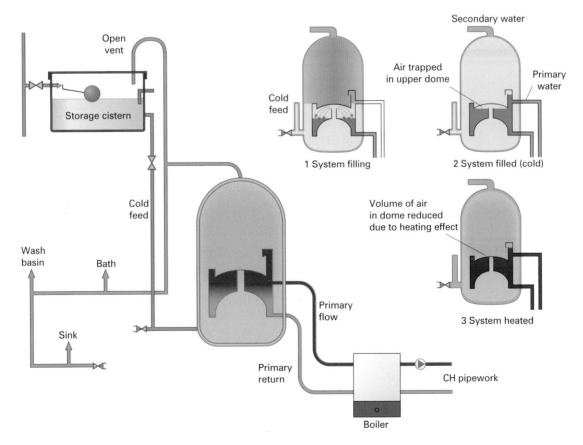

Figure 7.5: Single-feed vented hot-water storage system

- When the secondary supply is full, two air seals are formed and a seal is maintained. However, this seal could be lost if the water is heated too much and expands through this route (solid fuel appliances) or may be lost if the system volume is exceeded.
- Once the water is heated, expansion of the water in the primary circuit is taken up by forcing the air from the upper dome. This air is displaced, not lost.

Instantaneous water heaters can also be supplied indirectly and, again, should be fitted with a servicing valve. Indirect hot-water storage systems are where the cold-water supply is via the CWSC and installation details for these systems are covered in Unit 6.

Stratification

In a hot-water storage vessel, layers of water form from the top of the vessel where it is hottest to the base of the vessel where it is coolest.

Stratification must take place if the vessel is to function to its maximum efficiency. Manufacturers of storage vessels build stratification into the design. The following design rules enable stratification to take place:

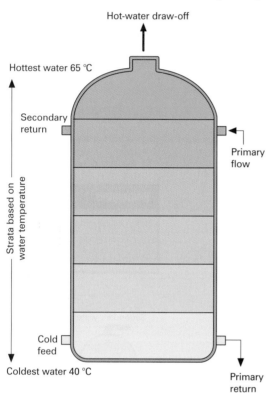

Figure 7.6: Stratification in a hot-water cylinder

- The vessel should be cylindrical in shape.
- The cylinder vessel should be installed in a vertical position rather than horizontal.
- The cold feed connection should be in a horizontal position.

Vented cylinder specification

Vented hot-water storage cylinders are available 'off the shelf' in a number of sizes to suit a particular circumstance, such as airing cupboard dimensions:

- standard widths vary from 300 mm to 600 mm
- standard heights vary from 675 mm to 1800 mm.

The most common cylinder size for a standard one-bathroom property is 450 mm × 900 mm, storing approximately 120 litres of water. Some manufacturers also produce 'one-off specials' made to the required dimensions for the job.

Cylinders are available in three grades. The grade of cylinder is based on the overall head of water generated by the cold-water storage cistern on the base of the hot-water storage cylinder:

- Grade 3 maximum head – 10 m
- Grade 2 maximum head – 15 m
- Grade 1 maximum head – 25 m.

Grade 3 cylinders are used in the majority of domestic properties.

Any new, or replacement, vented hot-water storage cylinders must comply with the requirements of the *Domestic Heating Compliance Guide*. The main points are as follows:

- A direct, indirect, single- or double-feed cylinder must be manufactured to the standards laid down in BS 1566
- The cylinder must be labelled as Building Regulations Part L compliant with the following information:
  - Type of vessel
  - Nominal capacity in litres
  - Standing heat loss in kWh/day
  - Heat exchanger performance in kW
  - Detail showing compliance with BS 1566.

In relation to energy efficiency, BS 1566 lays down minimum requirements for the insulation applied by manufacturers to vented cylinders, as well the surface area of the heat exchanger (coil) in an indirect cylinder. The greater the coil surface area the more energy efficient the cylinder tends to be.

The *Domestic Heating Compliance Guide* requires that all the pipework connecting to a hot-water cylinder should be insulated:

- to a point not less than 1 m from the cylinder, or
- to the point at which the pipework becomes concealed.

The storage capacity of a hot-water cylinder is based on an assessment of the consumption of hot water in the property. The recovery rate (the time needed for the cylinder to be re-heated) is also a factor that needs to be considered. BS 6700 outlines the methods of calculating the hot-water cylinder capacity. BS 6700 states that the minimum storage capacity in a small property should be no less than 100 litres.

If you fit an immersion heater into the cylinder, this must now be a dual thermostat control type. The first thermostat can be manually set by the installer normally between 60 and 65 °C. The second thermostat, known as the 'high limit', will be pre-set and if it operates will cut the system off until it is manually reset. However, if this does operate it will usually indicate that the normal control thermostat has failed and may need replacing.

Recently there were changes to the Building Regulations and document 'G3', which now covers vented hot-water cylinders as well as covered un-vented hot water. G3 states that a hot-water supply for any fixed bath must be designed and installed so that the temperature of the water cannot rise above 48 °C. This part of G3 applies to new-build or conversion properties and is a new requirement, intended to prevent scalding.

Combination storage systems (vented)

Under the requirements of the *Domestic Heating Compliance Guide,* combination type hot-water cylinders should be manufactured to BS 3198. This lays down the quality standards (including energy efficiency requirements) for this type of cylinder. The main feature of a combination type hot-water storage cylinder is that it contains a cold-water storage cistern integral to the unit. The base of the cold-water storage cistern must therefore be positioned at a higher level than the level of the highest water outlet (tap) in the property. This height (head) must also ensure that sufficient water flow rate is available at the outlets. These systems do not tend to be that widely used in newer installations as the relatively low head of water generated tends not to provide the required water flow rate at the outlets, unless the system is boosted with a water pump.

The advantages of a combination storage cylinder are:

- low installation costs
- they can be useful in flats, provided minimum water flow rates are achieved.

Disadvantages are that:

- they cannot normally be used with gravity-fed showers (unless the system is pumped)
- they tend to generate lower pressure and hence lower flow rate at outlets
- the cold-water storage space is often limited.

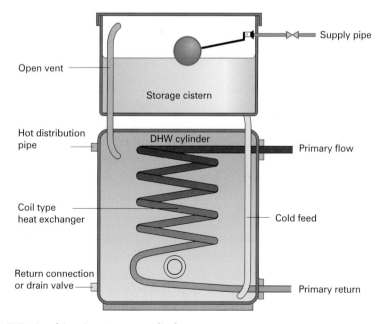

Figure 7.7: Combination storage cylinder

Unvented hot-water storage systems

This is a system type that is primarily covered at Level 3 and you need to hold an unvented hot-water certificate/card to allow you to install or maintain these systems. Below is an overview of the system.

An unvented hot-water storage system, as the name suggests, does not include an open-vent pipe or a cold-water storage cistern. Essentially it is a type of hot-water storage system that operates under the influence of mains/boosted cold water pressure, therefore providing higher water flow rates at taps (outlets).

As the hot-water storage cylinder operates as a closed vessel (at high internal pressure) the Building Regulations require:

- the cylinder to be fitted with a range of safety (temperature) controls to ensure that the stored hot water temperature never exceeds 100 °C
- the discharge pipes from the cylinder to be located in a safe place where they are not going to scald anyone
- the cylinder to be installed/maintained only by operatives who have passed an unvented systems competency test
- the cylinder installation to be notified to the local authority before installation or to be self-certified by a Competent Persons Scheme member company.

An unvented hot-water storage cylinder must be manufactured to the requirements laid down in BS EN 12897. This standard also details the insulation and heat exchanger performance requirements for the cylinder that are necessary to comply with Part L1 of the Building Regulations.

Under the requirements of the Water Regulations, the cylinder installation must include a range of functional controls preventing:

- over-pressurisation of the installation and therefore wastage of water
- contamination of the cold-water supply.

Figure 7.8 shows the typical layout of an unvented hot-water storage cylinder.

Table 7.1 outlines the purpose of the temperature-related control devices.

> **Did you know?**
>
> Companies who are in membership of an approved Competent Persons Scheme can self-certify certain types of work providing exemption from notification under the Building Regulations.

	Safety device	How this controls temperature
Safety item 1	Control thermostat	Maintains the water temperature in the cylinder between 60 and 65 °C
Safety item 2	High-limit thermostat (energy cut-out devices)	A non- self-resetting device that isolates the heat source at a temperature around 80–85 °C
Safety item 3	Temperature-relief valve	Discharges water from the cylinder at a temperature of 90–95 °C (water is dumped from the system and replaced by cooler water to prevent boiling)

Table 7.1: Temperature control safety devices

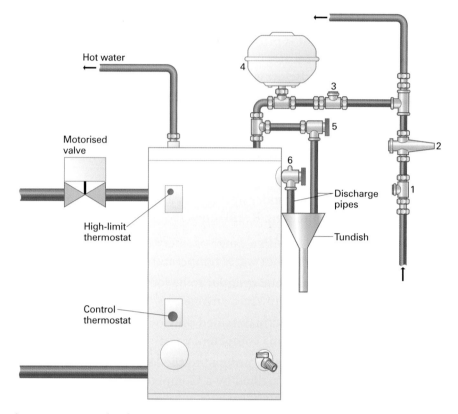

Figure 7.8: Typical indirect unvented hot-water storage cylinder

Table 7.2 outlines the purpose of the functional control devices fitted to the cylinder.

Functional control device	Tasks performed
Line strainer (1)	Prevents grit and debris entering the system from the water supply (causing the controls to malfunction)
Pressure-reducing valve; on older systems, this may be a pressure-limiting valve (2)	Gives a fixed maximum water temperature
Single-check valve (3)	Prevents stored hot water from entering the cold-water supply pipe (a contamination risk)
Expansion vessel or cylinder air gap (4)	Takes up the increase in water volume in the system due to the heating process
Expansion valve (5)	Operates if the pressure in the system rises above the design limits of the expansion device (i.e. if the cylinder air gap or the expansion vessel fail)
Temperature relief valve (6)	See notes under Table 7.1
Isolating (stop) valve (not shown in Figure 7.8)	Isolates the water supply from the system, for maintenance

Table 7.2: Functional control devices fitted to unvented hot-water storage cylinder

Water-jacketed heater

This is also known as a thermal storage system.

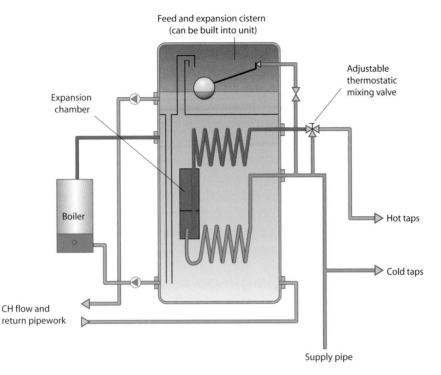

Figure 7.9: Water-jacketed tube heater

When the hot tap is turned on, cold water – from either the mains or storage cistern – passes through a heat exchanger, which is situated in a heat store of primary hot water. The size of this heat store will be calculated based on the volume and rate of flow that can be delivered without an unacceptable drop in temperature. The primary water flow from the boiler is programmed by the cylinder thermostat. Hot water is pumped to the radiator heating circuit and is returned to the heat store; the cooler water from the heat store is then returned to the boiler where it is reheated. This is similar to an indirect domestic hot-water system but in reverse.

Instantaneous hot-water systems

Instantaneous systems work by passing cold water from the service pipe through a heat source, which heats the water before it emerges at the application end. The heat source can be either gas, oil or electric.

The rate at which the water can be heated is limited, so the flow rate of the water needs to be controlled so it can be heated properly. Because of the reduced flow rate, it is not possible to supply a large number of outlet points all at once, so these systems are not installed in situations where there is high demand. For example, you might find a **multi-point** in a small property, or a single point in an office kitchen area or WC.

> **Did you know?**
>
> **Multi-point** refers to an instantaneous water heater that is capable of supplying water to more than one outlet.

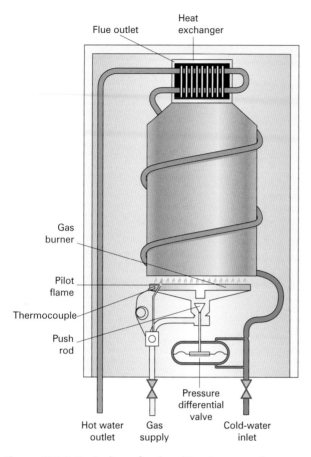

Figure 7.10: Typical gas fired multi-point water heater

Labels in figure:
Flue outlet
Heat exchanger
Gas burner
Pilot flame
Thermocouple
Push rod
Hot water outlet
Gas supply
Pressure differential valve
Cold-water inlet

Multi-point – gas fired

As you can see from Figure 7.10, this consists of a gas burner sited beneath the heat exchanger.

When the hot tap is opened, it allows water to pass through the heater. This causes the gas valve to open as a result of the drop in pressure in the differential valve. This drop in pressure is caused by water passing through the **venturi**, which creates a negative pressure as it sucks the water from the valve. The diaphragm is connected to a push rod and, as it lifts, it opens the gas line. The gas is then ignited by the pilot light. When the hot-water tap is turned off, the pressure in the differential valve is equalised, the diaphragm closes and the gas supply is turned off.

Single-point instantaneous water heater

This type of heater uses electricity or gas. Sited directly above the appliance, it is usually inlet controlled, with the hot water delivered via a swivel spout. The electric multi-point is a small tank of water with an electric heating element inside. Because of the low volume of water, it quickly heats up as it is drawn through the heater. The temperature at the outlet will be related to the water flow rate and the kW rating of the heater. Gas single-point heaters are now less common but you may cover them within the gas qualification.

You will find single-point heaters used in situations where a small number of hot water draw-offs are fed by individual heaters in a non-domestic type building, and where the use of a centralised hot-water system would be uneconomical. You might find them in the WC of small cafes, for example.

Instantaneous electric shower

The electric shower shown in Figure 7.11 is designed for mains connection, although some can be fed indirectly. The electrical rating can be in excess of 10 kW, so it is important that the supply is adequate and wired directly from the mains distribution unit (MDU). For a rating of 9.6 kW, the circuit protection device requirements would be 45 amp and the cable 10 mm². The shower should also be isolated with a switch; this should be located within easy access outside the shower room. Any pressure variations in the cold-water supply to the shower will be handled by the flow governor. Most electrical instantaneous showers are fitted with a flexible hose outlet – they will require a **check valve** to be fitted to protect against possible contamination of the supply, as required by the Water Regulations. This will usually be provided by the manufacturer.

Storage heaters

Outlet controlled

These are more common in large domestic or small commercial/ industrial buildings.

Figure 7.12 shows a typical storage heater. In this case, it is heated by gas, but electrical storage heaters are also available. A gas storage heater is basically a self-contained boiler and storage system. This system also includes an open flue, which must be terminated externally. Often referred to as pressure-controlled water heaters, they are usually designed to be fed by a cistern (indirect) or mains (direct) supply, in which case it would be included in an unvented system. This type of storage heater is classified as outlet controlled, as the supply is controlled at the appliance outlet such as a hot tap. It will also serve multiple outlets.

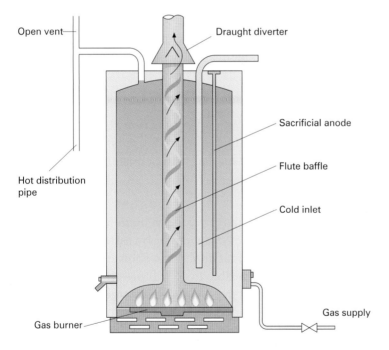

Figure 7.12: Direct-fired gas storage hot-water heater

Inlet controlled

As Figure 7.13 shows, these are generally seen as single-point heaters, fitted either above the appliance with a swivel outlet spout, or under the appliance.

Key term
Check valve – a device used to protect the water supply from any contaminants that may get into it.

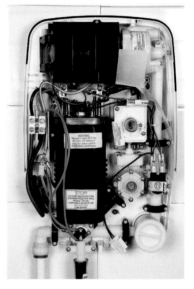

Figure 7.11: Instantaneous electric shower

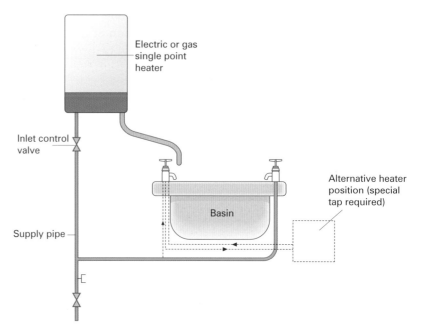

Figure 7.13: Small, inlet controlled single-point storage heater

Figure 7.14: Shower mixer

The heater is fed from the supply pipe, which has an inlet control. The outlet, and any connections made to it, must not be obstructed as the open outlet allows for expansion of the water on heating. If an under-sink model is used, then a special tap will be required to allow venting of the water heater. They can be heated by either gas or electricity.

Connection requirements for showers

This section looks at the layout and connection requirements of shower mixing valves. A shower can be fed by mains pressure, or it can be storage-fed by gravity or storage-fed using booster pumps.

Mains-fed showers

Water Regulations require that provision is made to ensure that backflow cannot occur. This is done either by installing a double-check valve, or by using a rigid connection to the showerhead. Alternatively, if a sliding rail is used, the flexible hose will pass through some form of device designed to prevent it from being submerged in water; this will prevent backflow. Mains connection can only be used on thermostatic mixing valves, using the manufacturer's valves designed specifically for this purpose.

Storage-fed showers

The hot and cold supplies need to be of equal pressure, and the showerhead needs to be a minimum of 1m below the bottom of the cold-water storage cistern to ensure that adequate pressure from the showerhead is achieved.

Boosted supply

In the system illustrated below, water pressure is increased using a booster pump. The diagram shows a single-impeller booster pump and a double-impeller booster pump. The pump increases the pressure, which means that the minimum static head of 1 m is no longer needed. However, a minimum of 150 mm head is necessary to allow the flow switch to open when the supply is turned on.

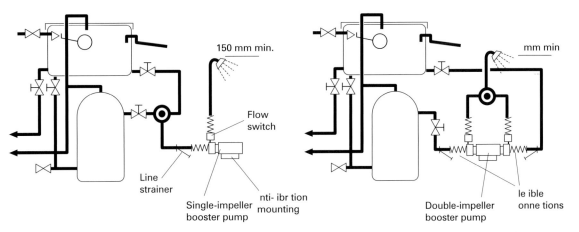

150 mm min.

Flow switch

Line strainer

Single-impeller booster pump

nti- ibr tion mounting

mm min

Double-impeller booster pump

le ible onne tions

Figure 7.15: Typical shower pump installations

Working life

Dilemma for Neeta

During alterations to a plumbing system on a house extension, Neeta, a final-year apprentice, has been given an ordinary shower mixer by the customer for fitting, although the cold water is mains fed. Neeta is aware that the shower should not be fitted directly to the mains, and she explains this to the customer. The customer accepts the explanation that it is not good practice and it would be against the Water Regulations to fit it, but insists that this is his house and that it is acceptable to him. 'Anyway,' he says, 'thermostatic showers are three times the price – no chance. Just fit it.'

- What should Neeta do now?
- Who would you go to for expert advice?

Bidet connection requirements

Bidets can be divided into two groupings:

- **ascending spray**, or over the rim taps, connected to a flexible hose
- over the rim with discharge via tap outlets.

Ascending spray bidets, or over-the-rim taps, with shower attachments should only be fed from a storage system that has dedicated hot/cold distributing pipework connecting to it. See Figure 7.17 for the pipework layout.

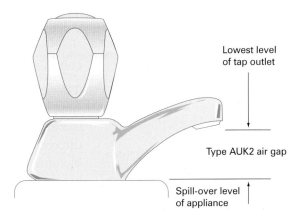

Lowest level of tap outlet

Type AUK2 air gap

Spill-over level of appliance

Figure 7.16: Type AUK2 air gap to over the rim bidet with no shower attachment fitted

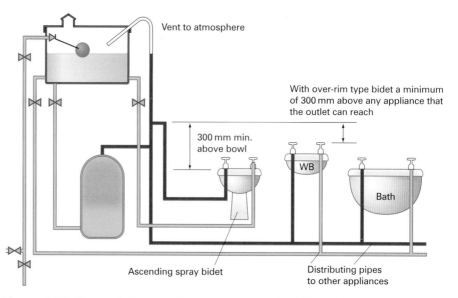

Vent to atmosphere

With over-rim type bidet a minimum of 300 mm above any appliance that the outlet can reach

300 mm min. above bowl

WB

Bath

Ascending spray bidet

Distributing pipes to other appliances

Figure 7.17: Pipework to ascending spray or over the bidet with shower attachment

Did you know?

Dedicated distributing pipework refers to a hot / cold pipework system that supplies water solely to an individual appliance.

Remember, these types of bidet cannot be directly fed with water from a combination boiler or an unvented hot-water system.

Over-the-rim taps with standard tap outlets can be supplied with cold water direct from the mains supply pipework, or from a storage cistern, without the need for **dedicated distributing pipework**.

Working principles of hot-water system components

Information on the working principles of cold-water system components is provided in Unit 6 on pages 270–80. Please refer to this section for more information.

Thermostatic mixing valves

Thermostatic mixing valves are generally used to prevent scalding and are used in most public buildings such as hospitals, schools and sheltered accommodation. However, they are now becoming more common in domestic properties.

The valve has a hot and cold supply to it and these are normally of similar pressure although, with the requirement of non-return valves to be fitted to prevent cross-contamination, it is not so important. If the pressures are considerably different it may be harder to set up the blending valve. You will need to check the manufacturer's specification before you install the valve to ensure it will operate with un-equal pressure supplies. The hot- and cold-water supplies are mixed within the valve to the required temperature that has been set and the water from its outlet will supply the fixtures with hot water at that temperature.

A valve can supply one or more fittings but the blended water pipe run should not be more than 2 m to prevent legionnaires' disease.

Working life

A couple lives in a two-bedroomed flat. The heating in the flat is taken care of using night storage heaters; the hot-water provision is a domestic hot-water combination storage cylinder (vented), which is heated by an immersion heater.

Adam has been called to the flat as the hot-water storage cylinder is leaking. He finds it is un-repairable, as the cylinder has started to corrode at its base. Adam informs the customer and says he will give an estimate to replace the faulty cylinder. The client asks if there is an alternative means of providing hot water as they are at work all day and the heated water tends to get wasted as the immersion has to be on but they never get the benefit. They consider this a waste of their money and, of course, energy.

Consider the following:

- For hot water demand, the flat has a kitchen and bathroom.
- A wet central heating provision will not be considered.
- Gas is available in the dwelling.
- This is a ground-floor flat.
- The cylinder cupboard is located in the bathroom on an outside wall.
- There is no shower as the system head is currently too low.
- Water pressure and flow are good.

What are the advantages and disadvantages regarding storage or instantaneous hot water in this scenario? What type of system or appliance would you recommend, taking into account the clients' wishes? Which type of fuel should be used and be most convenient? How long would the work take in terms of hours?

Progress check

1. What is a centralised storage system of hot water?
2. What is a localised system of hot water?
3. What is an instantaneous hot-water system?
4. What is storage system of hot water?
5. Explain the term 'gravity circulation'.
6. How can an immersion heater be controlled?
7. What is the minimum horizontal length of the draw-off pipe from the top of a storage vessel, to prevent one pipe circulation?
8. What is the maximum temperature that the thermostat controlling water in any hot-water storage cylinder should be set to?
9. Which grade of hot-water storage cylinder is the most common in domestic properties?
10. What does the *Domestic Heating Compliance Guide* identify?

2. Installation requirements of hot-water systems

The installation requirements of hot-water systems are very similar to those for cold-water systems. Please refer to the section covering this in Unit 6 on pages 286–91 for more information.

Insulation requirements for hot-water systems

This section outlines the key requirements laid down in the *Domestic Heating Compliance Guide* relating to the insulation of hot-water system pipework in order to conserve energy.

The requirements are as follows:

- Primary circulation pipework should be insulated wherever it passes outside the heated living space or through voids in a building, which are ventilated from unheated spaces.

- Primary circulation pipes for hot-water services should be insulated throughout their length, except where they need to penetrate joists or building structural elements.

- The pipes connected to a hot-water storage cylinder should be insulated up to 1m from the cylinder or to the point where they become concealed.

- If a secondary circulation system is used, the pipework included in this circuit should be fully insulated.

- Pipework in insulated areas should comply with the requirements of BS 5422.

- Insulation that complies with the *Domestic Heating Compliance Guide* must not have a greater heat loss than the figures quoted in Table 7.3.

Pipe diameter (mm)	Maximum permissible heat loss (W/m)
15	7.89
22	9.12
28	10.07

Table 7.3: Maximum permitted heat loss for insulation

Dead legs and the use of a secondary circulation system

Dead legs are long lengths of pipe from the hot-water storage vessel to the appliance (see Figure 7.20). Generally speaking, the length of pipe measured from the hot-water storage vessel or heater to the tap should be as short as possible.

Dead legs should be avoided for two reasons:

1. They waste water, because the user has to run off cold water before it turns hot.
2. They waste energy heating up the volume of water contained in the dead leg, which then cools.

One method of overcoming the problem of dead legs is **secondary circulation**. This can be demonstrated with the typical installation layout using a pumped secondary circulation system, shown in Figure 7.21. However, you would not be able to use secondary circulation if you had installed a combination boiler.

A flow-and-return loop is installed, which feeds all the appliances. The water is kept circulating, either by gravity or by a non-corrosive circulating pump (bronze manufacture).

The return pipe is connected in the top third of the cylinder. This prevents the cooler water lower down the cylinder from mixing with the hot water. The return leg of the cylinder normally includes a single-check valve to prevent reversal of the water flow in the return pipe when a tap is opened. The operation of the pump is normally controlled by a simple time clock, which is set to turn off the pump and prevent wasteful circulation of heated water during periods when the building is not in use.

Did you know?

The Water Regulations define a dead leg as 'a length of distribution pipe without secondary circulation'.

Remember

Trace heating can be used as an alternative to secondary circulation; the details of this process are covered in Unit 6 on page 291.

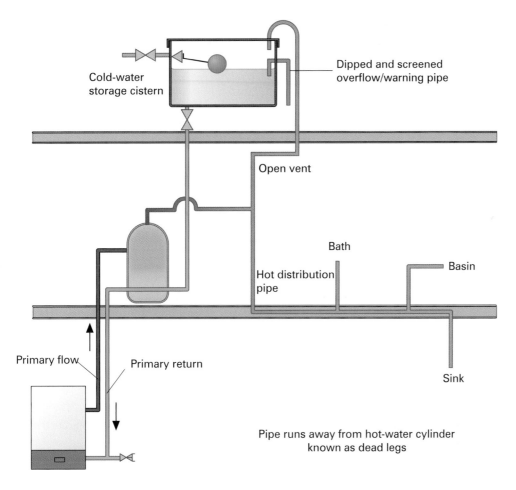

Cold-water
storage cistern

Dipped and screened
overflow/warning pipe

Open vent

Bath

Basin

Hot distribution
pipe

Primary flow

Primary return

Sink

Pipe runs away from hot-water cylinder
known as dead legs

Figure 7.18: Dead leg runs to individual plumbing appliances

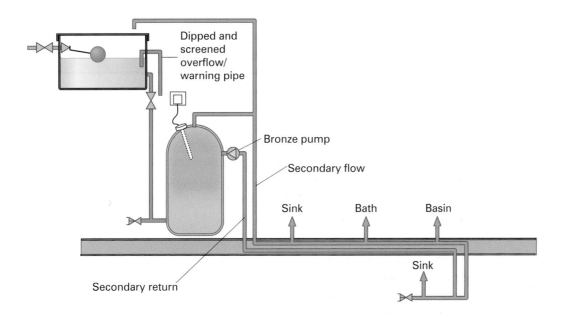

Dipped and
screened
overflow/
warning pipe

Bronze pump

Secondary flow

Sink

Bath

Basin

Sink

Secondary return

Figure 7.19: Secondary circulation system designed to overcome dead legs in systems

Sergio is a plumber who gained his qualifications in Portugal. He has worked in the UK for 14 years, and is experienced in general plumbing and heating work. However, he has not passed an approved unvented systems competency test.

He has recently been asked by an existing customer to replace a damaged vented cylinder system with an unvented hot-water storage cylinder, as the client would like to do away with the storage cistern and associated pipework in the loft.

Should Sergio complete the installation? What are Sergio's alternatives here? What are the legal requirements regarding the installation of unvented hot-water storage systems? Does Sergio have any exception from the regulations because unvented systems were not even a consideration when he qualified and because he qualified in another country?

Progress check

1. Briefly, what are the four requirements of an unvented hot-water system as laid down in the Building Regulations?
2. Under the requirements of the Water Regulations, an unvented cylinder installation must include a range of functional controls, which will prevent what?
3. Define the term 'multi-point'.
4. What is a check valve?
5. What is the function of a thermostatic mixing valve? In which buildings are these valves typically used?
6. All pipework connected to a hot-water storage cylinder should be insulated to what distance from the cylinder (unless the pipework becomes concealed before this)?
7. Why should 'dead legs' should be avoided in hot-water systems?
8. What is secondary circulation?
9. A circulating pump installed in a secondary circulation system should be made of bronze. Why is this?
10. What is the purpose of 'trace heating'?

Check your knowledge

1. Which British Standard covers hot water supply systems?
    a   BS 4571
    b   BS 6457
    c   BS 6700
    d   BS 7600

2. If an electrical fault is suspected on an immersion heater, what should be done?
    a   Locate and repair the fault
    b   Tell the customer
    c   Advise the supervisor
    d   Report the fault to the NICEIC

3. Which one of the following electrical devices is often fitted to a hot-water vessel to supplement the hot water supplied from the boiler?
    a   Diverter valve
    b   Heat pump
    c   Gas circulator
    d   Immersion heater

4. What is the purpose of a vent pipe?
    a   To fill the system
    b   To maintain atmospheric conditions
    c   To reduce atmospheric conditions
    d   To circulate the water, preventing dead legs

5. How many draw-offs or supplies can be taken off the cold feed to a cylinder?
    a   0
    b   1
    c   4
    d   As many as required

6. When installing an electric shower where should the service valve be located?
    a   After the stop valve
    b   On the outlet side of the shower
    c   As near as possible to the shower unit
    d   At least 300 mm from the heater

7. What is the main reason for installing a secondary circulation supply to a hot-water tap?
    a   To improve pressure
    b   To avoid fitting a pump
    c   To prevent dead legs
    d   To reduce pressure to the taps

8. What is the minimum head from the base of a cold-water storage cistern to the shower head of a storage-fed shower?
    a   300 mm
    b   500 mm
    c   750 mm
    d   1000 mm

9. How should the cold feed pipe in a domestic property be sized?
    a   In accordance with your supervisor
    b   In accordance with BS 6700
    c   In accordance with BS 1212
    d   According to the customer's requirements

10. What type of backflow prevention would be required for an over-the-rim bidet without a shower attachment?
    a   AUK1 air gap
    b   Double-check valve
    c   Single-check valve
    d   AUK2 air gap

Getting ready for assessment

The information contained in this unit of the book, as well as the continued practical assignments that you will carry out in your college or training centre, will help you with preparing for both your end-of-unit test and the diploma multiple-choice test. It will also support you in preparing for the practical assignments you will need to complete to demonstrateyour understanding in and enabling you to carry out and apply domestic hot-water system installation and maintenance techniques.

There are opportunities throughout the unit to test your progress in and understanding of the required underpinning knowledge; this will enhance your preparation for the forthcoming assessments, so make good use of them.

This unit will be assessed by the following assessment methods:

- externally set knowledge assessment
- externally set assignments.

With regard to the mechanical services industry, you will need to know:

- the types of hot water system and their layout requirements
- the site preparation techniques for hot water systems and components
- how to apply site preparation techniques for hot water systems and components
- the installation requirements of hot water systems and components
- how to install hot water systems and components
- the service and maintenance requirements of hot water systems and components
- how to service and maintain hot water systems and components

- the decommissioning requirements and how to apply them with regard to hot water systems and components
- the inspection and soundness testing requirements and how to apply them to hot water systems and components.

Always make sure that you are working safely throughout the tasks and assessments. Make sure you are working to all the safety requirements given throughout the assessments and wear all the appropriate personal protective equipment. When using tools, make sure you are using them correctly and safely.

Check it out – using your knowledge from this unit practice and test yourself on the following key terms:

Direct/indirect, conventional boiler, immersion heater, single point, instantaneous, multipoint, combination, demand/delivery, dead-leg, secondary circulation, bronze circulation pump, components, showers, thermostatic, backflow prevention, feed and expansion cistern, cold water storage cistern, directly/ indirectly heated storage, direct, single feed, double feed, water heaters, primary/secondary circuit, open vent, cold feed, thermostats, temperature/pressure relief, baths, wash basins, bidets, gravity fed, mains fed, statutory regulations, industry standards, manufactures instructions, vented/unvented, preparatory work, installation, commissioning/ decommissioning, tools/equipment, pipework/ fittings, access/exit, hazards, risk assessment, PPE, pressure/flow rate, fixings, expansion/contraction, clips/brackets, system layouts, cylinders/storage vessels, maintenance, visual inspection, soundness testing, flushing procedures, remedial work, insulation, frost protection, maintenance records.

Good luck!

Understand and apply domestic central heating system installation and maintenance techniques

Central heating systems are closely linked to the hot-water supply. One boiler is usually used to heat both the hot water and the central heating system. This unit focuses on the pipework systems, controls and components used in central heating systems, and the types of boilers available.

This unit will cover the following learning outcomes:

- Know the uses of central heating systems in dwellings
- Know the types of central heating system and their layout requirements
- Know the installation requirements of central heating systems and be able to install central heating systems and components

- Know, and be able to apply, the service, maintenance and decommissioning requirements of central heating systems and components.
- Know and be able to apply the site preparation techniques for central heating systems and components (see Unit 5 pages 231–39).

1. Know the uses of central heating systems in dwellings

Having central heating in dwellings is now considered part of normal life. However, many years ago it would have been considered a luxury. Central heating is installed to provide us with warmth during the colder months of the year, and during cold parts of the day. It can be classed either as full central heating, background heating or selective heating.

A full heating system will heat all habitable rooms within the dwelling to the normal design temperatures of 21 °C for the living and bathrooms, and 18 °C for all other rooms. This is based on an outside temperature of −1 °C. Background heating will also heat the rooms but to lower temperatures. Selective heating will allow a particular area, or number of areas, to be heated and this will normally be done by some form of control system.

The various types of central heating system are summarised in Figure 8.1.

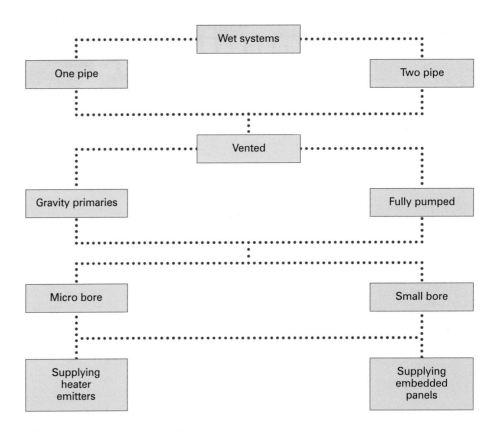

Figure 8.1: Types of central heating system

Sealed systems (unvented) or combination boilers are fully covered at Level 3, as are underfloor heating systems.

2. Know the types of central heating system and their layout requirements

The Building Regulations 2000 deal with the conservation of fuel and power in *Approved Documents L1A (new dwellings) and L1B (existing dwellings)*.

This document sets out requirements for:

- space-heating systems controls
- zone control
- timing controls
- boiler control interlocks
- hot-water systems
- alternative approaches for space heating and hot-water systems
- commissioning of heating and hot-water systems
- operating and maintenance instructions for heating and hot-water system controls.

The legislation sets out methods for improving energy conservation in dwellings by:

- improving the control of a system using temperature control to individual rooms
- permitting only the most energy-efficient boilers to be installed
- prohibiting the installation of gravity circulation systems in new properties.

Full-gravity heating systems

These are no longer installed, but you might come across an old system during maintenance work in older buildings such as village halls. They were sometimes referred to in domestic properties as background heating: one radiator, for example in the bathroom, was fed off the primary flow and return pipe from the boiler. As with any gravity system, it worked on the principle of having the pipework at the correct fall to aid circulation. Flow and return pipework to the radiator was usually 22 mm.

The key features of these systems were having large-diameter pipes and having no pump in the system for either heating or hot water.

> **Remember**
>
> Gravity systems were never really suitable for domestic properties because of the excessively large-diameter pipes that were required to get only the smallest amounts of heat to the radiators.

Pumped heating and gravity hot-water systems (semi-gravity)

Pumped heating and gravity hot-water systems can be either one-pipe or two-pipe systems.

One-pipe system

As with full-gravity systems, you are unlikely to install these from new, but you may come across them.

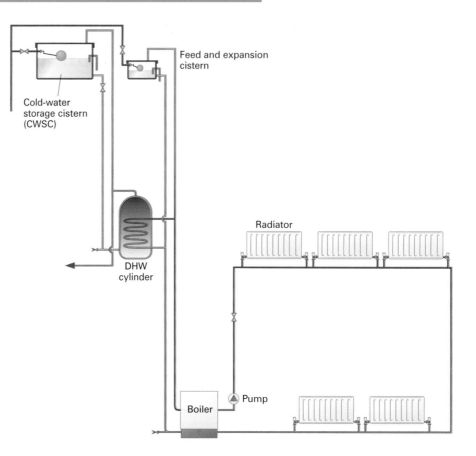

Figure 8.2: One-pipe semi-gravity system

This system works on the basis of a one-pipe run, with a flow and return from the boiler. The disadvantages of the one-pipe system tend to outweigh the advantages, so they are no longer installed in domestic properties (see Table 8.1).

Advantages	Disadvantages
Lower installation cost compared to a two-pipe system	The heat emitters on the system pass cooler water back into the circuit. This means that the heat emitters at the end of the system are cooler.
Quicker to install	The pump only forces water around the main circuit and not directly through the radiators. This means it is important to select radiators that allow minimum resistance to the flow of water.
Lower maintenance costs	The 'flow' side to the radiator is usually installed at high level to improve circulation, creating additional unsightly pipework.

Table 8.1: Advantages and disadvantages of a one-pipe system

Careful balancing of the radiators is a must and – for a given heat requirement in a room – the radiators at the end of the circuit will need to be larger than those at the beginning.

To ensure that radiators operate correctly they must be sited as close to the pipework ring as possible. If you try to cut a couple of connections 100 mm apart into the pipework and run it for 5 m to the radiator, it will not work. The full ring has to be extended to run under the radiator with short-tail connections onto it.

Two-pipe system (semi-gravity system)

This was a popular choice of system, particularly in the 1970s. These systems are no longer permitted on new properties (other than with solid fuel boilers) unless additional controls are installed. This type of system no longer meets the requirements of the Building Regulations for extension or boiler replacements to existing oil- or gas-fired systems.

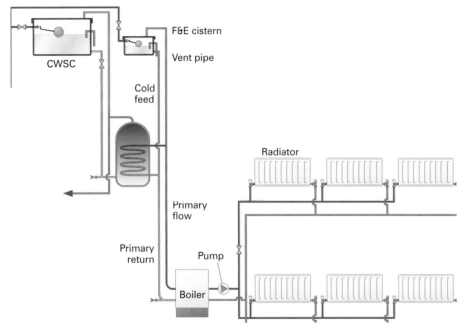

Figure 8.3: Two-pipe semi-gravity system

In a two-pipe system, water is pumped around both the circuit and the radiators. This improves the speed with which radiators heat up. The system can be balanced easily by adjusting the lock shield valve on each radiator.

Semi-gravity system with two-port valve

A semi-gravity system uses a two-port valve and provides independent temperature control of both the heating and hot-water circuits in a pumped heating and gravity domestic hot-water heating system. The pump and the boiler are switched off when space and hot-water temperature requirements are met.

Time control can be managed by either a time switch or a programmer. Thermostatic radiator valves (TRVs) can also be fitted to provide overriding temperature controls in individual rooms.

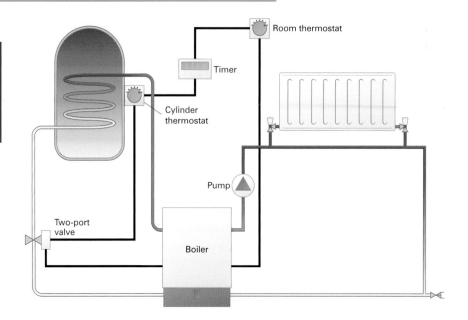

Figure 8.4: Semi-gravity system with two-port valve (image simplified for clarity)

The pump and circulation to the cylinder can be turned off independently by the thermostats, and there is an interlock to ensure that when the final circuit turns off so do the pump and boiler, thereby avoiding wasting energy.

Fully pumped systems

In this system, the hot water and the heating circuits are operated completely by the pump. Therefore, because there is no requirement

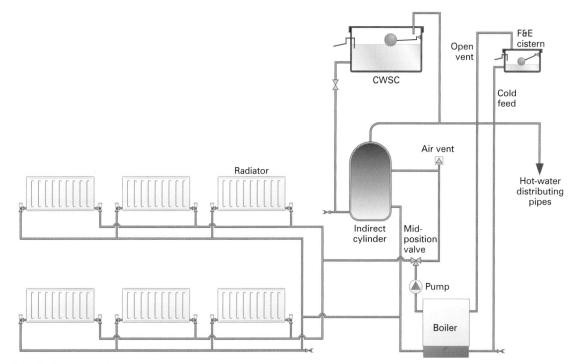

Figure 8.5: Fully pumped system

for gravity circulation, the boiler can be sited above the height of the cylinder, giving more design options.

Installations are controlled by motorised valves. There are a number of system designs incorporating two-port zone valves or three-port valves (two-position and mid-position) that meet the requirements of the Building Regulations *Approved Document L1*.

Fully pumped system using a two-position three-port diverter valve

This was one of the first fully pumped systems to be installed in domestic properties. However, it is no longer widely used.

This system is designed to provide independent temperature control of the heating and hot-water circuit in fully pumped heating systems. The design, when used with a programmer, satisfies the Building Regulations. The two-position three-port diverter valve is usually installed to give priority to the domestic hot-water circuit. That is, it can only feed either the hot water or the central heating system at any one time.

Because the system is a priority control system – to produce domestic hot water – it should not be used where there is likely to be a high hot-water demand during the heating season. The designed heating room temperature could drop below comfort level when the demand for hot water is high.

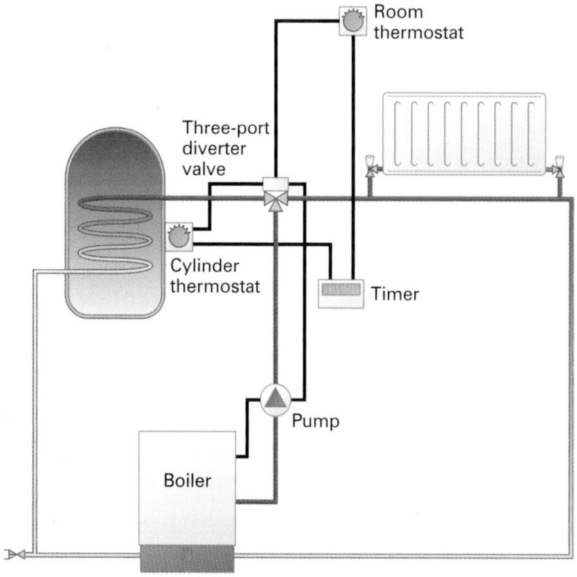

Figure 8.6: Fully pumped system with two-position diverter valve (image simplified for clarity)

Because the system feeds only one circuit at a time, you will find that this system is not widely installed.

Fully pumped system using three-port mid-position valve

This type of system is commonly installed in new domestic properties.

The system is designed to provide separate time and temperature control of the heating and domestic hot-water circuits. To fully meet the requirements of the Building Regulations, time control must be managed via a programmer and **TRVs**, and an automatic bypass valve must be fitted (where required).

The mid-position valve permits both hot-water and heating circuits to operate together.

Be careful when working on existing systems: the three-port diverter valve and mid-position valve system look similar. However, the three-port valve and timing device are completely different for each system.

This system is not suitable for dwellings with a floor area greater than 150 m².

Key term

TRVs – thermostatic radiator valves.

Fully pumped system using 2 × 2-port valves

This type of system is commonly installed in new domestic properties; particularly larger ones.

This system is recommended for use in dwellings with a floor area greater than 150 m². The main reason for this is the limited capacity of a three-port valve installation to satisfy the heat demands of a larger system. The use of the 2 × 2-port valves also gives greater flexibility in system design, with additional valves being added to the system to zone separate parts of the building. The system provides separate temperature control for both heating and hot-water circuits. Again, the features are similar to the other systems shown earlier.

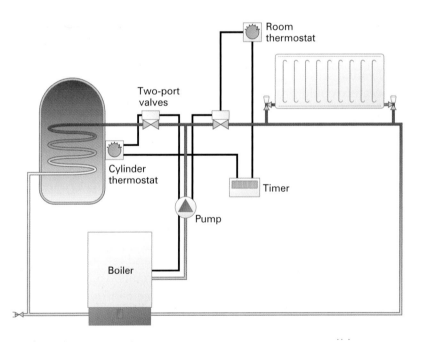

Figure 8.7: Fully pumped system with 2 × 2-port valves (image simplified for clarity)

System layout features for filling and venting systems

Sealed system

In a sealed system, the system is closed and is filled via a filling loop, which is connected to the mains cold water. The filling loop must always be disconnected after use to prevent any cross-contamination, although a non-return valve is also fitted. A pressure vessel then copes with expansion and contraction of the water. A safety valve is also used to relieve excess pressure. Some manufacturers have an arrangement integral to the boiler, which allows pressurisation without the need for an external filling loop.

It is common for this type of system to use a combination boiler although you can also use conventional boilers.

The main advantages of this system are that:

- you do not need to site a feed and expansion cistern in the roof space or at high level
- you usually cut down on pipe use as you do not require a open vent and cold feed.

Feed and expansion cistern

However, the feed and expansion (F&E) cistern is used on all open-vented central heating systems. While the cistern allows the system to be filled up, its main purpose is to allow water in the system to expand. The water level should therefore be set low in the cistern when filling the system. The cold feed to the system in an average domestic property is usually 15 mm minimum, and this pipe must not include any valves. This is to ensure that, in the event of overheating, there is a constant supply of cooler water to the system to prevent the dangerous condition of boiling. The servicing valve to the system should be located on the cold-water inlet pipework to the cistern. If a valve were fitted in the cold feed, it may be closed inadvertently: this could have disastrous consequences if the open vent also became blocked.

The F&E cistern is located at the highest point in the system, and it must not be affected by the position and head of the circulating pump. To avoid any problems with gravity primaries, a minimum height can be obtained by dividing the maximum head developed by the pump by three. In fully pumped systems, the level of water in the F&E cistern should be a minimum of 1 m above the pumped primary to the direct hot-water storage cylinder.

> **Remember**
>
> It is a Water Regulation that all float-operated valves must be fitted with a service valve.

Space for expansion of water

The cistern and float valve must also be capable of resisting a temperature of 100 °C. The system volume expands by about 4% when heated, so a system containing 100 litres would expand by 4 litres. Space must be allowed in the F&E cistern to take up the additional volume when heated.

Primary open safety vent

In a fully pumped system, the primary open safety vent should usually rise to a minimum height of 450 mm above the water level in the F&E cistern. This allows for any pressure-surge effects created by the pump. The open safety vent also helps to provide for the removal of any air which the system may collect, particularly on commissioning or re-filling.

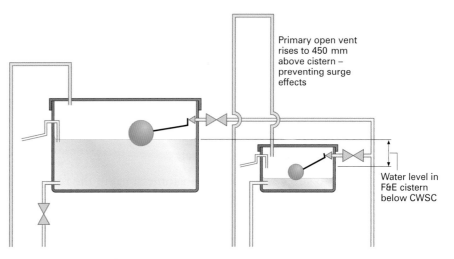

Primary open vent rises to 450 mm above cistern – preventing surge effects

Water level in F&E cistern below CWSC

Figure 8.8: Open safety vent to F&E cistern in a fully pumped system

The purpose of the primary open safety vent is to:

- provide a safety outlet should the system overheat due to component failure
- ensure that the system is kept safely at atmospheric pressure.

The minimum diameter of the safety vent is 22 mm and the pipe should never be valved.

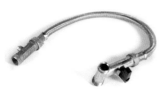

Figure 8.9: Filling loop with non-return valve

Figure 8.10: Typical pressure gauge

Figure 8.11: Combi boiler with built-in pressure gauge

Pressure gauge, pressure relief valve and fitting loop position

In a sealed system the feed and expansion cistern is replaced by an expansion vessel, pressure relief valve, pressure gauge and filling loop.

The filling loop is connected to a mains cold-water supply and has an isolation valve, enabling the user to isolate it when not required. The other end of the loop is connected into the heating system via a non-return valve. This ensures it complies with the Water Regulations. It should be noted that, when the filling of a system is not taking place, the filling loop should be removed and the ends capped off.

The filling loop takes the place of the cold water in the feed and expansion cistern. To show the person filling (charging) the system when the required pressure is reached, a pressure gauge needs to be installed within view of the filling loop. The pressure gauge is normally contained within the boiler panel on combi boilers or remotely on a system using a conventional boiler.

An expansion vessel is designed to take the expansion caused in the system as it heats up and will need to be sized to the particular system. In a combi boiler, this will often be in the appliance already. Again this replaces the space above the water level in the feed and expansion cistern of a vented system. An expansion vessel has a 4% expansion requirement.

A pressure-relief valve is also fitted as a further safety device, in case the expansion vessel fails or the system gets over pressurised when filling. This allows water to discharge outside the building in the event of excessive pressure within the system.

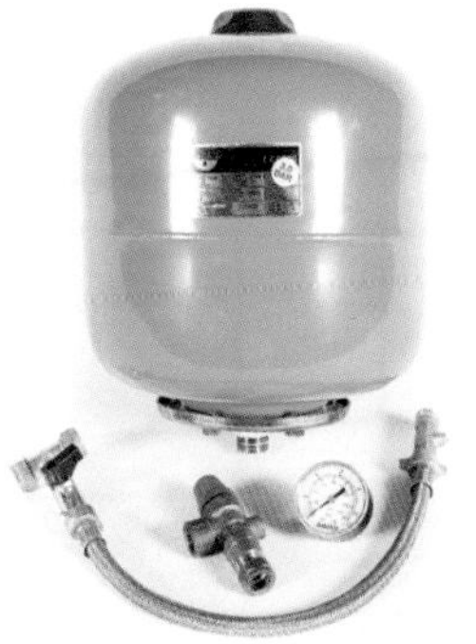

Figure 8.13: A typical pressure-relief valve

Figure 8.12: A combined expansion vessel; pressure gauge with filling loop

Domestic circulating pumps

When installed, domestic circulating pumps are fitted with isolation valves to permit service and maintenance.

Pumps are fairly simple. They consist of an electric motor, which drives a circular fluted wheel called an impeller; this 'accelerates' the flow of water by centrifugal force.

It is the pump's job to circulate water around the central heating system, ensuring that the water is delivered at the desired quantities throughout the system components. Most pumps have three settings, and pump manufacturers provide performance data for each, which shows flow rate in litres per second and pressure in kPa and metre (m) head.

The flow rate should not exceed 1 litre per second for small-bore systems and 1.5 litres per second for micro-bore systems; anything higher can create noise in the system. Most pumps deliver 5 m or 6 m head. This is usually enough to overcome the flow resistance of the whole heating circuit.

It is good practice to position the pump so that it gives a positive pressure within the circuit. This ensures that air is not drawn into the system through microscopic leaks.

> **Remember**
>
> Ensure the discharge pipe from the pressure-relief valve terminates where it will cause no harm to anyone or any damage to the structure of the building.

Figure 8.14: An A-rated domestic circulating pump

Figure 8.15: Ball-type pump valve

Figure 8.16: Gate-type pump valve

Figure 8.17: Pump stripped down to show the impellor

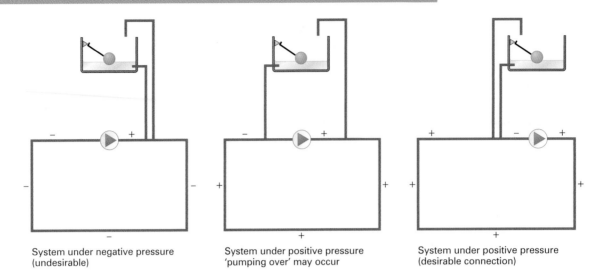

System under negative pressure (undesirable)

System under positive pressure 'pumping over' may occur

System under positive pressure (desirable connection)

Figure 8.18: Pump position in relation to cold feed and vent pipe

The pump position is even more critical in a fully pumped system because of its position in relation to the cold feed and vent pipe.

Air separators

The purpose of the air separator is to enable the cold feed and vent pipe to be joined closely together in the correct layout to serve the system.

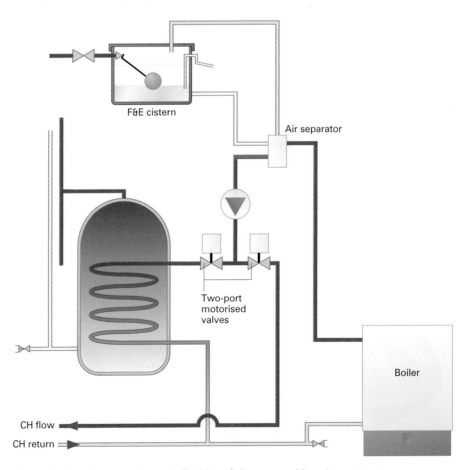

F&E cistern

Air separator

Two-port motorised valves

Boiler

CH flow

CH return

Figure 8.19: Air separator installed in a fully pumped heating system

The grouping of the connections inside the air separator causes turbulent water flow in the separator, which in turn removes air from the system. This reduces noise in the system and lowers the risk of corrosion.

Figure 8.19 shows a fully pumped system containing an air separator. This is also called a close-coupled method of feed and vent pipe connection, featuring only two pipes connected to the boiler.

Layout features for microbore systems

These systems use smaller than normal pipe diameters (8 mm and 10 mm) to feed the radiator circuits. These are sometimes referred to as radial circuits, as each radiator is supplied with an individual flow and return connection from the respective manifold.

Microbore systems were commonly used in the 1970s and early 1980s, and are still used in some cases today. It was based on the supply to the radiators provided from the heating mains via a number of manifolds, usually one downstairs and one upstairs. These in turn feed the radiators via individual 8 mm or 10 mm flow and return pipes.

Figure 8.20 shows a typical layout for the system, which could be installed with open-vented systems.

The manifold can be made up in sections to suit the number of connections required; as shown in Figure 8.20.

An alternative to using the manifold was to use a 'spider'; a device that does exactly the same job.

> **Did you know?**
>
> It is possible to mix 15 mm pipe runs to radiators and manifolds in the same system.

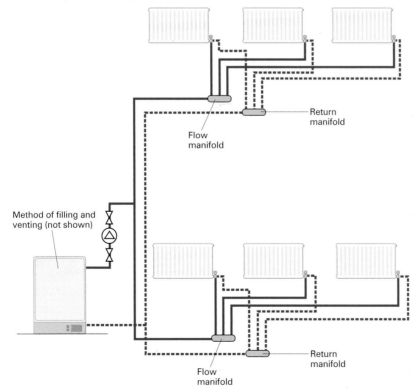

Figure 8.20: Layout of microbore system

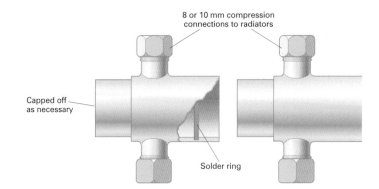

8 or 10 mm compression connections to radiators

Capped off as necessary

Solder ring

Figure 8.21: Manifold for microbore system

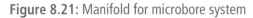

22 mm soldered connection

8 or 10 mm pipes brazed into 'spider' pipework connections made with straight couplings

Figure 8.22: Microbore manifold arrangement

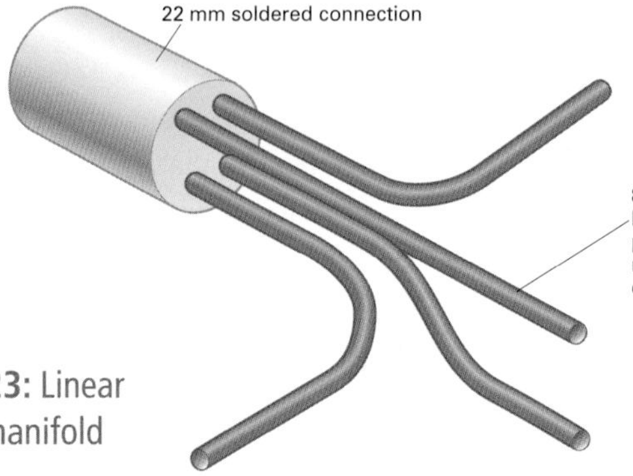

Figure 8.23: Linear ('spider') manifold

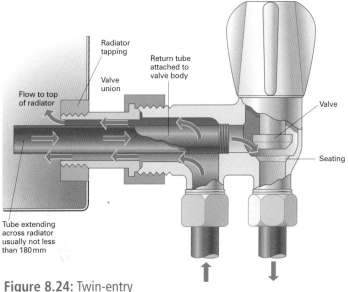

Radiator tapping

Return tube attached to valve body

Valve union

Flow to top of radiator

Valve

Seating

Tube extending across radiator usually not less than 180 mm

Figure 8.24: Twin-entry radiator valve

Twin-entry valves

The connections to the radiators may be different with this system, as twin-entry valves could be used.

The twin-entry valve, available as a manual or thermostatic valve, requires a tube to extend the flow connection across the base of the radiator. This ensures proper circulation across the radiator. With the introduction of radiators that have (largely) back-entry tappings, the twin-entry valve became redundant.

The main issue with the system layout, as shown in Figure 8.20, was that it was only suitable for smaller domestic properties. This is because of the possible longer pipe runs (greater frictional resistance), limited available flow rate and limited heat output (with 8 mm and 10 mm pipework). The type of property most suitable was two-bedroom starter homes.

There were some real advantages to the system if it was installed correctly as it saved time due to a quicker installation time. The system tended, however, to suffer from poor installation standards, including:

- pipe runs to radiators that were too long and overloaded
- unsupported pipe runs not laid to proper falls, causing severe airlocks.

However, 8 mm and 10 mm pipework have not disappeared. The tendency is to use these pipe sizes where appropriate on a job but to pipe up as a normal small-bore installation without the use of manifolds.

Remember that 8 mm and 10 mm pipework is supplied in soft copper rolls, which are malleable and easily flattened. You need to be careful about how you make the connections up to radiators. Installed as normal 15 mm pipework, they can be subject to flattening by vacuum cleaners, so the best approach is to rise up behind the skirting board and bring the pipe out to the radiator above 'knocking height'.

> **Working life**
>
> James is working on a property, which has a microbore central heating system. All radiators are connected from the manifolds using twin-entry valves. James commissions the system and it runs but the radiators only get lukewarm around the connection of the twin-entry valve into the radiator. The flow pipework is hot and the return pipework is cooler but still warm. The system is fully pumped, which rules out the possibility of a faulty circulation pump. All air has been successfully vented from the system.
>
> Discuss the probable cause of the radiators failing to heat.

State the general layout operating principles of solid fuel, oil-fired and gas-fired heat-producing appliances

> **Remember**
>
> In order to be competent at installing appliances you will need further qualifications, which will be covered at Level 3.

Unlike most other fuels, only limited controls can be used for solid fuels. Therefore, in a lot of cases, there are exceptions to some of the requirements of the Building Regulations. This section covers different types of heating appliances.

Solid fuel appliances fall into two general categories: hand feed and hopper feed.

Hand-feed boilers work on the principle that the fuel is loaded into the boiler and then ignited. After this fuel will need to be regularly hand fed into the boiler. This is like putting logs onto the fire to keep it burning.

Hopper feed boilers, on the other hand, have a supply of fuel that is either stored within the boiler or immediately external to the boiler. This stock is automatically fed into the boiler when required. Hopper feed contents are replaced less often than if you hand feed the boiler.

Once running, either type should burn with little or no control. You must ensure the installation can accommodate the dispersal of any heat without the aid of supplementary devices, such as a circulating pump.

Boilers will be covered in greater depth at Level 3.

Solid fuel boilers

In domestic installations you could come across:

- solid fuel boilers (room heater and independent free-standing boiler)
- cookers
- open fire with high output back boiler.

Figure 8.25: Convector-type solid fuel room heater

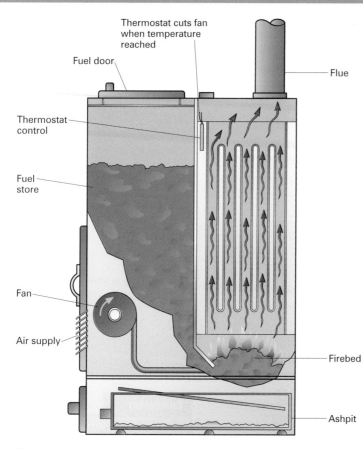

Figure 8.26: Section through an independent gravity-fed solid fuel boiler

Room heater

This type of appliance is available in a number of ratings, usually 10 kW and 13 kW. The 10 kW version will heat the room in which it is located, plus the domestic hot water, and can serve a total radiator surface of up to 13.8 m². The 13 kW version can serve up to 20.5 m² radiator surface. It fits a standard fireplace opening and is thermostatically controlled. This boiler requires a flue of 175 mm. It has two flow tappings and two returns for the domestic hot water and heating, all 1" BSP.

Independent free-standing solid-fuel boiler

The model shown in Figure 8.26 is a gravity-fed boiler.

This type of boiler is available in a range of outputs from 13.5 kW to 29 kW. The model shown has a range of 13.5 kW, 17.6 kW, 23.5 kW and 29.3 kW. There is a built-in fan to boost heat output on demand. The hopper is easy to fill from the top and a typical unit will hold enough fuel to burn continuously for 14 hours.

These boilers are designed for installation with a Class 1 flue system.

Figure 8.27: Typical solid fuel cooking range

Cooker

In some dwellings, especially in larger farmhouses, a cooking range is installed and often these are used to heat the hot water and possibly the bathroom radiator. This type of appliance is installed using gravity pipework as there is minimal control of its operation.

Open fire with high output back boiler

The high-output back boiler is installed in an open fire in place of the normal fire brick surround. The boiler is shaped the same but instead of being brick it has a steel water jacket. Again there is no automatic control to turn it on and off, and the hot-water and heating system will require some way of losing the heat. This would be done by using gravity for the hot water and at least one radiator.

Oil-fired boilers

The following ranges of oil-fired boilers are available:

- combination
- condensing
- traditional

- wall-mounted internal – room sealed or open flues
- wall-mounted externally sited
- floor-mounted internal – room sealed or open flues
- floor-mounted external.

The most common type of oil burner used on boilers today is the pressure jet. Oil is injected, via a nozzle, into an airstream created by a fan. The oil is then vaporised and ignited by a spark.

Older boilers used a vaporising system. The fuel pipe is coaxial with the air tube. The combustion air and the fuel delivery are infinitely variable. They can be controlled independently, through the turn of a knob, allowing excellent heat output and air/fuel mixture control. The fuel stream breaks up into droplets upon hitting the head of a cascade assembly. Most of the fuel burns completely without reaching the bottom of the combustion chamber. An annular groove in the base of the cascade assembly retains unburned fuel until it is vaporised, which usually occurs only during the starting period.

Oil boilers can be controlled in the same way as a gas boiler (see Table 8.2 for outputs). Where gas is not available oil is usually the next option chosen for heating and hot water. The boiler is supplied oil from a storage tank via a small-bore pipe, which is then ignited in the combustion chamber. The oil used in most domestic appliances is known as 28-second and this relates to its viscosity. The heat generated within the combustion chamber heats the water in the water jacket. Once the water has reached the required temperature, the thermostat will operate and cut off the boiler.

> **Did you know?**
>
> Oil boilers will be covered in greater detail if you follow the oil pathway at Level 3.

Type of boiler	Capacity	Comment
Combination	Available in a range of capacities	
Wall-mounted internal	20 kW, 22 mm flow and return connection	Suitable only for fully pumped systems
Wall-mounted external	14.6 kW and 19 kW, 22 mm flow and return connections	
Floor-mounted internal	Outputs up to 70 kW (240,000 Btu/h)	
Floor-mounted external	Outputs range from 14.7 kW to 33.7 kW (50,000 to 115,000 Btu/h)	

Table 8.2: Outputs of oil-fired boilers

Gas boilers

There are three basic types of gas boiler:

1. traditional wall mounted or floor standing
2. combination
3. condensing.

> **Did you know?**
>
> Gas boilers will be covered in greater detail if you follow the gas pathway at Level 3.

In all cases, new boilers have to meet minimum efficiency standards. These are set out as a **SEDBUK rating** (see Unit 3, page 113 for more information).

Traditional or regular boilers

Traditional or regular boilers include:

- boilers with cast-iron, high-alloy steel, copper or aluminium heat exchangers
- system boilers.

All of the above can be floor standing or wall mounted and are usually designed to fit in with kitchen unit installations. Gas-fired back boilers are also available; these are concealed in a chimney opening at the back of a gas fire.

Traditional or regular boilers supply hot water indirectly via a domestic hot-water storage cylinder, which is usually sited at first-floor level.

Cast-iron heat exchangers

The water in this type of boiler is heated by hot gases passing though the heat exchanger. The heat exchanger on a cast-iron boiler is a close network of waterways or, for the alloy type, a tube that passes through a series of fins. Boiler outputs for general domestic installations range from 9 kW to 29 kW.

Installation of traditional gas or oil fired boilers is not generally permitted in new and existing properties, where boilers should now be of the high-efficiency (condensing) type.

Condensing boilers

Under the Building Regulations, all boilers, whether being used in new systems or as replacements, are required to

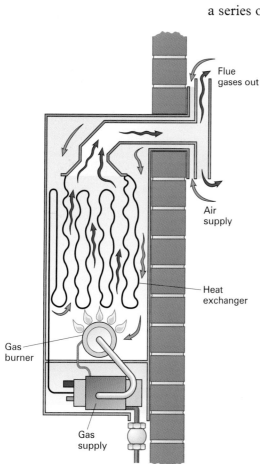

Figure 8.28: Section through a traditional wall-mounted boiler

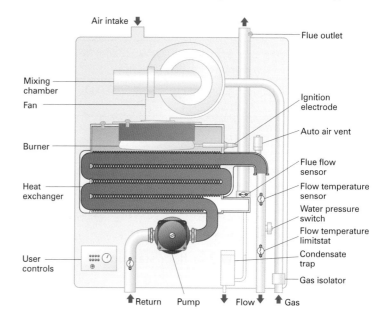

Figure 8.29: A condensing boiler

be of the condensing type. Key features of the condensing boiler are:

- being capable of extracting normally wasted heat from the flue gases to the point at which water condenses in the boiler and needs to be run safely to waste
- the condensate can be discharged to a soil pipe, waste pipe, gulley or a purpose-made soakaway (absorption point), used only as a last resort
- the condensate pipeline should normally be trapped, and the trap is often sited in the boiler.

System boilers

System boilers supply domestic hot water indirectly. The key system controls – pre-wired, pre-plumbed and pre-tested – are all built into the boiler unit, including the expansion vessel, the pump, and the system bypass.

Combination boilers

Combination boilers are designed to heat up a cold-water supply instantaneously for domestic hot water, so are classed as direct hot-water supply. When required, they will also supply hot water for the central heating system. Combination boilers are covered in greater detail at Level 3.

Figure 8.30: Gas-fired combination boiler

> **Working life**
>
> Gareth is a Level 3 apprentice who you work with sometimes. You are both asked to call at a job, which reports water being lost through the external discharge pipe on a combination boiler.
>
> On arrival, water is seen discharging from the pipe and has been doing so for some time. On inspection of the boiler, Gareth decides that it must be the pressure relief valve (PRV). He is confident as he has recently gained his unvented hot water certificate. While he is outside making a call to track one down, you notice that the filling loop is left connected and the pressure gauge is reading four bars. The boiler is not currently switched on.
>
> Other than a faulty pressure-relief valve (PRV) discuss what could be an alternative cause?

Operating principles of heat emitters

We use the term 'heat emitter' because it describes all types of devices used to heat the rooms that we live in. These include:

- cast-iron column radiators
- skirting heaters
- fan-assisted convector heaters
- panel radiators.

Cast-iron column radiators

Sometimes called hospital radiators, these are mostly found on older installations in such buildings as schools or village halls. However, some are now being installed in domestic properties as 'designer' decor.

Figure 8.31: Cast-iron column radiator

Panel radiators

Despite the name 'radiator', about 85% of the heat is given off by convection. The heat output of a standard panel radiator can be further improved by the addition of fins welded onto the back. These increase the surface area of the radiator, as they become part of its heated surface. The design of the fins will also help convection currents to flow.

Types of panel radiator

Radiator designs have developed dramatically over the last 60 years. Manufacturers aim to provide radiators that are efficient and also offer the maximum choice of styles. The most common types of steel panel radiators are shown in Figure 8.32.

Figure 8.32: Typical panel radiator

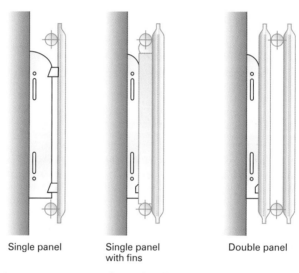

Single panel   Single panel with fins   Double panel   Double panel with fins

Figure 8.33: Types of panel radiator

> **Remember**
>
> When choosing radiators from catalogues, you should take note of manufacturers' fixing positions. It is often said that a radiator should be positioned beneath a window to reduce drafts. Curtains must finish 10 cm above the radiator.

Manufacturers will provide a range of at least four height options, from 300 mm to 700 mm. Width measurements are from 400 mm, with increments of 100–200 mm, to a maximum of around 3 m. The recommended height from the floor to the base of the radiator is 150 mm (depending on the height of the skirting board). This allows adequate clearance for heat circulation and valve installation.

Outputs will vary depending on design. You must ensure that the output will be sufficient to heat the room the radiator is going to be in.

Seamed top panel radiator

This is currently the market leader and is the most commonly fitted radiator in domestic installations. Top grilles are also available for this radiator.

Figure 8.34: Seamed top panel radiator

Compact radiators

These have all the benefits of steel panel radiators, with the addition of 'factory fitted' top grilles and side panels, making them more attractive to the consumer.

Rolled-top radiator

As the popularity of the compact has increased, the market for rolled-top radiators has declined. The method of manufacture means that some of the production seams can be seen following installation, making these radiators less attractive to the customer.

Combined radiator and towel rail

This product combines a towel rail and radiator in one unit. It allows towels to be warmed without affecting the convection current from the radiator. These are generally only installed in bathrooms.

Tubular towel rail

Often referred to as designer towel rails, these are available in a range of different designs and colours. They can also be supplied with an electrical element option, for use when the heating system is not required. They tend to be mounted vertically on the wall.

Low surface-temperature radiators (LSTs)

These were originally designed to conform to Health Authority requirements, where the surface temperature of radiators was not allowed to exceed 43 °C when the system was running at maximum. LSTs are now becoming popular in children's nurseries, bedrooms and playrooms, and in domestic properties where occupants are disabled.

> **Safety tip**
>
> You should not wedge the back of radiators over skirting boards or tightly under window sills, as this will prevent air circulation across the back panel; reducing the output. Radiator shelves have a similar effect.

Panel radiator accessories

Tappings

Radiators are usually provided with ½" BSP tappings, into which you attach the radiator valves, and a plug and air-bleed valve. Domestic installations have the valves placed at the bottom of the radiator at opposite ends. Alternatively, the connections could be placed top and bottom at opposite ends if a person is elderly or disabled, making access easier.

Radiator brackets

A number of brackets and fixing styles are available and are usually provided with the radiator. The bracket shown in Figure 8.40 has deep

Figure 8.35: Compact radiator

Figure 8.36: Rolled-top radiator

Figure 8.37: Combined radiator and towel rail

Figure 8.38: Tubular towel rail

Figure 8.39: Radiator plug and manual air vent

Figure 8.40: Typical radiator bracket

hanging slots and corresponding lug positions. These provide greater stability. Plastic inserts are used to seat the radiator precisely and help to minimise expansion and contraction noises.

Skirting heaters

These work on the principle of natural convection; the fins provide a large surface area for heat output. The radiator is heated up by conduction from the heating pipe. Cool air then passes through it and is heated; this heated air rises and passes from the panel via louvres at the top. Skirting heating is no longer widely used in domestic properties, owing to the restrictions placed on output from the heater. However, you may still come across it in some older properties.

Figure 8.41: Skirting heater

Fan-assisted convector heaters

These work by forcing cooler air through the heating fins (heat exchanger) using an electrically controlled fan. They therefore require connection to the central heating control system.

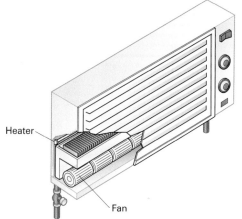

Figure 8.42: Fan-assisted convector heater

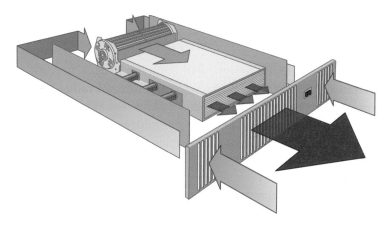

Figure 8.43: Kick space heater

A client asks you to replace a radiator on her existing system, which is leaking. She explains that several radiators over the last few years have all had to be replaced. When draining the system it is obvious that the problem is corrosion due to the amount of sludge in the system. You explain that the system will require power flushing and the addition of a rust inhibitor on completion. She informs you that the system had rust inhibitor added last time the system was drained 12 months earlier. The client also asks you to look at the float-operated valve in the roof space as the warning pipe often runs water onto the yard at the rear of the property.

The system is an open-vented, fully pumped type. Explain what the cause of the corrosion problem could be.

Operating principles of central heating control components

Some of the components you will use for central heating have been covered earlier in this book. Information about these can be found on the following pages:

- hot-water storage cylinders (pages 308–316)
- circulating pumps (pages 339–40)
- drain valves (pages 272–75)
- devices designed to reduce sediment (pages 277–79).

Radiator valves

There is a wide selection of valves available from manufacturers. Building Regulations require thermostatic radiator valves (TRVs) to be installed on new systems unless there is another means of controlling an individual room's temperature. A thermostat and motorised valve may be used but these are usually too expensive to include in all rooms. You may be required to install manual radiator valves on repair jobs.

Wheel-head radiator valves

These enable the occupier of the building to control the temperature of the radiator manually by turning it on or off.

Rotating the plastic 'wheel-head' anticlockwise will raise the spindle through the body of the valve, lifting the valve and opening the flow to the radiator.

Figure 8.44: Wheel-head radiator valve

Lock-shield radiator valves

These valves are intended to be operated by the plumber; not by the occupier of the building. The plastic cap conceals a lock-shield head, which can only be operated with a special key or pliers. The plumber will use this valve to isolate the supply if removing the radiator, or to balance the system when commissioning. The valve illustrated in Figure 8.45 shows an added feature: an in-built drain-off facility.

Figure 8.45: Lock-shield radiator valve with drain-off facility

> **Remember**
>
> The valve head of a TRV needs to have good air circulation around it in order to sense the room temperature properly. If the head is going to be covered by a curtain, for example, then a special TRV with a remote sensing element needs to be installed.

Thermostatic radiator valves (TRVs)

These control the heat output from the radiator by controlling the rate of water flowing through it. The user adjusts the valve to maintain a

Figure 8.46: Thermostatic radiator valve

desired temperature in the room; the valve then works automatically to maintain that set temperature.

TRVs are fitted with a built-in sensor, which opens and closes the valve in response to room temperature. Liquid, wax or gas expands into the bellows chamber as the sensor heats up. As the bellows expand they push the pin down, closing the valve. There are a number of settings on the head of the valve to enable a range of room temperatures to be selected.

You need to install a bypass valve to prevent the boiler and pump working against a closed system should all the TRVs close down and the water flow rate in the system fall to a low level. This component and its positioning in the system will be covered fully later in this unit.

The bypass must be fitted in accordance with the boiler manufacturer's instructions.

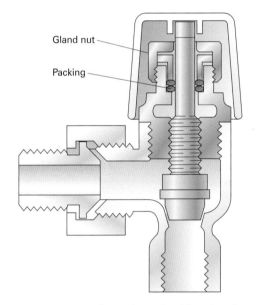

Gland nut

Packing

Figure 8.47: Section through a wheel-head radiator valve

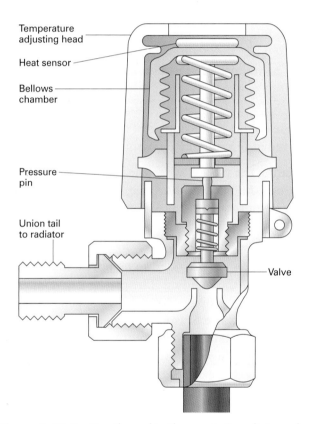

Temperature adjusting head

Heat sensor

Bellows chamber

Pressure pin

Union tail to radiator

Valve

Figure 8.48: Section through a thermostatic radiator valve

Automatic air vents

These are used on central heating systems in order to remove air from the system automatically. A vacuum break on the bottom of the valve prevents an airlock forming and encourages air to be released from the water.

Vents should always be installed on the positive side of the system and positioned where air is likely to get trapped.

Figure 8.49: Automatic air vent

Motorised valves

The type of valve used will depend on the system design, but the following are available:

- Three-port diverter valve
- Three-port mid-position valve
- Two-port motorised zone valve.

All motorised valves must provide the correct flow direction and priority to match the system.

Three-port diverter valve

This valve controls the flow of water on fully pumped central heating and hot-water systems on a selective priority basis, which is normally used for the domestic hot water.

Three-port mid-position valve

This valve looks very similar to the diverter valve. It is used on fully pumped hot-water and central heating systems, in conjunction with a room and cylinder thermostat. It provides full temperature control of both the hot-water and heating circuits, which can operate independently of each other or both at the same time.

Figure 8.50: Three-port diverter valve

Two-port motorised zone valve

A single-zone valve is used on gravity domestic hot-water and pumped central heating systems to enable separate temperature control of both the heating and hot-water circuits. Motorised valves are also used in fully pumped systems to provide separate control of both heating and hot-water circuits. They can be used to zone different parts of a building, such as the upstairs and downstairs.

Figure 8.51: Two-port motorised zone valve

Automatic bypass valves

These are mechanical devices used to make sure water can flow through the boiler to maintain a minimum water flow rate should a system using zone valves or TRVs become closed. Once set, the valve opens automatically as the TRVs or zone valves close and the system pressure goes over the pre-set limit.

Use of the bypass valve reduces system noise and increases pump life by preventing the pump from working against a 'dead head'.

Time switch

A time switch is an electrical switch operated by a clock to control space heating or hot water, or both together.

Full programmer

A full programmer can be used on gravity-primary and fully pumped systems. It allows the time settings for space heating and hot water to

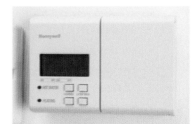

Figure 8.52: Central heating programmer

be fully independent. More sophisticated programmers can include separate time controls for multiple space heating circuits.

Room thermostat

Figure 8.53: Room thermostat

This device measures the air temperature within the building or room where it is sited and controls the operation of the heating circuit (depending on whether the timing device is calling for heat). It can be used directly to switch on a circulating pump or boiler, or to operate a motorised valve.

Programmable room thermostat

Figure 8.54: Programmable room thermostat

This is a combined time switch and room thermostat. It enables the user to set target temperatures for different periods for the central heating. This can be done on a daily or weekly cycle: for example, a higher temperature may be required for a certain period in the evening or at weekends.

Cylinder thermostat

The cylinder thermostat measures the temperature of the water stored in the hot-water storage vessel and, when it reaches a pre-set temperature, switches off the hot-water circuit to the vessel. This is done by connecting to the circulating pump or boiler, or by the operation of a motorised valve.

Frost thermostat

Figure 8.55: Frost thermostat

Frost thermostats give automatic frost protection to boilers and pipework that have to be located in areas at risk from severe cold. External boilers are a typical example.

Progress check

1. Identify the design temperatures for the following rooms:
    - living rooms and bathrooms
    - all other rooms.
2. Answer the following questions regarding the Approved Document L1.
    - Where is the approved document located?
    - In which year did the current regulations become legislation?
    - What is the full title of the Approved Document L1?
    - List the eight requirements set out in the Approved Document L1.
3. Which central heating systems are no longer permitted by the approved document.
4. All new domestic heating systems must be fully pumped. What is the one exception?
5. Briefly explain the operation of the following diverter/zone valves:
    - two position three-port diverter valve
    - three port-mid-position diverter valve
    - two-port valves.
6. How much allowance for expansion should there be in an open-vented feed and expansion cistern?
7. In a fully pumped heating system, what is the minimum height of the open vent above the water level in a feed and expansion cistern?
8. What is the minimum height referred to in Question 7 (above) designed to prevent?
9. What does a filling loop on an unvented hot-water heating system replace in an open-vented system?

3. Know the installation requirements of central heating systems and be able to install central heating systems and components

Central heating system controls

Everyone has a moral obligation to do what they can to conserve energy. Energy conservation is also a high priority for the government. For this reason, Building Regulations have been amended to improve energy efficiency in buildings.

The revised Regulations have had the effect of enforcing fixed specifications for system design and, in particular, its control. This section looks at the mechanical and electrical controls used on central heating systems, beginning with the basic system controls.

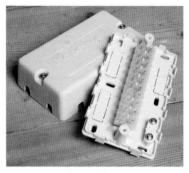

Figure 8.56: Wiring centre

Junction box or wiring centre

Junction boxes and wiring centres provide the connections between the electrical system components and the mains electricity supply.

Most manufacturers supply their controls in packs, and these can include a wiring centre, which is designed to simplify the wiring of a particular system pack. Packs include control valves, programmers and thermostats. All the terminal connections are clearly marked and full instructions are included with the wiring centre.

Boiler control interlocks

This is a term used in Building Regulations. It is not an actual control device but an interconnection of controls – such as room and cylinder thermostats, programmers, and zone valves – designed to ensure that the boiler does not fire up when there is no demand for heat.

In a system with a traditional boiler, this would require the correct wiring of the room **stat**, cylinder stat and motorised valve(s). It may also be achieved by more advanced controls such as a boiler energy manager.

How the controls work together

So far, we have looked at system layouts, components and controls. We now need to think about how they work together. We will focus on the following system types, as they are the ones commonly installed.

Fully pumped systems are the only option for new installations and the preferred option for any existing system modifications or upgrades. There are several types:

- Fully pumped system using three-port mid-position valve
- Fully pumped system using 2 × 2-port zone valves.

> **Remember**
>
> Building Regulations *Approved Document L1* is the government publication that deals with the conservation of fuel and power in domestic premises and sets out the regulations for the installation of domestic heating systems

> **Key term**
>
> **Stat** – a widely used abbreviation for 'thermostat'.

Fully pumped system using three-port mid-position valve

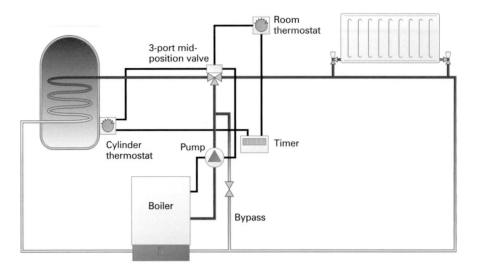

Figure 8.57: Components of a fully pumped system using a three-port mid-position valve

Installation:

- three-port valve:
  - water flow must be into port AB
  - port A is connected to the heating circuit
  - port B is connected to the hot-water storage vessel primary circuit
  - **room and cylinder stat** as for the boiler control interlocks (page 355)
- **time control** must be provided by a programmer that permits the selection of heating without hot water
- **bypass:** an automatic bypass may be installed immediately after the pump, between the flow and return. Pipework should be 22 mm and flow should be in the direction of the arrow marked on the valve body. However, most installers do not fit an automatic bypass, as this valve would always leave an open circuit.

How this installation works

- **Heating only:** when the room stat calls for heat, the valve is activated so that the central heating port only is opened and the pump and boiler are switched on.
- **Hot water only:** when the cylinder stat calls for heat, the valve remains open to the hot-water storage vessel; the pump and boiler are switched on.

- **Heating and hot water:** when both stats call for heat, the valve plug is positioned so that both ports are opened. The pump and boiler are switched on; they're switched off once the stats are satisfied. The valve remains in the last position while the programmer is in the 'on' position.

Fully pumped system using 2 × 2-port zone valves

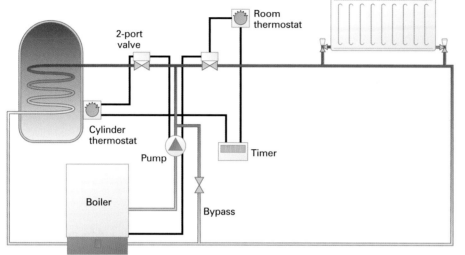

Figure 8.58: Components of a fully pumped system using 2 × 2 port valves

Installation

- Two-port valves: water flow must be in the direction of the arrow on the valve body.
- All other components are the same as the previous system (fully pumped system with three-port mid-position valve, see page 356). A programmer must be used to satisfy Building Regulations.

This system provides greater flexibility than the others. Additional zone valves and thermostats can be added to separate the heating circuit into several zones. This improves temperature control and efficiency.

How this installation works

When either stat calls for heat, the respective motorised valve will be opened. Just before it is fully opened, the auxiliary switch sited in the motorised valve head will be closed and will switch on both pump and boiler. Once both stats are satisfied, the valves are closed and the pump and boiler switched off.

Working life

Working life

Jarred is plumber who has been called to a fault on a fully pumped domestic hot-water heating system using 2 × 2-port valves. The client has complained that there is no domestic hot water but the central heating system is operating satisfactorily. Jarred checked both motorised valves and found that the central heating circuit valve was open but the hot-water circuit valve would only open if the spring return

was operated manually. Jarred decided to call in a qualified electrician, Makszi, as he felt sure the problem was due to an electrical fault. On checking the circuits, Makszi informed Jarred that there was no electrical fault as the circuits were reacting to the demands of the system.

Discuss what the probable cause of the fault could be.

Progress check

1. Name the three most common fuel sources used to run central heating systems.

2. What does the abbreviation SEDBUK stand for?

3. What term is used to define all new boilers whether being fitted in new installations or as a replacement in an existing system, as defined In Part L1 of the Building Regulations?

4. What is the key difference between combination boilers and system boilers?

5. In a domestic property, where would a room thermostat usually be fitted? Give a reason for your answer.

4. Know and be able to apply the service, maintenance and decommissioning requirements of central heating systems and components

Fault diagnosis will be fully covered at Level 3. At Level 2 you are required to understand the principles of maintaining the components within the system; excluding the boiler. Level 2 also requires you to be able to change a defective radiator valve and central heating pump, so this section focuses on these two components.

By now, you should have realised how important energy efficiency is. Maintenance plays an important role here: it ensures that the system components continue to operate as they are designed to. For example, a pump that is not performing to its specification could make the boiler work harder, thus wasting energy.

Maintenance procedures consist of:

- finding out what the fault is
- informing the customer of your action
- isolating the supply
- stripping the component (if possible)
- repairing or replacing the defective part

- reassembling components (if applicable)
- turning on supplies and testing the component for correct operation.

Soundness testing procedures

The testing requirements for hot and cold systems have already been covered, and these aspects apply equally to central heating systems. This section covers flushing central heating system pipework and using additives to protect against corrosion.

More information about the procedures used in soundness testing can be found on pages 249–50 in Unit 5.

Visual inspection

In the same way as for hot- and cold-water pipework, testing includes making sure that all pipework and fittings are thoroughly inspected to ensure that:

- they are fully supported, including F&E cisterns and hot-water storage cylinders
- they are free from jointing compound and flux
- all connections are tight
- in-line valves and radiator valves have been properly tightened and are closed to allow stage filling
- the inside of the F&E cistern is clean
- all the air vents are closed.

Before filling, it is a good idea to remove the pump and replace it with a section of pipe. This will prevent any system debris entering the pump's workings.

Remember

Advise the customer or other site workers that soundness testing is about to commence.

Working life

Anthony is an experienced plumber who has an apprentice, Lyndon. Their employer has asked Anthony to brief Lyndon on the visual inspection procedure for a property, which has an open-vented hot-water heating system.

Discuss a checklist of visual checks which need to be carried out prior to the testing for water soundness.

Pressure testing

On larger jobs, testing could be part of the contract specification. It is usual to test the system to 1½ times the normal working pressure or 3 bar, whichever is the greater, for a period of one hour. The test pressure is achieved by using hydraulic test equipment.

Figure 8.59: Two-port motorised valve showing safety cover removed

Remember

Sometimes contractors receive calls from consumers who think their system is faulty, when in fact they do not understand how to work it properly. It is vitally important to make sure that your customer fully understands how to operate the system you install.

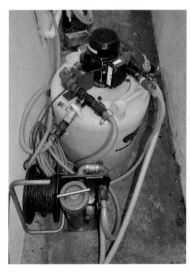

Figure 8.60: Power flushing kit

Remember

Again, remember to include a system additive when refilling.

Hydraulic test equipment

A safe mains connection (one that will not contaminate the supply) or a hydraulic test kit can be used to pressure test the system. All open ends in the system must be temporarily sealed prior to the test and a test point provided to connect the hydraulic kit or mains supply. The system is then increased to 1½ times its normal working pressure and left to stand for one hour. Larger systems must be tested section by section, e.g. first floor, second floor, or by zones.

Final system checks

During the process of filling and testing the heating system, the system should be balanced and the boiler fully tested and adjusted. These procedures are fully detailed at Level 3.

Routine checks and dealing with defects

Prevention is better than cure, and more and more plumbing firms offer their customers planned maintenance packages for the full domestic system; not just heating. This will include:

- servicing of boiler and other heating appliances at least once a year (a legal requirement for gas appliances in tenanted properties)
- regular servicing of taps and valves
- cleaning and flushing of central heating systems.

Maintenance contractors report that the majority of maintenance and repair jobs in domestic properties are caused by:

- failure of central heating controls, boiler parts, motorised valves and thermostats
- leaks or burst pipes
- leaks caused by householders themselves, a typical example being the removal and replacement of radiators for decorating.

Motorised valves are usually very reliable, but will eventually come to the end of their working life. At one time this meant draining the system. However, it is now possible to replace defective motors or the actual power head.

On older systems, it may be necessary to flush out pipework when carrying out maintenance work. A power flushing kit is designed for this purpose.

Approved cleaning agents should be used with this equipment, which comes complete with hoses and valved connections. Always follow manufacturer's instructions when power flushing, as procedures vary from manufacturer to manufacturer.

Effective operation of thermostats

In its simplest form, a thermostat is nothing more than a switch, but one that automatically reacts to changes in temperature. There are various types of thermostats and although bi-metallic thermostats are still manufactured, electronic versions are more likely to be found on newer systems.

In the case of the bi-metallic strip, temperature change causes mechanical movement. This is because it is actually made up of two joined strips of different metal that are rolled into a coil. The two metals have a different coefficient of expansion and therefore expand at different rates as they are heated. These different expansions force the strip to bend one way if heated, and the other way if cooled below its initial temperature.

Consequently, as the room heats up or cools down, the metal reacts to the change in temperature. Once the thermostat reaches a specific set level, it sends a signal to the central heating boiler to switch it on or off. These versions were used extensively because of low cost, but could be slow to react to temperature change.

In the case of an electronic thermostat, it uses a device known as a thermistor (a resistor whose electrical resistance changes with temperature) to sense room temperature and a micro-controller (a micro-computer on a single chip) that measures the resistance change and converts that number to a temperature reading. Once the temperature in the room moves above or below the set temperature, the thermostat sends an electric signal to the boiler to switch it on or off.

Testing for leaks

To test for leaks, you need to carry out the following checks:

- Turn on the stop tap if the installation was for a complete cold-water, domestic hot-water and central heating system, or the service valve to the F&E cistern if only the central heating circuit has been installed.
- Allow the system to fill.
- Turn on the radiator valves fully and bleed each radiator.
- Visually check all the joints for signs of leaks.
- Drain down the system, flushing out all debris, wire wool, flux, etc.
- Reinstall the pump.
- Refill and test for leaks again.
- Make sure the water level in the F&E cistern is at a level that allows for expansion of water in the system when hot.
- Heat system and flush while hot.

Working life

On page 359, you discussed a checklist of visual checks for Anthony. Anthony is now confident that Lyndon has a good understanding of what is required regarding visual inspection and so he decides to carry out a similar exercise, with regard to testing for leaks and commissioning on the same system described above.

Make a checklist of criteria, which needs to be carried out when testing for water soundness, including preparation for handing over to the client.

Figure 8.61: Hydraulic test kit

System cleaning

This part of the process is vital as it prevents corrosion in the heating system and the build-up of deposits that could lead to system failure.

The cold flush

Once you've checked and tested for leaks, it is time for the cold flush. You will need to drain the system from the drain-off points as quickly as possible. First turn off the water supply, then drain off with a hose pipe; this allows air into the system at the various bleed points.

The cleansing process

BS 7593 details the requirements for the treatment of water in domestic heating systems. These guidelines state that all systems, as part of the commissioning process, should be treated with a proprietary system cleaner to ensure the removal of foreign matter and bodies that may cause corrosion, such as flux residues.

The cleansing is done by introducing a system cleaner into the system. The amount of cleaner required is specified by the manufacturer, and based on the system's volume in litres. The system cleaner must be added via the F&E cistern as it is refilling. Adding the system cleaner into the cistern after it has refilled will not allow it to work properly.

Once the cleaner has been introduced the system can be refilled and vented.

The hot flush

Most manufacturers recommend a two-hour period for the cleaning operation, after which the system cleaner can be emptied out of the system. Turn off the system electrics so that the controls are isolated. Open the drain valves full bore and empty the system contents, being sure to fully open any bleed valves and motorised valves. Once the system contents have been drained, close all the valves and introduce the required amount of corrosion inhibitor. This is done prior to refilling, at the same fill point at which you introduced the cleaner. Refill the system and vent the air. Run the system back up to temperature, venting air as required and ensuring all circuits are operating correctly.

Replacing a defective radiator valve

Valves can be one of three types:

1. TRVs
2. Lock shield
3. Manual radiator valve.

Replacement of any of these valves will follow the same procedure. There are four ways to do this job:

1. Create a vacuum in the system to hold the water while replacing the valve – this is usually the preferred option as it is quicker.

2. Drain down the central heating system – this is usually a last resort.

3. Use pipe-freezing equipment to isolate the supply to the radiator valve.

4. Some manufacturers have produced specialist servicing tools and valve inserts for both manual valves and TRVs. These enable the valve to be serviced without draining down.

The first two methods are the most commonly used, so the next section will focus on them.

Creating a vacuum in the system

This is a quick and simple solution to changing a component such as a radiator valve as it saves draining down the system. It works by completely plugging and closing all outlets to the system such as vent pipes and automatic air vents. It only works where you are working on one open pipe end (you cannot use this procedure for cutting into pipework) and it can only be used for vertical sections of pipework – making it ideal for radiator-valve replacement.

Safety tip

When creating a vacuum to access a radiator valve, there must be no existing leaks in the system.

Replacing a defective radiator valve

Step 1: Close both valves on the radiator

Step 2: Drain the radiator

Step 3: Relieve system pressure

Step 4: Saw through compression ring (optional)

Step 5: Replace valve

Step 6: Finish the job

Isolate the system

Make sure the electrical supply to the heating system is isolated. Turn off the boiler. Remove the fuse from the spur outlet to the wiring centre or junction box for the controls. Advise the customer of what you are doing and tell them not to touch the controls. Only work on the system when the water has cooled. If you speak to the customer beforehand to

Safety tip

Be careful to support the valve fully when opening the union so that you do not strain the pipework.

arrange the job, it is a good idea to ask them to turn off the system so that it is cold when you get there.

Close all outlets

Turn off the service valve to the F&E cistern; close any automatic air vents. The open vent pipe and cold feed now need plugging – manufacturer devices are available to carry this out.

Drain down the radiator

Before progressing any further, the radiator that is serviced by the valve being replaced needs to be drained down. This is carried out by isolating both radiator valves and loosening the union nut on the radiator valve being replaced. Catch the water from the radiator in a small bowl, and have a bucket handy to empty the bowl from time to time. Air needs to be let into the radiator through its air vent to ensure that the water in the radiator fully drains out.

Relieving the system pressure

The radiator valve being replaced should now be 'cracked open' at its control mechanism. A small amount of water should be discharged into the bowl until the pressure in the system reduces to atmospheric pressure. After a short period, the water flow should stop. If it does not, or the flow is excessive, this suggests that you have not closed all of the outlets or there is a leak in the system. If there is a leak, there is no option but to try another method of isolation, such as draining down. When the flow stops, you can move on to replacing the valve.

Replacing the valve

Key term

Allen key – a hexagonal tool designed to fit into hexagonally recessed nuts and allow them to be turned.

Check that the new valve body will fit the existing valve tail screwed into the radiator. If it does not fit, change the tail. If it is the type with an external nut, this can be done with an adjustable spanner. If not, then you will need a radiator spanner/**Allen key**, which goes inside the tail to remove it. When fitting the new tail, wrap PTFE tape around the threads five or six times in the direction of the thread.

At this point you will need to establish whether modifications to the existing pipe are required. If you are fitting a valve by the same manufacturer, it will probably be a straight swap; otherwise, if there is some lift in the pipe, you may be able to remove the old compression ring and nut by sawing and lifting the pipe to the required height of the new valve. It is also possible that the length of pipe may have to be extended. Once the pipe has been prepared, the valve should be installed and all the nuts tightened.

Finishing the job

You are now ready to put the system back into operation. Remove the plugs from the cold feed and vent pipe, and open any automatic air vents in the system. Open the radiator valves and refill the radiator, checking for any leaks to the replacement valve. Replace the fuse and operate the system. Advise the customer that the work has been completed.

Draining down the system

This turns the relatively simple task of changing a defective radiator valve into a long job. It is not the draining down that takes time but the refilling and making sure all the system is working correctly.

Isolate the system

Make sure the electrical supply to the heating system is isolated. Turn off the boiler. Remove the fuse from the spur outlet to the wiring centre or junction box for the controls. Advise the customer of what you are doing and tell them not to touch the controls. Only work on the system when the water has cooled. If you speak to the customer beforehand to arrange the job, it is a good idea to ask them to turn off the system so that it is cold when you get there.

Drain down the radiator

Locate the drain valve at the lowest point below the radiator valve to be changed. If you are working on an upstairs radiator, drain only the first-floor level, not the whole system. You can do this by slightly loosening the connection from the valve to the radiator (have a bowl and cloth ready) and checking it until the water has stopped dripping. Drain-off is done by connecting a hose pipe to the drain valve and running the other end to an outside drain.

Fitting the valve

Once the system has been drained, one of two situations will exist:

1. The valve can be fitted without altering the pipework. This is because the valves are the same size, or there is sufficient 'play' in the pipe to move it up or down to fit.
2. The valve dimensions are different, so the pipework has to be altered.

If the valve can be changed without altering the pipework, remove the valve, leaving the existing nut and compression ring on the pipe.

Check that the new valve body will fit the existing valve tail screwed into the radiator. If it does not fit, change the tail. If it is the type with an external nut, this can be done with an adjustable spanner. If not, then you will need a radiator spanner/Allen key, which goes inside the tail, to remove it. When fitting the new tail, wrap PTFE tape around the threads five or six times in the direction of the thread.

Remove the nut and compression ring from the new valve and place the new radiator valve body in position. If it is a TRV, remove the thermostatic head first. Because the compression ring has been used, some plumbers wrap PTFE tape or jointing compound around the joint between the compression ring and the valve body.

If you have to alter the pipework – either shortening by cutting, or extending by use of a fitting – complete the process, fit the new nut and compression ring to the pipe, and follow the next steps.

Once the valve has been secured to the pipe, fit the valve body to the radiator tail and tighten using adjustable spanners/grips. Make sure you use opposite force on the valve body to stop the valve twisting and possibly affecting the nut and compression ring joint.

Finishing the job

Turn off the drain tap, turn on the supply to the F&E cistern and begin the air-release process. This will involve:

- bleeding all the radiators, so do not forget your air key
- bleeding the pump
- bleeding the air valve (if manual type) on the pumped primary on fully pumped systems.

Using pipe-freezing equipment

There are a number of kits on the market. They vary in specification and cost, and you will have to decide which kit is best suited to your needs and your pocket. The kits illustrated can generally be used on copper and plastic and all are suitable for use on central heating systems.

Firstly, you need to make sure you can use your kit competently. Only carry out this process on cold pipework with no water flow. If you try to use a pipe-freezing kit on warm pipes, you cannot predict how long the plug of ice will last, so you may get wet and the house may be flooded.

Then, follow the steps below:

- Make sure the supply to the opposite end of the radiator is turned off.
- Remove the floorboards from above the pipe run to allow plenty of access to the pipe to be frozen.
- Following the manufacturer's instructions, place the freezer head or jacket at the required distance from the valve.
- Most kits will freeze the supply within 2 minutes and will last for 30 to 45 minutes.

Remember

Don't forget to add corrosion inhibitor at the end, and check your handiwork for water tightness.

Figure 8.62: Pipe freezing

- Removing and replacing the valve is the same as the draining-down job, but remember that here you will have to drain the excess water out of the radiator. When you're doing this, loosen the air vent to allow the water to drain more quickly. Have a bowl and cloth available to deal with any excess water.
- Once the process of replacing is complete, fill the radiator and test.

Replacing a defective heating pump

In the same way as valve replacement, advise the customer of what you are doing, both in advance and during the actual job. On maintenance jobs, a pump replacement could be needed on any type of central heating system and the pump could be located almost anywhere in the building. However, it is more than likely that it will be located close to the domestic hot-water storage vessel. On solid fuel systems or gravity primaries, pumps can be located near the boiler. On system boilers, pumps are located inside the boiler casing.

Carry out some preliminary checks:

- First, make sure that the pump really is faulty. The fault could be in the electrical control, or it could be something as simple as a blown fuse on the fused spur to the pump.
- Make a note of the manufacturer's codes, size and type of pump to ensure that a correct or suitable alternative pump can be identified.
- Check to see whether the replacement pump will fit without any pipework alterations. If it does not, go and get a pump that will! Check that the pump valves are in good condition (which they rarely are). If the answer is yes to both these checks, you will not have to drain the system down. If the answer is no to either, then you will.

Isolate the supply

Depending on the type and age of the system, there may be a fused spur to the pump, or a fused isolating switch. The pump could also be part of a system control package, in which case you will have to isolate the fused spur to the wiring centre or junction box.

In all cases, remove the fuse to isolate the supply. If a wiring centre or junction box has not been used (i.e. older systems), turn off the boiler as well. If you need to alter the pipework, you will have to drain down the system as when replacing a radiator valve.

Disconnect the pump cable (Step 1)

You need to disconnect the electrical supply. First remove the plastic cover. This will reveal wiring that is the same as on an ordinary plug.

Remove the plastic cable retainer and disconnect the cables, making sure there is enough bare wire ready for reconnection to the pump fittings, and that the wires are in good condition. Cut back the wire and re-strip if not, or replace the cable, and wire the new pump before it is put in position. This makes the job easier.

Remember

Always read the manufacturer's instructions properly.

You may have to alter pipework, which will involve heat. Allow it to cool before freezing.

Safety tip

Always follow the safety instructions that come with the equipment.

You have only got 30–45 minutes, so get moving!

Always use a temporary continuity bond before breaking any pipework connections.

Remember

Because the electrical system is involved in this job, the safe isolation procedure must be followed at all times. Don't forget: some manufacturers produce pump extensions that allow you to avoid draining the system, so check for these first.

Replacing a defective pump

Step 1: After safe isolation disconnect the cable and close the valves

Step 2: Undo union connections

Step 3: Remove old pump

Step 4: Clean the union surfaces

Step 5: Fit the new pump

Step 6: Complete electrical connections

Remember

Remember your plug wiring: blue to neutral, yellow and green to earth and brown to live. On older systems the wiring could be red to live, black to neutral and green to earth.

Remove the old pump (Steps 2 and 3)

Assuming that you do not have to drain down, turn off the isolating valves and remove the pump. If it has been in place for a long time, this may be difficult. You may have to apply heat to the pump valve nuts to loosen them. You will also need large **stillsons** or grips on both valve nuts to stop the pump rotating. Be careful when doing this not to disturb the joint between the pipe and the pump valve.

Cleaning the union surfaces (Step 4)

Once you have got the pump out, make sure the pump valve faces are free of any left-over valve washers.

Fitting the new pump (Step 5)

Fit the new pump in accordance with the manufacturer's instructions and make sure the new washers are fitted. If you altered the pipework to fit the new pump, ensure the pump and the new valves are fitted. Turn the water supply back on. If you drained the system, refill it as you did for the radiator valve replacement. If not, turn the pump valves back on. Bleed the pump through the slotted head in the body of the pump or pump bleed screw.

Pumps have a directional arrow, which you must follow when installing.

Complete electrical connections (Step 6)

Reconnect the pump cable to the cable terminals, carry out any electrical checks to the electrical supply and test the pump. Advise the customer that the system is back in service and complete any commissioning records that may be required.

Poor heat emission from a heat emitter

Radiators may not heat up as they should. There are a number of reasons why this may occur. Firstly check the valves are open. If the radiator has a thermostatic valve, ensure it is not stuck closed by removing it and testing.

Once you have confirmed the valves are open and working, ensure the system is not operating and then check there is no air in the top of the radiator. Air can be released using the air vent located at one end of the top of the radiator. Open the vent until there is a flow of water and then reclose the valve.

> **Remember**
>
> When venting the system, the circulating pump and system must always be turned off.

If the system is a sealed system you may need to top up the pressure before continuing (this will be done automatically in an open-vented system via the cold feed and expansion cistern). If, after this, the radiator is still not working, check the circulating pump is operating correctly, especially if the radiator is at the furthest point in the system. This could be done by closing off other radiators. If the radiator still does not operate correctly you may have to remove the radiator and see if it needs flushing out.

Decommissioning central heating systems

As with the decommissioning of hot- and cold-water systems covered earlier, this could include the following situations:

- An old system is going to be completely stripped out of a domestic property and replaced with a new system, such as a gravity solid fuel system being replaced with a fully pumped condensing boiler system.
- A system is going to be stripped out permanently, such as before the demolition of a building.

Make sure you:

- keep customers or other site workers informed of the work being carried out
- make sure all supplies are isolated and capped off (where necessary) and that electrical supplies are effectively isolated
- use notices on taps or valves, such as 'not in use' and 'do not use'.

Working life

Ben has to replace a faulty thermostatic radiator valve (TRV). The valve pin, which is operated when the actuator head pushes on it, is damaged as someone has bent the pin trying to release it from its closed position.

Discuss Ben's options for replacing the whole valve in the system successfully. When you have listed these options, select the one that would create the least disruption and save time. Then discuss the procedure for this to be carried out.

Progress check

1. The valve pin in thermostatic radiator valve heads (TRVs) often become stuck in the closed position. Why does this occur? What can be done to reduce the risk of this occurring on installations?

2. Before carrying out any maintenance work on a domestic hot-water system or components, what is the first thing to do in terms of health and safety?

3. How can you work out which direction to install a replacement circulation pump into an existing system, if the old pump has been removed by someone else?

4. After installing a replacement circulation pump and turning on the valves, how is air removed from pump?

5. What should be done to the system before opening the air-release valve found at the top of a radiator panel?

Check your knowledge

1. If a two-pipe system is being installed in a dwelling, which of the following statements will be true?
   a Water in the flow and return pipework will be mixed.
   b The return water will be at the same temperature as the flow.
   c The water in the flow and return pipework must be kept separate.
   d The temperature of the return water will be higher than the flow.

2. Which of the following systems requires a temporary initial filling arrangement?
   a Fully pumped system of hot water and heating (vented)
   b Sealed central heating system with vented hot water
   c Combined hot-water and heating system with vented primary pipework
   d Small-bore heating system with vented primaries and auxiliary heating

3. What is a zone valve in a pumped heating circuit designed to do?
   a Isolate the circuit from the heated water supply
   b Permit the system to be heated to temperatures above 100 °C
   c Stop the pump vibrating on its fixed mounting
   d Prevent the mixing of water from the flow and return pipes

4. What is the main purpose of the vent pipe on a heating circuit?
   a To provide hot water
   b To introduce air into the system
   c To provide protection in the event of overheating
   d To allow filling up of the system

5. What is the purpose of a thermostatic radiator valve?
   a To control individual temperatures of radiators
   b To control the temperature of the boiler
   c To control the temperature of different zones
   d To control the output from a gas fire

6. Which of the following is fitted to a hot-water storage cylinder to control the temperature of the water?
   a Cylinder thermostat
   b Cylinder thermometer
   c Programmer
   d Thermostatic radiator valve (TRV)

7. What is the normal size of pipe used in micro-bore heating systems?
   a 8 and 10 mm
   b 8 and 15 mm
   c 10 and 15 mm
   d 15 and 22 mm

8. What are combination boilers designed to do?
   a Provide hot water and heating
   b Work with oil and gas at the same time
   c Provide hot water from a storage cylinder
   d Pressurise cold water

9. What steps should be taken before decommissioning a central heating system?
   a Check for isolation and remove the pump before draining down
   b Apply a temporary earth bond to the stop tap
   c Check safe isolation, apply temporary continuity bonding across the system and then drain down
   d Check safe isolation and save the inhibitor to put back in the system

10. What may cause the upstairs radiators to work correctly while the downstairs radiators are not working?
    a Faulty gas valve
    b Faulty pump
    c Faulty boiler
    d Faulty cylinder

11. What is the main reason for allowing a 150 mm gap between the floor and the base of a radiator?
    a So the customer can vacuum under the radiator
    b So the skirting board can be painted
    c To allow adequate clearance for heat circulation
    d To allow the joiner to remove the skirting board if necessary

Getting ready for assessment

The information contained in this unit of the book, as well as the continued practical assignments that you will carry out in your college or training centre, will help you with preparing for both your end-of-unit test and the diploma multiple-choice test. It will also support you in preparing for the practical assignments you will need to complete to demonstrate your understanding in and enabling you to carry out and apply domestic central heating system installation and maintenance requirements.

There are opportunities throughout the unit for you to test your progress in and understanding of the required underpinning knowledge; this will enhance your preparation for the forthcoming assessments, so make good use of them.

This unit will be assessed by the following assessment methods:

- externally set knowledge assessment

- externally set assignments.

With regard to the mechanical services industry, you will need to know:

- the uses and types of central heating systems and their layouts requirements in dwellings

- how to carry out and apply site preparation techniques for central heating systems and components

- the installation requirements and be able to install central heating systems and components

- the service and maintenance requirements of central heating systems and components and be able carry them out

- the decommissioning requirements central heating systems and components and be able to carry them out

- how to carry out and apply inspection and soundness testing requirements of central heating systems and components.

Check it out – using your knowledge from this unit practice and test yourself on the following key terms:

Access/exit, air relief, anti-gravity valve, automatic bypass valve, background, British standards, circulating pump, clips/brackets, clocks/programmers, Codes of practice, column, combination, combination boiler, commissioning/decommissioning, components, condensing, cylinder/room thermostats, cylinders, defects, drain valve, expansion vessel, expansion/contraction, F&E cistern, fan assisted, fan convector, filling and venting, filling loop, flow, flushing procedures, freestanding, frost protection, full, fully pumped, gas, gravity, hazards, industry standards, inhibitor, installation, insulation, isolation of fuel/electricity supplies, low surface temperature, maintenance records, maintenance, manufactures instructions, mid-position valve, oil, one and two pipe systems, open flued, open vented, panel, pipework/fittings, PPE, preparatory work, pumped with gravity hot water, purpose, radiator valves, remedial work, return, risk assessment, room sealed, sealed system, sediment build up, selective, solid fuel, soundness testing, statutory regulations, system boiler, temporary de-commissioning, thermo-mechanical control valve, thermostatic radiator valve, tools/equipment, towel warmer, traditional, two-port valve, vaporising, visual inspection, wall mounted and warning notices.

Good luck!

Understand and apply domestic rainwater system installation and maintenance techniques

Guttering catches rainwater running off roofs and is available in different shapes, sizes and materials. The most common guttering for domestic dwellings is plastic. This is either half-round- or square-shaped, and brown, white or grey in colour. The water is taken from the guttering to low level via a downpipe. The downpipe connects to the drainage system, soakaway or a rainwater harvesting system. Without guttering, the water running off a roof would cause damage to the brickwork or other problems such as rising damp.

This unit will cover the following learning outcomes:

- Know the general principles of gravity rainwater systems
- Know the layout requirements of gravity rainwater systems
- Know, and be able to apply, the installation requirements of gravity rainwater systems
- Know, and be able to apply, the service and maintenance requirements of gravity rainwater systems
- Know, and be able to apply, the inspection and testing requirements of gravity rainwater systems

Information about knowing and applying site preparation techniques is given in Unit 5.

1. Know the general principles of gravity rainwater systems

Gutters are used to collect rainwater that falls onto the roof of a building. It flows down the rainwater pipe and into surface-water drains, combined drainage systems or soakaways. If gutters and rainwater pipes were not installed, the rainwater would damage gardens due to water running off the roof, staining on brickwork and getting into houses, leading to such problems as rising damp.

However more and more people are looking at using this water to help the environment and reduce their water costs. This process is known as rainwater harvesting and will be covered at Level 3.

Gutter and rainwater systems

Gutters are specified by their shape in cross section.

Generally, rainwater pipes are either square or round. Guttering is supplied in 2 m and 4 m lengths, and can be obtained in a range of diameter sizes from 75 mm to 150 mm. The most common diameter in domestic dwellings is 112 mm, with rainwater pipes in 2.5 m, 4 m and 5.5 m lengths.

From the guttering there is a downpipe, which takes the water from high level down to the low-level drain. The downpipe is normally 65 mm or 68 mm in diameter.

Guttering of 150 mm and over is not normally fitted to a domestic dwelling. It is common on larger industrial and commercial buildings, which have larger roof areas. This is normally known as high-capacity guttering, as the amount of water running off a large roof is considerably greater than that from an average domestic dwelling.

Figure 9.1: High capacity guttering: internal stop end

Figure 9.2 shows the different types of guttering available for buildings, including round and square sections.

Standard half-round   Square   Ogee

Figure 9.2: Types of guttering

Common gravity rainwater system materials

Gutters and rainwater pipework systems installed on new domestic properties tend to be plastic. Plastic piping has the advantages of being:

- light
- cheap
- flexible
- low maintenance.

It is available in white, black, brown and various shades of grey. Because PVC gutters and rainwater pipes have a smooth internal surface, they provide a better flow rate than other materials.

Plastic has a high expansion rate, so it can be affected by the variations of winter and summer temperatures. Allowance for expansion and contraction is made in the fittings, so it is important that manufacturers' fitting and fixing instructions are followed.

Other materials used

You are quite likely to come across cast-iron and asbestos cement guttering on maintenance jobs.

Cast iron

This is a very strong material, but it requires constant painting to prevent rusting. Leaking joints also take longer to repair/replace than their plastic counterparts, as it takes time to remove the gutter bolt fixings, which tend to be corroded and difficult to unscrew. More often than not they have to be cut off using a hacksaw.

Cast iron may still be specified on jobs where its strength is needed, or on listed buildings where original materials have to be used.

Asbestos cement

Asbestos cement gutters were quite popular before the introduction of PVC. They are no longer available because of the dangers of working with asbestos.

If you are called to a maintenance job, you must follow the procedures as outlined in Unit 1 (pages 14–15). Asbestos guttering should not be repaired but removed and replaced with plastic, with all safety procedures being followed.

> **Did you know?**
>
> Plastic gutters and rainwater pipes do not require painting, making them very low maintenance.

Figure 9.3 Guttering fittings

Progress check

1. What size guttering is usually found on larger roof areas, such as commercial buildings?
2. List the advantages of plastic guttering.
3. Why do plastic gutters offer a better flow rate than other materials?
4. When fitting plastic guttering, where is the allowance for expansion and contraction made?
5. Which materials, other than plastic, might you come across when installing and maintaining guttering?

2. Know the layout requirements of gravity rainwater systems

Factors determining type of gutter system

When installing guttering, there are several factors to consider prior to installation. These include:

- customer preference
- roof area
- outlet positions
- fall and any changes in direction.

If you are installing replacement guttering you will only need to consider some of these factors. This unit covers all the factors so that you are aware of them all.

If it is a new installation you may be working to a specification. If so, these factors will have been decided before you started work. Most manufacturers have a design department or a calculator service you can use online. To use this you will need the following information:

- Roof size, height, length and width
- Angle of the roof (normally under 25, between 25 and 45, between 45 and 60 or over 60 degrees)
- Number of outlets and their location (e.g. is there just one in the middle or are there two with one at each end?).

Customer preference

A customer may have preference for a certain types of guttering such as colour, shape and material. As the installer you will need to know, or find out from manufacturers, if this preference is available and if it is suitable to cope with the amount of water that will run off the roof. If this is not the case then you will need to advise the customer and offer an alternative explaining why they will need to use this instead.

Roof area/rainfall intensity

Areas with a high intensity of rainfall will also collect more water and require more water to be drained off.

High-capacity guttering is usually larger and is designed to cope with larger amounts of water that will run off larger roofs.

Remember

The roof area is very important as the larger the roof and the greater the pitch, the more water will run off. BS EN 12056–3:2000 uses a factor of 0.021 ltrs/sec per square metre of roof.

Outlet positioning

Another important factor is the number of downpipes and their position. If it is an existing property the positions will already be determined. You will need to ensure these can cope with the amount of water; the size of the outlet will affect the capability of the guttering so as not to overflow.

Likewise, any change of direction will also cause some resistance to the flow of water within the guttering. If guttering is fitted level, the flow rate will be less than if it is installed with a fall. The normal fall for guttering is 1:600 (1 mm fall for every 600 mm).

Purpose of components used in eaves gutters

You will encounter a number of different components when working with eaves gutters. You will need to be familiar with these components to make the correct decisions about the construction of new guttering.

Running outlets

These are used to join two lengths of guttering together and provide an outlet in the middle for a downpipe. The running outlet is normally fixed by two screws either to a fascia board or to the brickwork.

Gutter angles

These are used when a change of direction is required in the guttering. Most manufacturers produce set angles of 90° and 135° for internal or external corners.

Union brackets

Gutter unions are used to join lengths of guttering together. In most cases they incorporate a bracket for fixing the lengths together. Union brackets are sometimes also known as gutter unions. To provide extra support, these normally have two fixing points like the running outlet for screwing to either the fascia or brickwork.

Stop ends and stop end outlets

Stop ends can be either internal or external and are used at the end of the guttering to seal off the ends. Stop end outlets incorporate an outlet for a downpipe.

Fascia brackets

Fascia brackets hold the guttering in place and are screwed to the **fascia board** of the building. If there is no fascia board, other alternatives would be required such as drive-in or rafter brackets.

Basic fascia bracket

Union bracket

Angle

Running outlet

External stop end

Internal stop end

Figure 9.4: Guttering components

Figure 9.5: Fascia bracket

Brackets should be installed no more than 1 m apart unless manufacturers state otherwise. When carrying out installations, the manufacturers' recommendations should always be followed.

Specialist unions between different gutter materials

These are used when joining new plastic to existing cast iron, metal or other material used for guttering. Different manufacturers have a selection of fittings available, depending on your requirements.

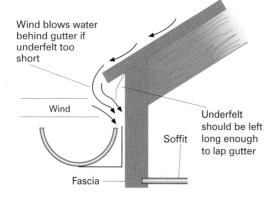

Wind blows water behind gutter if underfelt too short

Wind

Underfelt should be left long enough to lap gutter

Soffit

Fascia

Figure 9.6: Fascia fixed to rafter

Fitting rainwater systems

On new work, the gutter will be fixed to fascia boards using a fascia bracket. A fascia bracket is also shown.

On some older properties, fascia boards may not be fitted. On these jobs, metal rafter brackets similar to the ones shown in the following illustrations can be used for metal and plastic gutters.

Find out

A customer has asked you to call and give a quotation for new guttering on their dwelling. Find three different manufacturers and styles of guttering you could show them to give them a choice.

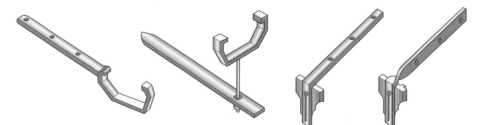

Figure 9.7: Gutter brackets

Components used in rainwater downpipes

Offsets

Offsets are used to return the outlet from the guttering back to the building to allow clipping of the downpipe. This can be purchased as a unit but it is more commonly made up using bends.

Angles and bends

These are supplied in a variety of angles enabling the downpipe to either be returned from the guttering to the building or to overcome obstructions. The ends are called socket and spigot. They need to be assembled with the socket end upper to ensure water will not leak out.

Figure 9.8: Fixed or manufactured offset

Figure 9.9: Angles and bends (note, socket at top and spigot at bottom)

Figure 9.10: Branch

The socket ends are larger than the spigot ends as shown in Figure 9.9. These can also be used to make offsets.

Branches

Branches are inserted into the downpipe to allow a connection from another downpipe. This could possibly be from a lower guttering on a porch.

Hopper heads

Hopper heads serve several purposes. They can be used as a collection box for multiple gutter downpipes to empty into and funnel these into one downpipe leading down to the ground. They can also mix air into the conductor to help eliminate back pressure or vacuum and promote heavy rain water flow through the downspouts.

Decoratively and visually hopper heads also help break up lengths of downpipes, providing architectural interest, elegance or panache. These will have holes in the back to allow fixing to the fabric of the building using two screws.

Shoes

Shoes look like the angles and bends previously described but they only have a socket end for the downpipe to be inserted. This is used at the bottom of the downpipe to direct water into a drain or divert it away from the building. Rainwater shoes should be secured to the fabric of the building using a down pipe clip.

Specialist connectors to the drainage system

A specialist connector may be used to connect the downpipe to the underground drainage pipe. It is fitted at ground level where the downpipe connects directly to the underground system. The connector in Figure 9.13 runs from square downpipe to underground drainage but you can also get them for round downpipes.

Figure 9.11: Hopper head

Figure 9.12: Shoe

Figure 9.13: Specialist connector

Find out

Search online to find out what types of specialist connectors are available and describe when you may use them.

Square to round adaptor

Clients may want square guttering with round downpipes or visa versa. If you have square guttering then the chances are the outlet will be square so you will need to be able to fit round downpipes to the guttering. To do this you will need to fit an adaptor between the guttering and the downpipe.

Clips

Clips are used to support the downpipe and keep it attached to the fabric of the dwelling. Normal spacing for these is a maximum of 1.5 m and they need to be placed where a joint occurs. These are fixed to the fabric of the building using two screws.

Figure 9.14: Clip

Working life

Teresa qualified as a plumber a few years ago. After several years of 'working on the tools', the company she worked for invited her to apply for the position of plumbing estimator. She was successful and now visits sites and individual properties measuring, advising clients and preparing estimates. She also sees the contract through by ordering materials, supervising work and preparing final accounts.

She has been asked to prepare an estimate for a replacement guttering installation at a dwelling, which is a bungalow. At the property she measures up and records the following.

- The fascia is 18 m long.
- The measurement of the return from fascia to the brickwork is 250 mm.
- There is one connection to the drainage system at ground level, which is located in the middle of the dwelling. This is connected to a rainwater-harvesting tank.
- The measurement from ground level to the bottom of the fascia board is 2.4 m.
- The existing wooden fascia is in a poor condition, showing several areas that are rotten.

What alternatives are available to replace the decaying fascia board and who should Teresa contact regarding the work? Which components will Teresa need to include? Which tools and equipment will be required to complete the installation?

Use the information discussed to produce an estimate with regard to the materials costs (excluding the new fascia boarding).

Jointing procedures for gutter systems and rainwater piping

PVC-U

Most manufacturers use a rubber seal together with a clip-in system for the jointing of the guttering, and a spigot and socket method for the downpipe as shown on page 379. This is a quick and simple system, which is one of the reasons that plastic guttering is most commonly used on domestic properties.

The rubber seal is located in the fittings and the guttering pipe is laid inside the fitting normally by placing the back edge in first and then pressing in the front. Remember to check that the rubber seal is located correctly before inserting the guttering. The guttering can be taken apart in reverse order and in most cases the rubber seals can be replaced if required.

Aluminium

Aluminium guttering is available and the fittings have the same names as those used in either plastic or cast iron. However the jointing is carried out differently from that used on plastic. Joints with aluminium are normally made by using a non-setting mastic and either rivets or nuts and bolts as shown in Figure 9.15.

Downpipes are assembled in the same way as plastic with the socket and spigot method. However, some systems use self-tapping screws to hold the fittings together.

Figure 9.15: Aluminium joint using a non-setting mastic

Cast iron

The fittings have the same names as for other types of guttering. However, most cast-iron guttering comes with a spigot end where either a rubber joint can be inserted or a non-setting mastic applied. Either way, a bolt is placed through to hold this in place.

Gutters can be cut using a hacksaw. The blade should be tungsten tipped with 50 teeth per inch. A powered saw or disc cutter can be used but health and safety must be adhered to. After cutting, edges must be sealed and this can be done by painting any bare metal using an approved product. Downpipes also have a socket and spigot system.

The sockets are normally quite large and allow for either a complete seal to be used or, in some cases, a number of wedges, normally made from lead, to stop the downpipes rattling and keep them central in the joint.

Progress check

1. Which factors need to be considered before the installation of a guttering system?
2. You are planning to use an online guttering/down pipe calculator. What information will you need?
3. What is the factor laid out in BS EN 12056–3:2000 regarding the run-off rainwater from a pitched roof?
4. What is the normal fall (ratio) of rainwater guttering?
5. What is another term sometimes used to describe a union bracket?
6. On domestic properties, how far apart should gutter brackets be (i.e. what are the maximum centres)?
7. What is a fascia? Where would it be found and what materials could it be made from?
8. On domestic properties, how far apart should downpipe clips be spaced (what are the maximum centres)?
9. How are the joints sealed and secured in place on plastic guttering?
10. How are joints sealed and secured on aluminum guttering?

3. Know, and be able to apply, the installation requirements of gravity rainwater systems

The components you will encounter when working with guttering were covered on pages 377–80, which also looked at their positioning and fixing requirements. Installation requirements are covered below.

Installing guttering

The first job is to set out the gutter brackets. Gutters can be laid level or to a slight fall. The fall should be 1 in 600, which for a length of 6 m gutter would be 10 mm. This is hardly noticeable once the gutter is fixed in position.

Once you have worked out the fall over the total length of the installation, the first bracket is fixed at the highest level on the run and the last bracket is fixed to give the amount of fall required.

A string line is fixed between the two brackets. This is used to position the rest of the brackets, by offering each bracket up to touch the string line and so ensure an adequate fall is allowed through the brackets which are usually spaced at 1 m centres unless the manufacturer states otherwise.

The gutter can then be positioned in the brackets. On the first length, depending on the roof shape, you would fit the stop end (or corner bracket) and union bracket.

The next joint would be dealt with in position.

Fixing a gutter

Step 1: Locate back edge of gutter union bracket.

Step 2: Pull front edge down until level with lip of bracket.

Step 3: Snap the gutter under the lip of the bracket.

Expansion and contraction

Make sure you allow for thermal movement. On a plastic gutter, a raised fixing mark is usually found on the inside joint. If expansion marks are not on the gutter joints, you should allow 3 mm for every metre run. Expansion takes place due to temperature differences and therefore cannot be avoided so it must be allowed for in an installation. After the guttering has expanded to its full extent and it starts to cool, it will then go back to its original length. This is known as contraction. If allowance is not made the guttering could expand and force joints to fracture.

Installing downpipes

Once the gutter system is complete, you can work on the downpipe. The brackets are marked out using a plumb line. This is dropped from a masonry nail driven into the nearest mortar joint beneath the running outlet. The brackets are centred with the plumb line and the wall marked through the fixing holes.

You usually have to install an offset (known as a swan neck), to clear the width of the fascia and soffit. The easiest way to do this is to install the top section of downpipe with the first offset on, placing the second offset against the outlet of the running outlet, and taking the measurements in position.

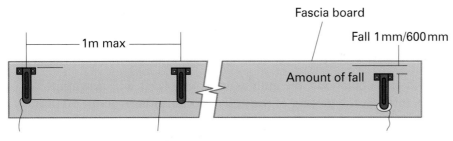

Figure 9.16: Positioning the brackets

The fall pipe is finished off at the base with a rainwater shoe or connection to the drain.

Clip spacings should be every 1.5 m and allowance should be made for thermal movement: 6–8 mm should be left from the end of the pipe to the inside shoulder of the fitting. Clips should also be placed where pipes are jointed.

All screws used for fixing should be alloy or stainless steel so that they do not rust.

Cast-iron downpipes and gutters should be cut using an angle grinder.

Methods for making new rainwater pipework connections to the drainage system

The connection of the downpipe to the underground system can be made using a number of different methods. It is either connected directly and sealed or it can be un-sealed if using a gully. If un-sealed then care needs to be taken to ensure that water cannot leak elsewhere. Leaking can cause problems, such as freezing water on paths in winter, which create a slip hazard.

Discharge to gully using a shoe

The gully finishes at ground level. The downpipe will be run to low level just above the gully and terminate with a shoe. The shoe will throw

Remember

Ensure that the gully has a surround to catch the water and direct it down into the gully.

water away from the building. It is good practice to ensure that the gully has a surround to catch the water and direct it down into the gully. Without this, some water may shoot out onto the surrounding area. This method is not recommended for rainwater harvesting systems and is normally used for soakaway and mains drainage systems.

Discharge connection to a drainage bend

Again the drainage bend or pipe will terminate at ground level and could go to the main drainage, rainwater harvesting or a soakaway system. Any of these will then require a specialist connector (as shown on page 379). The type of connector used will depend on the type of downpipe used and the size of the underground pipe. This will allow the downpipe to be sealed to the underground system without the risk of any leakage from the system.

Direct connection to the gulley

As with all installations, the gully is at ground level. The downpipe passes through the grating of the gully and terminates about 50 mm inside. This reduces the chance of any water shooting out. The downpipe is not sealed to the underground system so there is still a possibility of water overflowing out of the gully should it get blocked. Again, this is not a recommended method for rainwater harvesting systems but is acceptable for all others.

Working life

Josh is a Level 2 apprentice who has been sent to carry out the replacement of the rainwater system to a two storey semi-detached house. He notices that the existing guttering and downpipes are made from asbestos cement. His employer had not been informed of this. He also notices that the gutter is shared between the two properties (semi-detached), with one central downpipe in the middle of them.

Josh has also been told that he must get the job finished that day. He is needed on a new contract starting the next day, which will last for several weeks.

What should Josh do? His employer has pressurised him to get the job competed urgently. Is there a procedure that should be adopted? If so, what would this entail?

When the work can be started, will both properties have to have new plastic guttering system fitted due the difference between the profile of plastic and asbestos cement?

Should Josh replace the adjoining property's gutter anyway, without the residents being asked?

Progress check

1. What is the purpose of a string line, which is fastened to the first and last fascia brackets when installing a rainwater gutter?

2. In a plastic guttering system, what is the allowance for expansion per 1 metre run?

3. What is another term sometimes used to refer to an 'offset'?

4. What allowance for expansion/contraction should be made in the joints of a rainwater downpipe?

5. Screws are used to secure rainwater guttering downpipes. What should these screws be made from?

4 & 5. Know, and be able to apply, the service and maintenance, inspection and testing requirements for domestic rainwater systems

Domestic rainwater systems share many of the same requirements as above-ground drainage systems. More information on these can be found in Unit 10 on pages 389–436.

Issues relating directly to rainwater systems are covered below.

Basic maintenance and cleaning

Access covers to soil and vent pipes should be checked to make sure they are operating properly, and a visual inspection should be made of waste traps and fittings for signs of leakage.

Blocked pipes and drains

In the event of a more serious blockage a drain auger may be used (see Figure 9.19). Simple drain augers may be manually operated. More sophisticated devices, primarily for soil and drain runs, may be electrically powered. If there is a serious blockage, high pressure jetting of the system may be undertaken. This job is usually carried out by a specialist contractor who has been fully trained in using the equipment required. The equipment involved is expensive.

Gutters should be checked regularly and cleaned as required. They collect silt residue from roof tiles, leaves and bird droppings. The silt can build up and becomes a garden for seeds carried by the wind. Plastic gutters can be stripped in lengths between joints, emptied into a bucket and wiped clean. Cast-iron gutters should be similarly cleaned out and then painted to prevent rusting.

Leaking joints on cast-iron stacks can be cleaned out and re-made using non-setting pipe-jointing compounds. Leaking cast-iron gutter joints can be repaired by removing the old joint, thoroughly cleaning the jointing surfaces of the gutter and fitting, applying jointing compound, and fitting a new gutter bolt.

To test an installed rainwater system, you normally pour water into the guttering and check all joints for leaks; ensuring all the water flows away down the outlets. Common causes of plastic guttering leaking are rubber sealing strips becoming dislodged. Downpipe joints will also leak if installed the wrong way round.

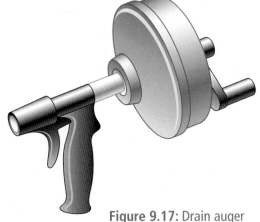

Figure 9.17: Drain auger

Basic maintenance and cleaning

Gutters and downpipes need to be kept clear of any blockage. Otherwise, damage could occur to the fabric of the dwelling. This will depend on the location of the dwelling in relation to trees and so on.

Maintenance should consist of:

- periodic internal cleaning of guttering
- flushing of downpipes
- a visual inspection of all joints.

Where there are signs of leakage the joint should be taken apart and checked. If you are unsure then the component should be replaced. Plastic guttering can be washed with warm soapy water and cast-iron gutters should be painted to prevent rusting.

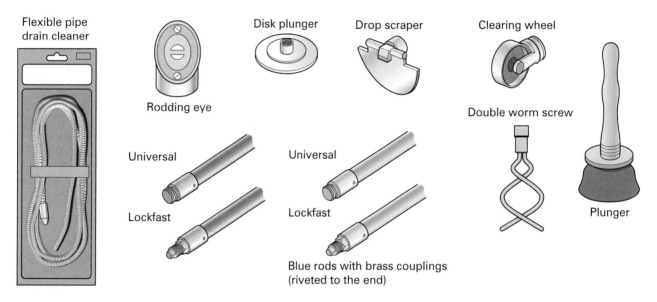

Flexible pipe drain cleaner

Disk plunger

Drop scraper

Clearing wheel

Rodding eye

Double worm screw

Universal

Universal

Lockfast

Lockfast

Blue rods with brass couplings (riveted to the end)

Plunger

Figure 9.18: Equipment for unblocking pipes and drains

Check your knowledge

1. Which of the following is a type of plastic guttering?
   a Half-round
   b Half channel
   c Elsan
   d Mansard

2. It there is no fascia board on the dwelling, what should be used to install guttering?
   a Purlin brackets
   b Stand-off brackets
   c Rafter brackets
   d Eaves brackets

3. What is the maximum distance between brackets on a plastic gutter system?
   a 900 mm
   b 1000 mm
   c 1100 mm
   d 1500 mm

4. What is the recommended fall to which plastic guttering should be installed?
   a 1:600
   b 1:1200
   c Fit underneath the tiles, very tightly
   d Fit from the top to the bottom of the fascia

5. What is the allowance to allow for expansion, that should be included per metre run of guttering?
   a 1 mm
   b 2 mm
   c 3 mm
   d 4 mm

6. Where would a stop end outlet be fitted on a rainwater guttering system?
   a In the downpipe
   b In the middle of the guttering
   c At ground level
   d At the end of the guttering

7. If a rubber jointing seal has dried out what would be used to lubricate it?
   a Oil
   b Grease
   c Silicone lubricant
   d Washing up liquid

8. What would determine the maintenance frequency of a guttering system?
   a The height of the dwelling
   b The age of the guttering
   c The type of guttering
   d The location of the property

9. What is the name of the pipe that connects the guttering to the underground drainage system?
   a Downpipe
   b Up-pipe
   c Connector pipe
   d Water outlet pipe

10. What is the recommended spacing for downpipe brackets?
    a 1600 mm
    b 1800 mm
    c 3000 mm
    d 1500 mm

11. Which of the following is a type of gutter fitting?
    a Boot
    b Shoe
    c Slipper
    d Trainer

Getting ready for assessment

The information contained in this unit of the book, as well as the continued practical assignments that you will carry out in your college or training centre, will help you with preparing for both your end-of-unit test and the diploma multiple-choice test. It will also support you in preparing for the practical assignments you will need to complete to demonstrate your understanding in and enabling you to carry out and apply domestic rainwater system installation and maintenance techniques.

There are opportunities throughout the unit for you to test your progress in and understanding of the required underpinning knowledge; this will enhance your preparation for the forthcoming assessments, so make good use of them.

This unit will be assessed by the following assessment methods:

- externally set knowledge assessment
- externally set assignments.

With regard to the mechanical services industry, you will need to know:

- the general principles and layout requirements of gravity rainwater systems
- the site preparation techniques of gravity rainwater systems and how to apply them
- the installation requirements of gravity rainwater systems and how to install them
- the service and maintenance requirements of gravity rainwater systems and be able to apply them

- the inspection and testing requirements of gravity rainwater systems and be able to apply them.

Check it out – using your knowledge from this unit practice and test yourself on the following key terms:

1:600, 3 mm per metre, angles, angles, asbestos cement, blockages, branches, British standards, cast iron, Codes of practice, commissioning/decommissioning, components, connection to drainage system, downpipe, expansion gap, expansion/contraction, exposed rafters, extruded aluminium, fascia brackets, gutter fall, half round, hazards, high capacity, hopper head, industry standards, installation, jointing, maintenance, manufactures instructions, masonry fixing brackets, offsets, ogee, positioning of outlets, PPE, preparatory work, protection of customers property, purpose, PVC-u, rainfall intensity, rainwater damage to buildings, rainwater harvesting, rainwater shoe, remedial work, replacement/renewal defects, risk assessment, roof area, round downpipe, running outlets, safe access/exit, safe handling of materials which may be contaminated, soakaways, soundness testing, specialist connectors, specialist unions, square downpipe, square gutter, statutory regulations, stop ends, stop-end outlets, tools/equipment, unions, visual inspection and working principles.

Good luck!

Understand and apply domestic above-ground drainage system installation and maintenance techniques

Above-ground discharge systems (AGDS) include sanitary appliances, sanitary pipework and fittings, and rainwater systems. This unit covers the various appliances on the market, looking at how to fit them and pipe them up, and how they are tested and maintained. AGDS are essential to keep our environment clean and hygienic.

This unit will cover the following learning outcomes:

- Know the uses of sanitary appliances and their operating principles

- Know the types of sanitary pipework system and system layout requirements

- Know, and be able to apply, the site preparation techniques for sanitary appliances and connecting pipework systems

- Know, and be able to apply, the installation requirements of sanitary appliances and connecting pipework systems

- Know, and be able to apply, the service and maintenance requirements of sanitary appliances and connecting pipework systems

- Know, and be able to apply, the decommissioning requirements of sanitary appliances and connecting pipework systems

- Know, and be able to apply, the inspection and soundness testing requirements of sanitary appliances and connecting pipework systems

1. Know the uses of sanitary appliances and their operating principles

Sanitary appliances can be divided into two main groups: those used for washing purposes and those used for the removal of human waste.

Sanitary appliances play a very important role in our lives: they mean we can keep ourselves and our eating utensils clean, and we can use the toilet in privacy and comfort.

Although this unit focuses on domestic installations, it also covers urinals and flushing cisterns, which you may come across when working in public toilets or larger buildings.

British Standards and Regulations

British Standards

British Standard 8000 is the Code of Practice covering workmanship on building sites. Part 13 covers above-ground drainage and sanitary appliances, including:

- materials handling
- site storage of components
- the preparation of work, materials and components
- the installation of sanitary appliances
- inspection and testing.

BS 8313 gives guidance on the holes, chases and ducts required for pipework.

Water Regulations

Water Regulations do not refer specifically to sanitary appliances. However, they do deal with how sanitary appliances perform, by covering undue consumption, waste of water and protection against backflow. While you will not study the Water Regulations in great detail until later in your qualification, you do need to have an understanding of the requirements at this stage.

The regulations governing WC cisterns, urinals and automatic flushing cisterns are limited to the capacity of water for flushing.

Backflow prevention is achieved either through mechanical methods (using valves) or non-mechanical means (using air gaps between outlet and appliance water levels) to prevent the possibility of water being drawn back into the supply pipework and into the main. This could occur if an open tap outlet were allowed to become submerged in a sink or bath, and if the main outside stop valve were turned off and water drained from the system.

Remember

Backflow prevention has a bearing on sanitary appliances, which are categorised by the Water Regulations in terms of risk. This covers WC suites, washbasins, bidets, baths, shower trays and sinks.

Siphonic action uses differences in the atmospheric pressure in two parts of a pipe to draw water through it. Refer to Unit 4 if you cannot remember how this works.

As the water drained down from the system, water from the sink or bath would be drawn into the pipework due to siphonic action. Water Regulations will be covered in greater detail at Level 3.

Working principles of sanitary appliances

A vast range of styles and designs of sanitary appliances is available. However, British Standards ensure that there is standardisation in terms of dimensions, including the size of waste outlets.

Water closets and cisterns

There are many different types and designs of water closets (WCs) to choose from. They are categorised as:

- back-to-the-wall
- close-coupled
- low-level
- high-level
- concealed.

In the main, WCs are manufactured from **vitreous china** conforming to BS 3402. Vitreous china is made from a mixture of white burning clays and finely ground materials. These are fired at high temperatures, and even before glazing the material cannot be contaminated by bacteria, so it is totally hygienic.

Vitreous china is coated with an impervious non-crazing vitreous glaze in either a white or coloured finish. The material is stain proof, burn proof, rot proof, rust free, non-fading and resistant to acids and alkalis.

WCs used in public places may also be manufactured in stainless steel. This material is more resistant to vandalism than vitreous china.

WC cisterns

Prior to 1993, the capacity of a WC flushing cistern was 9 litres. The Water Regulations brought this down to 7.5 litres and, from January 2001, reduced it further to 6 litres. In addition, Water Regulations also permit the use of dual-flush cisterns, which deliver 6 litres for a full flush, and 4 litres for a lesser flush. The dual-flush cistern is now specified more often that the single flush. To meet the requirements of the Water Regulations, a dual flush should have a label on it, stating that it is a dual-flush cistern.

There are two types of cistern:

1. Siphonic
2. Dual-flush valve (often known as a drop valve).

With the siphon type (see Figure 10.1), when the lever is pressed the water in the bell of the siphon (1) is lifted by a disc (or diaphragm) up

> **Remember**
>
> Water can also be drawn back into the supply pipework and into the main through other causes. This could include a drop of pressure, including bursts on the water main or from peak demand periods.

> **Key term**
>
> **Vitreous china** – produced from a solution called slip, or casting clay. Slip has the consistency of pouring cream and contains ball clay, china clay, sand, fixing agent and water.

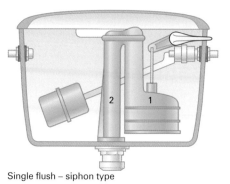
Single flush – siphon type

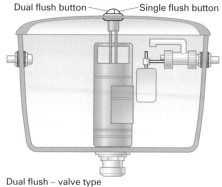
Dual flush button — Single flush button

Dual flush – valve type

Figure 10.1: Different types of cistern

and over into the leg of the siphon (2). This creates the siphonic effect, which continues until the water level in the cistern has dropped to a level which allows air to enter the bell.

The dual-flush valve-type cistern is operated by pressing either the full flush or lesser flush buttons, which have to be clearly marked on the cistern. These operate a valve that releases the water into the WC pan.

Another type of fitting (see Figure 10.2) is also available for 6-litre-capacity cisterns. This works on the principle of lifting a hinged flap and is suitable for full flush only.

Cisterns of 9 litres and 7.5 litres are still available. Water Regulations permit these to be installed as replacements for existing cisterns, because in these cases, the WC pan was designed to work on those capacities.

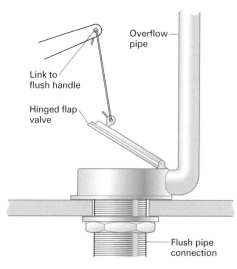
Overflow pipe

Link to flush handle

Hinged flap valve

Flush pipe connection

Figure 10.2: Cistern fitted with hinged flap mechanism

The overflow pipe can now discharge into the WC pan using an integral overflow, eliminating the need to provide an external overflow pipe. This pattern of cistern is now supplied with most WCs.

WC pans

There are a number of designs for WC pans, but there are only two main types:

1. Wash-down WC pan

2. Siphonic pan.

The wash-down pan, which tends to be the most common type, uses the force of the water from the cistern to clear the bowl. The siphonic pan creates a negative pressure below the trap seal.

Flush pipe

Figure 10.3: Wash-down pan

With the single-trap siphonic pan, this is done by restricting the flow from the cistern and is achieved by the design of the pan.

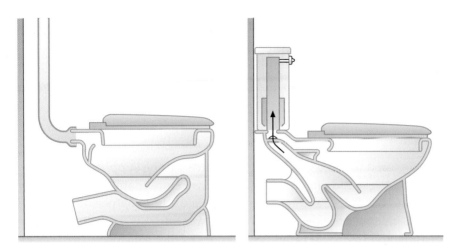

Figure 10.4: Single- and double-trap siphonic pan

Safety tip

When handling used sanitary appliances, suitable PPE should be used.

The double-trap close-coupled pan uses a pressure-reducing device between the cistern and the pan. As the water is released into the second trap, it has the effect of drawing air from the void between the two traps, and siphons the content from the bowl.

Figure 10.6 shows what the pressure-reducing valve looks like between the cistern and pan.

Popular WC types include:

- **concealed** – where the cistern is hidden
- **close-coupled** – where cistern and pan are joined in one unit
- **low-level** – where the cistern is no more than 1 metre above the pan
- **high-level** – where the cistern is more than 1 metre above the pan necessitating flushing by a chain.

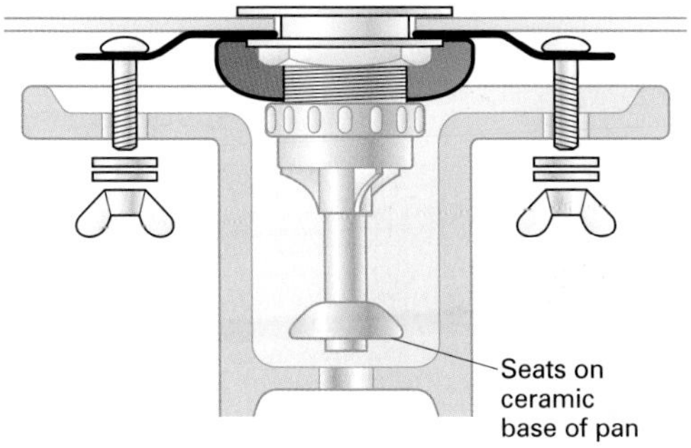

Seats on ceramic base of pan

Figure 10.5: Typical pressure-reducing valve between cistern and pan

The choice of WC suite depends on a number of factors:

- **Cost** – the customer's budget
- **Location** – for public toilets, factors such as durability and ease of cleaning are a consideration. For example, stainless steel units are often used because of their resistance to vandalism
- **Aesthetics** (how good it looks) – Victorian versions of the high-level cistern are now considered very desirable.

On larger housing contracts, or in public buildings, the choice of WC may already have been made by the client's architect as part of the contract specification document.

Most WC pans manufactured today are 'P' trap and use either a straight or bent pan connector depending on the soil stack location.

Some cisterns are reversible – that is, the handles, overflow outlet and water inlets can be either left- or right-handed. In some cases, the cistern is plastic rather than vitreous china.

Kits are usually supplied for the soil outlets, and will be either a 'P' trap or an 'S' trap.

Baths

Manufactured to BS 4305 (EN 198), baths can be supplied in acrylic sheet, heavy-gauged enamelled steel and vitreous enamelled cast iron. Victorian-style free-standing designs give a traditional 'period look'. Vitreous enamel provides a smooth hardwearing surface that resists corrosion and is easy to clean.

Baths come in a huge variety of styles, and taps and waste can be either end-mounted or centre-mounted. Taps can even be mounted on risers from the floor, external to the bath itself. Space-saving styles include corner baths and compact baths.

More expensive options include enamelled baths manufactured from heavy-gauged porcelain enamelled 3.5 mm thick onto heavy gauge steel. Steel sheet is pressed into shape and then coated in layers of vitreous enamel, which is fired at high temperatures. The enamel and steel bond made during firing gives the enamel a glass-like high-gloss finish. This type of bath is particularly suitable for installation in hotels, or in housing association and local-authority housing where durability is important.

The jet-system bath is a popular choice in the 'luxury' market. It is like a standard bath but contains a number of jets (generally eight) in the side of the bath to provide a massage effect. In some cases additional outlets are sited in the floor for foot massages.

The water is circulated through pipework located around the outside of the bath and forced through the jet nozzles by a pump.

Great care needs to be taken with the electrical connections to this type of bath as the bath is a high risk-appliance containing water. The

Did you know?

Acrylic sheet baths are vacuum-formed to the desired shape.

Figure 10.6: Combined waste and overflow fitting

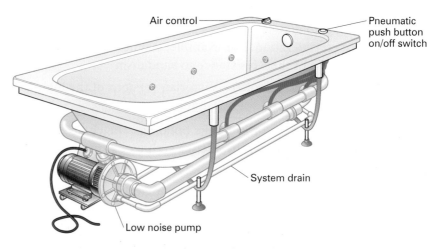

Air control — — Pneumatic push button on/off switch

System drain

Low noise pump

Figure 10.7: Typical jet-system bath

electrical safety precautions that must be employed are significantly higher. A qualified electrician will usually deal with the connection of such appliances.

Baths are manufactured with two holes for pillar taps, or single holes for monobloc fixings. Most come with an overflow hole, and waste hole that will take a 380 mm (1½") threaded waste. Combined overflows and waste fittings are also provided, as shown in Figure 10.6.

Bidets

Made from vitreous china, there are two types of bidet:

1. Over-the-rim supply
2. Ascending spray.

The ascending spray bidet has to be piped up correctly to avoid risk of contamination to the supply. It cannot be used on the mains supply, such as with combination boilers and unvented hot-water systems. This is a requirement of the Water Regulations. The bidet can be supplied for use with pillar taps, monobloc fittings and pop-up waste.

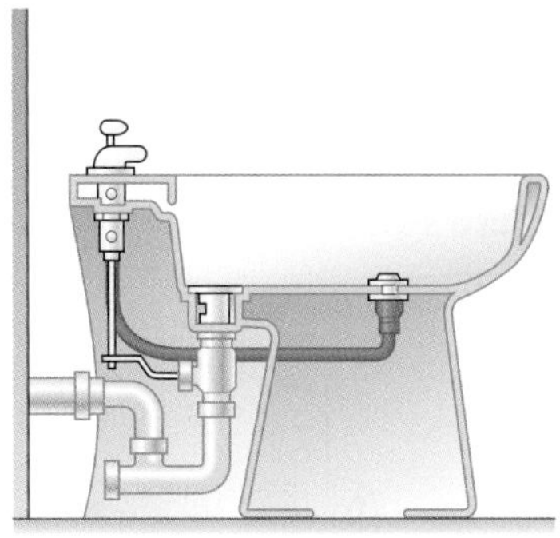

An over-the-rim bidet presents less of a water contamination risk and may be connected to supply pipework fed directly from the mains, provided that the Type AUK2 tap gap (non-mechanical) is maintained above the spill-over level of the sanitary appliance.

If the over-the-rim bidet includes a flexible hose, it should be treated as though it is an ascending spray type, because the flexible hose and spray head may be dropped into the contaminated water in the bidet bowl. An ascending spray bidet or an over-the-rim bidet with flexible hose must be fed from storage and must have its own dedicated piping system in order to avoid potentially serious contamination of the water supply.

Figure 10.8: Ascending spray bidet

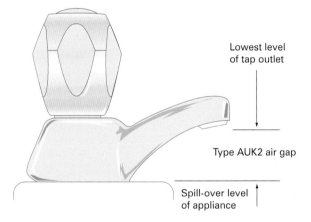

Figure 10.9: Air gap requirement for over-the-rim bidet

Washbasins

There are two main types of washbasin:

1. Fixed to the wall
2. Countertop basins.

All basins are available with either single tap holes for a monobloc mixer tap, or dual tap holes for pillar taps. Basins are also available with three holes for fittings with an independent spout.

Fixed to the wall

These are secured by using brackets or are supported by a pedestal. Larger washbasins, fixed using brackets, are mostly found in

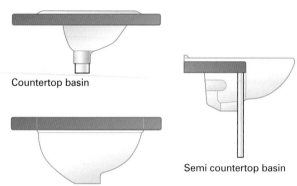

Countertop basin

Under countertop basin

Semi countertop basin

Figure 10.10: Countertop

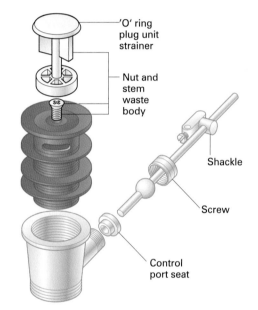

'O' ring
plug unit
strainer

Nut and
stem
waste
body

Shackle

Screw

Control
port seat

Figure 10.11: Pop-up waste fitting

Figure 10.12: Standard 32 mm slotted basin waste, plug and chain

non-domestic installations. The pedestal type is probably the most popular choice in homes. Appearance is the main factor as all the pipework and fittings can be seen with the bracket-mounted basin.

Countertop basins

There are three types of countertop basin:

1. Countertop – sits proud of the work surface
2. Semi-countertop – used where space is tight as the front of the basin projects clear of the top
3. Under the countertop – has the flange surface on top of the basin.

All basins are available with either single tap holes for a monobloc mixer tap, or dual tap holes for pillar taps. Basins are also available with three holes for fittings with an independent spout.

Countertops are installed in all sorts of buildings, from homes to pubs and restaurants. They are manufactured from either vitreous china or a high-impact plastic called acetyl.

The basin outlet is designed to take a 32 mm (1¼-inch) slotted waste fitting. Waste fittings can be the standard slotted waste with a plug and chain or a pop-up type, which is operated by lifting and pressing a knob on the top of the basin.

Hand-rinse basins

Called hand-rinse basins due to their size, these are used in cloakrooms with WCs. They can be a corner design or traditional design, either countertop, pedestal or bracket-mounted.

There are three types:

1. Standard countertop – this fits into a pre-cut hole and is secured by fixing brackets that secure the underside of the basin to the countertop; the joint is sealed using waterproof sealant
2. Semi-countertop – this can be made from vitreous china or acetyl
3. Under countertop – this appliance is made from acetyl.

Shower trays

Shower appliances were covered in Units 6 and 7. This section concentrates on shower trays.

As with other sanitary appliances, there are a number of designs. Trays are available in various materials: reinforced cast acrylic sheet, fire clay

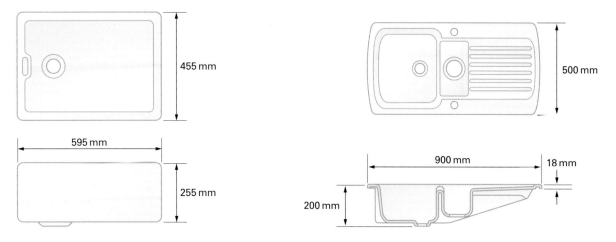

Figure 10.13: Section through a Belfast sink   **Figure 10.14:** Section through a modern acetyl plastic kitchen sink

for heavy-duty applications or resin-bonded. The outlet is designed to take a 38 mm (1½-inch) threaded waste fitting. Corner trays are used where space is limited.

Sinks

Belfast sinks and London sinks

Belfast sinks are used in both domestic and commercial situations. They are usually fed by bib taps but can also be supplied by pillar taps or monobloc taps in domestic installations.

London sinks are similar but without an overflow. Both types are made of heavy-duty fire clay, which is manufactured in a similar way to vitreous china. They are available in white only and are designed to take a 38 mm (1½-inch) threaded waste.

Kitchen sinks

A vast range of designs is available. Sinks are produced in stainless steel, plastic-coated pressed steel, fire clay and plastic in the form of acetyl. They can be single drainer, double drainer, single with basket (suitable for connection to a small waste-disposal unit), or double sink with single drainer. They are designed to take a 38 mm (1½-inch) slotted waste fitting. This can be slotted for use with a sink overflow, or plain. Taps can be monobloc or pillar.

Urinals

Urinals are normally installed in non-domestic buildings, such as public toilets and schools. They can be manufactured from vitreous china, plastic, stainless steel or fire clay. Urinals can be single or multiple bowl, and of trough or slab type, but all will need to have an automatic flushing cistern.

> **Safety tip**
>
> Care should be taken when handling stainless steel sinks as the edges are razor sharp.

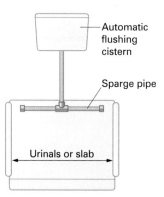

Figure 10.15: A slab-type urinal

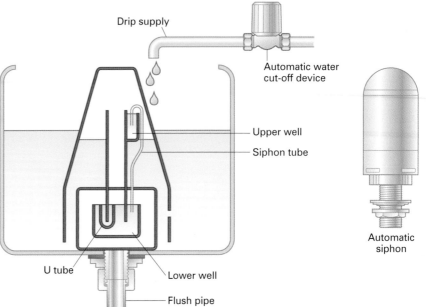

Figure 10.16: Automatic flushing cistern

Urinals are fitted with an automatic flushing cistern. Its automatic siphon operation is very simple:

- As the cistern fills, and the water rises, the air inside the dome of the automatic siphon is compressed.
- The increased pressure forces water out of the U tube, which reduces the pressure in the dome.
- The reduction in pressure causes siphonic action to take place, flushing the cistern.
- When the cistern has emptied, the water in the upper well is siphoned into the lower well.

Water Regulations state that auto-flushing cisterns must not exceed the following maximum volumes:

- 10 litres per hour for a single urinal bowl, or stall
- 7.5 litres per hour, per urinal position, for a cistern servicing two or more urinal bowls, urinal stalls or per 700 mm of slab.

The flow rate can be achieved using urinal flush-control valves, which allow a small amount of water to pass into the system.

Timed flow control valves can also be used. These have the additional advantage of being switched off when the building is not in use (e.g. evenings, weekends, factory shutdowns or school holidays). They can be set for 'one-off' hygiene flushes, such as once every 24 hours. Many public buildings now control the supply of water to the automatic flushing cistern using infra-red sensors that detect motion (similar to those used in some burglar alarms). When motion is detected, a solenoid valve is opened permitting a flow of water to the automatic flushing cistern.

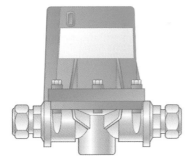

Figure 10.17: Hydraulic flush-control valve

Figure 10.19 shows a typical example of a urinal set-up with individual wall-fixed urinals (bowls). These are made of vitreous china. The picture also shows the cistern and flush-pipe arrangements. On slab urinals, the flush pipe is often mounted on the face of the urinal and has a number of holes, which are used to wash the face of the slab and the channel. These are called **sparge pipes**.

Working principles of sanitary appliances

The main working principles that you need to consider with regard to sanitary appliances apply, on the whole, to WC pans and cisterns.

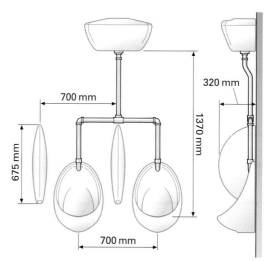

Figure 10.18: Typical two-bowl urinal layout

Some WC pans clear their waste by the wash-down action of the water as it is released into the bowl. Other designs use siphonic action: siphonic WC pans, siphons in cisterns and automatic siphons. The scientific aspect of siphonage is covered in Unit 4 (pages 175–76), and pages 392–393 looked at how siphonage is used.

> ### Key term
>
> **Sparge pipe** – a flush pipe mounted on the face of a urinal connected to a number of holes (urinal spreaders), used to wash the face of the slab and channel.

Progress check

1. According to the current Water Regulations, what is maximum capacity of a WC flushing cistern?
2. Why is it still permitted to buy and install WC cisterns with the older regulation capacities of 7.5 and 9 litres?
3. How are WC soil outlets referred to regarding their 'trapping' arrangements?
4. Which two types of bidet are available?
5. What are the two main types of washbasin?
6. Generally, why is a slotted waste fitting required with an appliance such as washbasin?
7. What is the difference between a hand-rinse basin and a washbasin?
8. What is the difference between the sink patterns known as Belfast and London?
9. How is the water supply to urinals controlled?
10. What is the maximum volume of water that the Water Regulations allow for an automatic flushing cistern, which should not be exceeded for a single urinal bowl or stall?
11. What is a sparge pipe?

2. Know the types of sanitary pipework system and system layout requirements

Sanitary pipework systems

At one time, pipework systems were two-pipe; you may still come across this type of system on maintenance jobs. In a two-pipe system, the pipework from the WC was separate from that of the washbasin, bath and sink, and the wastewater only joined together once it entered the drain (see Figure 10.20).

Today, all pipework systems used on domestic dwellings are one-pipe systems, of which there are four types:

1. Primary ventilated stack system
2. Ventilated discharge branch system
3. Secondary modified ventilated stack system
4. Stub stack system.

Soil and vent pipe systems can be installed inside or outside the building. Whichever system is used, the design and installation must comply with Building Regulations Part H1. Generally speaking, Part H1 states that the foul water system must:

- convey the flow of water to a foul water outfall (a foul or combined sewer, a cesspool, septic tank or settlement tank)
- minimise the risk of blockage or leakage

> **Safety tip**
>
> Remember to wear the correct PPE when handling used sanitary pipework.

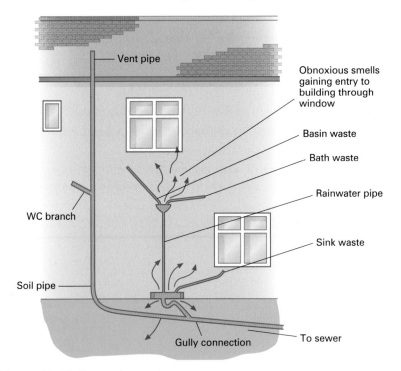

Figure 10.19: External two-pipe system

- prevent foul air from the drainage system from entering a building under working conditions
- be ventilated
- be accessible for cleaning blockages.

The other important document regarding sanitary pipework is BS EN 12056 Part 2. This gives guidance on the minimum standards of work and materials to be used.

The primary ventilated stack system is the most commonly used.

The main aim of the following design principles is to prevent the removal of the trap seal, since loss of this seal would allow obnoxious smells to enter the building.

Primary ventilated stack system

Sometimes called the single-stack system, this type of system is often specified for domestic dwellings because it does not need a separate ventilating pipe like other one-pipe systems. This means that it costs less, in terms of materials and installation time, which is good for both business and the environment.

Building Regulations Part H1 sets out a number of rules about the design of this system. The rules governing pipe diameter and the minimum depth from the lowest connection above the invert of the drain of 450 mm are shown in Figure 10.20.

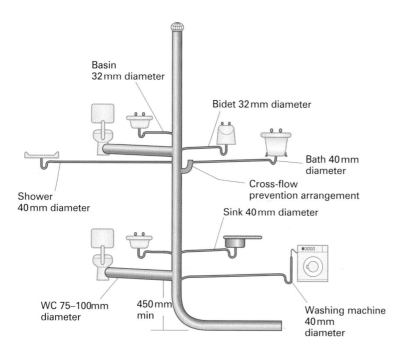

Basin 32 mm diameter

Bidet 32 mm diameter

Bath 40 mm diameter

Cross-flow prevention arrangement

Shower 40 mm diameter

Sink 40 mm diameter

WC 75–100 mm diameter

450 mm min

Washing machine 40 mm diameter

Figure 10.20: Primary ventilated stack, with full range of domestic appliances

In addition, there are limits to the maximum lengths of the branch connections and their gradients. These are shown in Table 10.1.

	Pipe size (mm)	Maximum length (m)	Minimum slope (mm fall per metre run)
Basin	32	1.7	18–20
Bath	40	3.0	18–90
Shower	40	3.0	18–90
WC	100	6.0	18

Table 10.1: Limits to length and slope of pipes

In the primary ventilated stack system it is necessary to have the appliances grouped closely together. There is some flexibility, however: for example, if you did install a shower with a 50-mm waste fitting it could be located up to 4 m away from the stack, as opposed to 3 m if using 40 mm pipe. The size of the branch pipes should always be at least the same diameter as the trap. The maximum pipework lengths and gradients may be exceeded but only when additional ventilation is provided within the system using a device such as an anti-siphon trap or a self-sealing valve.

Branch connections

The location of a branch pipe in a stack should not cause **cross-flow** into another branch pipe. Cross-flow can be prevented by working to the details shown in Figure 10.21. This will prevent trap seal loss.

It is, however, permissible to have connections from two WCs in opposing positions.

You may find that on some installations it is easier to run the kitchen-sink waste pipe into a gully rather than a pipe to the stack. This is permissible as long as the pipe end finishes between the grating or sealing plate and the top of the water seal.

Where these conditions cannot be met, separate ventilation will need to be installed.

This can be done in two ways:

1. Ventilating each appliance into a second stack – the ventilated discharge branch system
2. Directly ventilating the waste stack – secondary ventilated stack system.

The trap sizes and seals apply as described in the section on traps on page 403. Figure 10.22 shows what this looks like.

Branch ventilating pipes

A way of overcoming excessively long pipe lengths or steep gradients is to ventilate the pipework system using separate ventilating pipework. This does not happen on many new installations, as anti-siphon traps

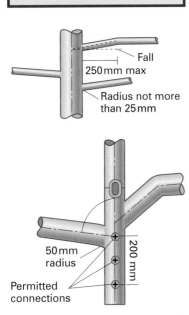

Fall
250 mm max
Radius not more than 25 mm

50 mm radius

200 mm

Permitted connections

Figure 10.21: Cross-flow

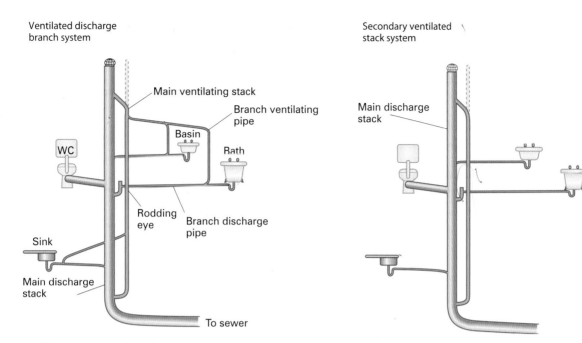

Figure 10.22: Ventilated discharge system and secondary stack system

or self-sealing valves are now commonly used as a modern alternative, reducing material costs and installation times.

There are some factors to consider:

- The branch vent pipe must not be connected to the discharge stack below the spill-over level of the highest fitting served.
- The minimum size of vent pipe to a single appliance should be 25 mm. If it is longer than a 15-m run or serves more than one appliance, then it must be 32 mm minimum.
- The main venting stack should be at least 75 mm. This also applies to the 'dry part' of the primary vented stack system.

Stub stacks

These reduce the amount of ventilating pipework on a ventilated discharge stack. They also avoid the need for weatherings on internal stacks. The highest waste connection allowed is 2 m above the invert of the drain and 1.3 m to the base of the WC. The length of the branch drain from the stack is 6 m for a single appliance connection and 12 m if there is more than one. The stub stack is commonly used to connect the ground floor appliances in a property containing more than one bathroom, where the high point in the drainage run includes a ventilating pipe to fully ventilate the drainage system.

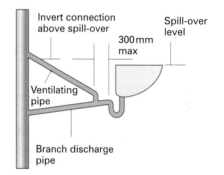

Figure 10.23: Branch ventilating pipes

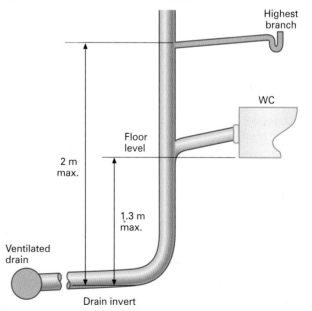

Figure 10.24: The stub stack system

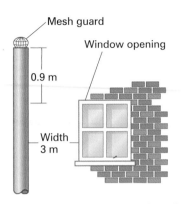

Figure 10.25: An external stack must be terminated as shown

Figure 10.26: A typical air admittance valve

These dimensions can be exceeded when an air admittance valve is used to allow air to enter the system. The valve, which operates on negative pressure (the open valve lets air in) and positive pressure (the closed valve contains the smells), can be located in the roof space or the pipe-boxing arrangement to the WC, subject to certain requirements.

General discharge stack requirements

All stacks should have access for cleaning and clearing blockages. Rodding points and access fittings should be placed to give access to any length of pipe that cannot be reached from any other part of the system. All systems pipework should be easy to get to in case of repair.

Figure 10.27: A terminal mesh guard should be fitted to prevent the possibility of birds nesting

Figure 10.28: A vent cowl could be fitted where the stack is sited in exposed windy conditions

Traps and trap seal loss

Traps are used on above-ground discharge system pipework and appliances. They are mainly manufactured in plastic (polypropylene to BS EN 274), although they are also available in brass for use on copper pipework, where a more robust installation is required, and can be chromium plated to provide a pleasing appearance. Table 10.2 shows minimum trap sizes and seal depths.

Find out

Why do you think it is necessary to install a trap?

Appliance	Diameter of trap (mm)	Depth of seal (mm of water or equivalent)
Washbasin[1] Bidet	32	75
Bath[2] Shower[2]	40	50
Food waste disposal unit Urinal bowl Sink Washing machine[3] Dishwashing machine[3]	40	75
WC pan – outlet <80 mm WC pan – outlet >80 mm	75 100	50 50

1 The depth of seal may be reduced to 50 mm only with flush grated wastes without plugs on spray tap basins.
2 Where these appliances discharge directly to a gully, the depth of seal may be reduced to no less than 38 mm.
3 Traps used on appliances with flat bottom (trailing waste discharge) and discharging to a gully with a grating may have a reduced water seal of no less than 38 mm.

Table 10.2: Minimum trap sizes and seal depths

The purpose of a trap is to retain a 'plug' of water to prevent foul air from the sanitation and drainage pipework entering the room. The depth of this 'plug' of water will depend on where it is to be used. Part H of the Building Regulations includes a table detailing the minimum trap diameter and hence connecting pipe diameter for a range of sanitary appliances. It also includes the requirements for minimum depth of seal.

You must ensure that the minimum trap seal depths are maintained. If you don't follow these requirements then noxious smells and unsanitary conditions may be created in the building.

The depth of the trap seal is measured as shown in Figure 10.30.

P and S traps

You can see why these are called P or S traps from their shape (note that they are turned in Figure 10.29 to show how they fit into the appliance). They can be bought in tubular design, or with a joint connection, which allows a few more options when installing pipework and fittings. The traps can also include a cleaning eye for cleaning and maintenance.

P traps are often used where the waste pipe is installed directly through a wall from the appliance and into a drain or directly into a stack. The S trap is used where the pipe has to go vertically from the trap through a floor or into another horizontal waste pipe from another appliance (with other constraints being used to avoid induced siphonage, of course). P traps and bottle traps can be converted to S traps using swivel elbows like the one shown in Figure 10.31. S traps can be a real problem as the fall of pipe from the basin is generally too steep, which can lead to trap seal loss and obnoxious smells entering the building.

Bottle traps

Bottle traps are often used because of their neat appearance and because they are easier to install in tight situations. Figure 10.33 shows what the trap looks like inside and how the depth of seal is measured. They

Figure 10.29: A selection of plastic traps

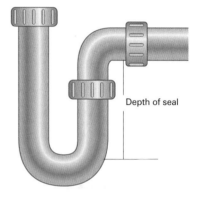

Figure 10.30: Depth of seal

Depth of seal

Figure 10.31: P trap

Figure 10.32: Bottle trap

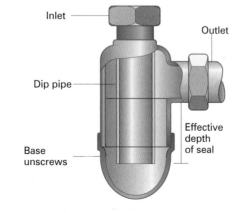

Inlet

Outlet

Dip pipe

Effective depth of seal

Base unscrews

Figure 10.33: Cross-section of bottle trap

Figure 10.34: Swivel elbow

should be avoided on sinks and, to a lesser extent, on baths and shower trays; here, they can be prone to blockage due to the accumulation of food or soap deposits.

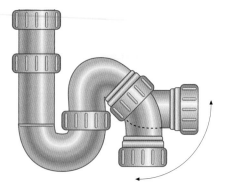

Figure 10.35: Tubular swivel trap

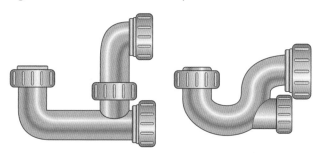

Figure 10.36: Low-level bath taps

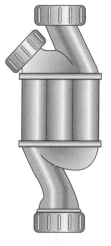

Figure 10.37: Straight-through trap

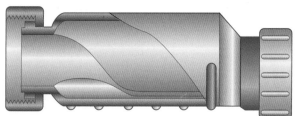

Figure 10.38: Self-sealing valve

Tubular swivel traps

These are particularly useful on appliance replacement jobs, as they allow more options for connecting to an existing waste pipe without using extra fittings or altering the pipework. On new jobs they are often used on sinks with multiple bowls, again because of their multi-position options.

Low-level bath traps

These are designed so that they can fit into tight spaces under baths and shower trays. This type of trap is available with a 50 mm seal depth making it suitable for connection into a soil and vent stack. Be careful – the 38 mm seal depth version is only suitable for waste connection into a ground floor gulley.

Straight-through trap

These are used as an alternative to an S trap where space is limited. They are also easier to hide behind pedestal basins. The main problem with this design is the two tight bends, which slow down the flow of water.

The self-sealing valve

The self-sealing valve can be used as an alternative to a trap. It works on the simple principle of using an internal membrane as a seal. The membrane allows water to flow through it when the water is released, then closes to prevent foul air from entering the building. The valve can be used on systems meeting BS EN 12056: Part 2. It is ideal for fitting behind pedestals and under baths and showers, and is supplied with a range of adaptors so that it can be used in various situations. The valve has the potential to revolutionise the installation of above-ground systems, the requirements of which will be covered below. System design procedures for the straight-through trap will be covered fully at Level 3.

Running traps

You may see running traps being used in public toilets where one running trap is used for a range of un-trapped washbasins. In domestic installations it could be used where a P or S trap arrangement is not possible. Running traps are sometimes used with a

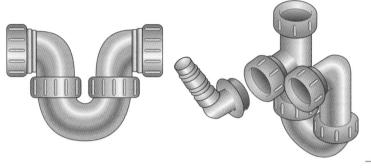

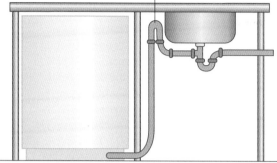

Waste pipe looped above sink overflow

Figure 10.39: Running traps   **Figure 10.40:** Combined sink trap fittings   **Figure 10.41:** Combined sink/washing-machine trap

washing machine waste outlet or for dishwashers, although specialist traps are available for these appliances.

Resealing and anti-siphon traps

If an above-ground discharge system is designed and installed correctly, loss of trap seal should be prevented. Some of the reasons for loss of trap seals are covered next, but these particular traps are designed to prevent seal loss due to the effects of siphonage. The Hepworth valve provides a very similar function although it is less complicated. These types of traps could be specified or fitted in situations where normal installation requirements cannot be met.

The anti-siphon trap uses an anti-vacuum valve located on the top of the trap. If the pressure drops inside the pipework, the valve is activated, allowing air to enter the system to equalise the pressure.

The resealing trap has a bypass within the body of the bottle. The dip pipe allows air to enter the trap via the bypass arrangement. As the seal is lost due to siphonage, air is allowed into the trap, breaking the siphonic effect.

Access

Whatever type of trap is fitted, it is important that you can access the trap for cleaning. Some traps have cleaning eyes; others can be split at their swivel joints to enable a section of the trap to be removed.

Trap-seal loss

Trap seals can be lost as a result of poor practice and by not following the regulations; the detail of which we have just covered.

There are three main ways in which trap seal may be lost due to poor practice:

1. Self-siphonage
2. Induced siphonage
3. Compression.

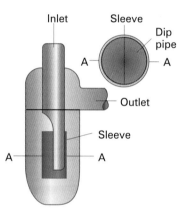

Inlet   Sleeve   Dip pipe

A — A

Outlet

Sleeve

A — A

Figure 10.42: Resealing and anti-siphon traps

Figure 10.43: Anti-siphon trap

Self-siphonage

This is most common in washbasins, as their shape allows water to escape quickly. As the water discharges, a plug of water is formed, creating a partial vacuum (negative pressure) in the pipe between the water plug and the basin. This is enough to siphon the water out of the trap. Ensuring that the waste pipe is within the lengths allowed and to the correct fall, or that it is ventilated, should prevent self-siphonage. Resealing traps would also avoid this problem.

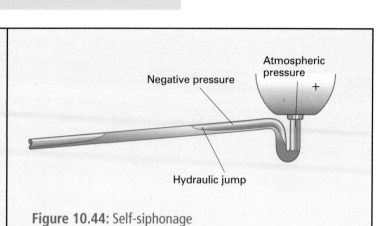

Figure 10.44: Self-siphonage

Induced siphonage

This is caused by a discharge of water from an appliance that is connected to the same waste pipe as other appliances. As the water plug flows past the second appliance connection, negative pressure is created between the pipe and the appliance that siphons the water out of the trap. This arrangement is not acceptable on a primary ventilated stack. In Figure 10.46, fitting a branch-ventilating pipe between the two traps would solve the problem (as would fitting a resealing trap).

Figure 10.45: Induced siphonage

Compression

As the water is discharged from an appliance into the main stack (usually a WC at first-floor level), it compresses at the base of the stack, causing back pressure. This can be enough to force the water out of the trap, thus losing the seal.

The use of large-radius bends and minimum 450 mm length between the invert of the drain and lowest branch pipe are defined in the regulations in order to prevent this.

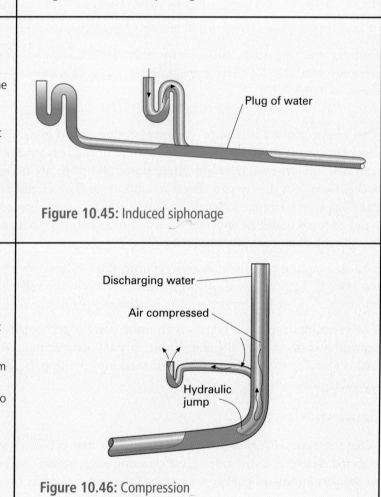

Figure 10.46: Compression

Why traps lose their seal

The main reasons for traps losing their seal are bad pipework design and poor installation. Table 10.3 shows why a trap may lose its seal through natural causes.

Cause	Effect
Capillary action (see Figure 4.13)	Only occurs in 'S' trap arrangements and is not a regular occurrence. If a thread of material becomes lodged as shown in the illustration in this table, water could be drawn from the trap by the effect of capillary attraction.
Momentum	The seal of a trap is removed by force of water. This can happen if you pour a bucket of water into a sink or toilet.
'Wavering'	Wind pressure across the top of the soil and vent pipe, particularly in exposed locations, causes the water in the trap to produce a wave movement and wash over the weir of the trap. This does not happen often, and can be avoided by fitting a 90° bend or a cowl to the top of the vent pipe.
Evaporation	The most common form of natural seal loss. Occurs in very warm, dry weather when the water in the trap simply evaporates. It is unlikely to occur in traps with a 75 mm seal.

Table 10.3: Why traps lose their seal

Connections to drains

Plumbers very rarely work on below-ground drainage systems. This job tends to be done by the builder on smaller projects or ground workers on large contracts.

However, you need to know the basics of the systems and how to connect the soil and vent pipe to the drain, because that **is** a plumber's job. Essentially, you must ensure that **foul drainage** cannot enter the surface water system.

This is because with certain types of system, **surface water** is discharged cheaply via separate pipework to streams, etc. Foul water drainage should never be discharged to such pipework, as it could be a potential health hazard.

There is a simple rule here: always know what you are discharging into – then you can't go wrong.

There are three main types of drainage system:

1. Combined system
2. Separate system
3. Partially separate system.

The combined system

In this system, the water from the sanitary appliances (foul water) and the rainwater all go into one sewer.

The separate system

Here, the foul water runs into one sewer, and the rainwater runs into a separate sewer for surface water.

> **Key terms**
>
> **Foul drainage** – anything discharged from a sanitary appliance such as a WC, bath, basin, or sink.
>
> **Surface water** – the term used for water collected via the rainwater system

svp = soil vent pipe
rwp = rainwater pipe
fwg = foulwater gully

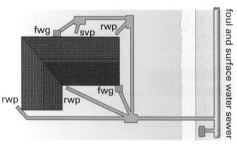

Figure 10.47: Combined drainage system

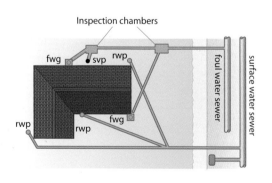

Figure 10.48: Separate drainage system

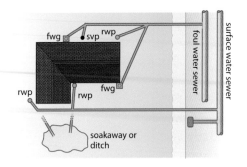

Figure 10.49: Partially separate system

Partially separate system

This system still uses two pipes, but some of the surface water is discharged into a watercourse, soakaway or drainage ditch.

The advantages and disadvantages of each system are shown in Table 10.4.

System	Advantages	Disadvantages
Combined	Cheap and easy to install. Gets a good flush-out during periods of heavy rain.	More costly to treat the water at the sewage works. At times of heavy rain, inadequately sized drains could overflow. All gullies have to be trapped.
Separate	No need for water treatment of surface water. No need for trapped gullies on surface water drains.	Danger of cross-connections – foul to surface water.
Partially separate	Greater flexibility with the system design.	Danger of cross-connection.

Table 10.4: Advantages and disadvantages of different drainage systems

Remember

Remember about access. Check whether an access pipe fitting will be needed.

Figure 10.50: Drain connector

Connecting the soil and vent pipe

On a new job, such as on a large building contract, this information will be on the drawings and specification. When you go into the property, it is obvious because all the outlets to the drain will be in place for the soil and vent pipe or downstairs WC.

On a replacement job, the connection will be the same as the existing soil and vent pipe, so you will use that.

The soil and vent pipe are likely to be connected to a drainage pipe. Fittings are made to connect the two. Some drainpipe terminations, particularly on older systems, are finished with a collar. Drain connectors, such as the one illustrated in Figure 10.50, are inserted into the collar and a joint is made using a sealing compound. This is finished off with a sand and cement joint of a 1 to 4 mix.

Working life

Gary is a Level 3 plumber who has been sent by his boss to a complaint. When he arrives he discovers that the complaint is from the tenant of a house. His landlord had employed a plumber six months previously to install a new bathroom due to the state of the old one.

The tenant had complained time after time to the landlord that things weren't right with the new bathroom and there must be a blockage as there were quite often smells in the room, and when the washbasin was emptied waste water appeared in the bath.

Gary had a quick look round and soon identified the problem. It is nothing to do with a blockage.

Discuss the following points:

- What is the probable cause of the problem if it is not a blockage?
- What term is used to identify this type of trap seal loss?
- Should Gary carry on and repair the problem?
- What are the consequences if he does continue and repair the problem?
- Who would be responsible for paying for the work to be carried out?

Progress check

1. In a primary ventilated stack system, what is the minimum depth allowed between the lowest connection and the invert of the drain?

2. How can the maximum recommended pipework lengths and gradients be exceeded in a primary ventilated stack system?

3. What is the important requirement if a kitchen-sink waste pipe is terminated into a gully rather than the stack pipe?

4. Other than installing anti-siphon traps or self-sealing valves, what other measure can you take if excessive pipe lengths and/or steep gradients cannot be avoided?

5. Subject to certain requirements, where can air admittance valves be located internally?

6. Why should bottle traps be avoided when connecting to sinks and, to a lesser extent, baths and shower trays?

7. When is it permissible to install a 38-mm seal bath trap?

8. How does a self-sealing valve connected to a soil and waste system prevent smells entering the building?

9. In public toilets, a running trap may be found. What is it used for?

10. What type of trap seal loss is most common in washbasins?

11. What measures are likely to prevent trap seal loss by compression?

12. What is a combined system of below-ground drainage?

13. What is a separate system of below-ground drainage?

14. What is a partially separate system of below-ground drainage?

3. Know, and be able to apply, the site preparation techniques for sanitary appliances and connecting pipework systems

It is important to prepare properly before you start installing any appliance. This should ensure there are no hold-ups later owing to missing fittings, or pipe chases not having been prepared as part of the specification.

Site preparation was covered in Unit 5, pages 231–39. This section looks at some issues that relate specifically to working with sanitary appliances.

Storage of materials on site

This is particularly important on larger jobs where you might take delivery of a number of items of sanitary ware. However, some aspects also apply to one-off jobs. Sanitary appliances are expensive items, so it is important that they are stored properly to prevent damage or theft.

Here is a storage checklist:

- Before a delivery takes place, make sure there is somewhere suitable to store the appliances – a lockable materials cabin, with adequate room, is ideal. The storage surface should be raised off the ground of the cabin (use pallets or similar), and it should be clean and dry.

- Before accepting delivery, check the appliance carefully for signs of damage. This does not mean removing the wrappings, which should be left on as long as possible. Production defects are rare, but keep an eye out for these as well.

- Check all traps, taps, wastes and plugs, brackets and seats are with the appliances. There is nothing worse than starting a job and then finding that a waste fitting or bracket is missing.

- When you are happy that everything is complete, check off all the items ordered against the delivery note. If you have the authority, sign for the delivery. You will receive a customer's copy of the delivery note. If anything is not correct, note the missing items and contact the supplier immediately. If you are desperate for the delivery as it stands, sign for the items and get the person making the delivery to countersign that some items are missing.

- Take care when handling the material. Some sanitary items are heavy: use the lifting technique from Unit 1 on health and safety. Most pieces of sanitary ware are very fragile and can break into sharp pieces, so always wear gloves, especially when cleaning up accidents.

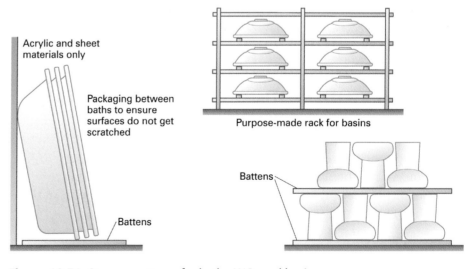

Figure 10.51: Storage patterns for baths, WCs and basins

Checklist

Work site

- Check access to the work area, and check that the work area itself is free from obstructions and potential hazards. This should ensure that the risk of accidents is kept to a minimum.

- Make sure you have any job details or specifications/ drawings showing the fixing positions of the sanitary appliances and any specific fixing requirements. You will also need the manufacturer's fixing instructions.

- If you find any installation problems, or defects in the appliance as you are working, report them to your supervisor.

- On some jobs you will cut holes or chases or install ducts for pipework. On other jobs, this preparatory work is done by other trades in advance, so you will need to check that it has been done. If not, report back to your boss or the site supervisor.

- When you're at the stage of fixing the appliances, the first-fix pipework must be installed. This may have been done by you, particularly on a single-dwelling job, or it could have been done by another plumber on a large housing contract.

- Make sure servicing valves have been installed where required. Good practice would see them put on every appliance that requires regular service and maintenance.

- On first-fix jobs, open-ended pipes should have been capped. This is to prevent debris getting into the system.

- On jobs where the appliances are being fixed against tiles, baths and showers should be installed before tiling, and concealed bidets, washbasins and WCs after tiling. This makes the job easier for the tiler as they won't have to cut around the shape of the appliance.

- Edges of baths should be set into plaster or plasterboard. Careful planning and liaison with other trades can ensure that plasterwork is not taken to full-floor level on the bath-edge wall, and in the case of stud-partitioned walls, extra struts can be fixed to support the edges of the plaster board.

- Finally, as you are carrying out the work, avoid standing on any of the appliances. This is sometimes unavoidable with showers, or baths with a shower appliance installed above it. In each case, make sure the surface of the appliance is fully protected with dust sheets and cardboard packaging.

- Store the materials in the materials cabin, which should be clean and dry. The appliances should be stored away from plaster and concrete, and should be kept away from areas where other materials could fall onto or into them (e.g. brick stacks).

The illustrated ways of storing appliances (see Figure 10.51) are simple but effective. WC pans should be stacked no more than four high. Make sure the storage surfaces (battens/pallets) are clean: battens are used for all sorts of things on site, and if they are covered in grit, storing materials on them will be no better than storing them on the ground.

> **Safety tip**
>
> Always use the correct lifting and carrying techniques when transferring materials from the store to the job. Ensure that your route is planned and safe.

> **Remember**
>
> It could be a few days between first and second fix, and in the meantime other trades, such as plasterers, may have been in. That is why you should always cap off pipework.

4. Know, and be able to apply, the installation requirements of sanitary appliances and connecting pipework systems

Positioning and fixing requirements of sanitary appliances

This section works through the installation of a typical new bathroom suite, including the:

- bath
- pedestal washbasin
- close-coupled WC – wash-down pattern
- shower tray and enclosure
- bidet.

There are no set rules for the sequence you should use to install the appliances. Plumbers will work out the best sequence based on the position of the appliances and the size and layout of the bathroom. You should base your decision on what you think will be easiest and quickest.

On jobs where you are replacing a suite and the dwelling has only one WC, the first appliance you install will be the WC suite, to keep any inconvenience to a minimum.

It is not very often that you will install a bath and shower tray in the same room, unless it is a large property. Most baths have shower units installed over the appliance, so a shower tray may be used in an en-suite room, particularly where space is tight, or the occupier has difficulty in using a bath due to age or illness.

Generally speaking it is preferable to install the bath before the shower.

Preparing the appliances

This is often referred to by plumbers as 'dressing the suite' – it means installing the taps, wastes and, in the case of the bath, the cradle frame or feet. It will also include installation of float valves, overflows, siphons or flushing valves, and the handle assembly to the WC cistern.

Fitting the taps

Fitting the taps is a relatively simple process. Tap manufacturers' designs have to comply with regulations, so your installation will also satisfy the regulations. Remove just enough of the protective coating to make sure a clean joint can be achieved between the fittings and the surface of the appliance. Also, double-check to make sure that the surface is clean and dry. You will not have to use any other jointing compound: the washers supplied by the manufacturer are all you need. There may be slight differences in the washer kits supplied by manufacturers, but the illustrations here show typical details.

It is good practice to 'break' the tap headgear away from the tap body on new taps. This enables the installer to check for any faults, like damaged washers. More importantly, it enables the removal of the headgear from the body at a later date (possibly several years later), and potentially avoids damaging the appliance.

Pillar tap to a basin

The grip washer goes on the underside of the tap. Sometimes a thicker washer is also supplied, which goes between the back nut and the underside of the basin. Tighten the back nut using a purpose-made basin wrench. You can test if it is tight enough by checking by hand that the tap does not move. Make sure both taps are pointing outwards in parallel; some plumbers turn the taps slightly inward.

Monobloc taps

These are often supplied with an 'O' ring, which should be fitted between the base of the tap and the basin. The washer goes underneath as shown in Figure 10.54. The assembly is completed using a metal washer held in place by a fixing nut.

Tap and spout assembly: three-hole

These are installed in three-hole basins and ascending spray bidets, as follows:

* Remove all the parts that go above the basin.
* Fit the spout with its washers.
* Loosely connect the tubes between the body and side valves.
* Raise the assembly into position from below, making sure the sealing washers are in place.
* Fit the body seal, washer and lock nut loosely, then fit the valve flanges on top of the basin, aligning them with the top of the side valve as in Figure 10.55.

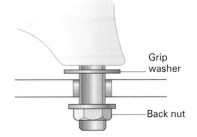

Figure 10.52: Seal to basin tap

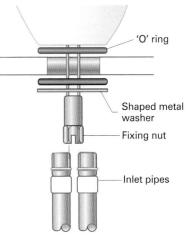

Figure 10.53: Monobloc tap connections

> **Remember**
> The correct tap conventions are: hot on the left, cold on the right.

> **Remember**
> For all tap installations make sure nuts are not cross-threaded.

> **Safety tip**
> Never use linseed oil compounds on plastic wastes as they degrade the plastic causing it to fail; it can ruin the bath and shower tray as well.

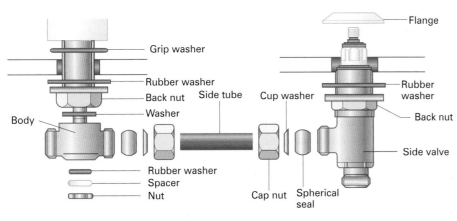

Figure 10.54: Three tap-hole basin

- Carefully tighten all the components, and then finish by fitting the valve head gear, any shields, drive inserts and hand wheels.

Other appliances

The same principles apply to installing taps on other appliances. Remember to make sure that you use the washers supplied by the manufacturer.

Waste fittings

For washbasins, these include slotted waste fittings, moulded seals and pop-up wastes.

Slotted waste fittings

Slotted waste fittings can be sealed using non-setting plumbers' compound or silicone sealant. This is done by moulding a thin strip of compound around the flange of the waste (Step 1) and then placing the waste into the basin, making sure that the slots in the waste line up with the overflow slots of the basin (Step 2). Clean off excess sealant or compound with a soft dry cloth. Also check to see that the compound or silicone have not obstructed the overflow slots in the basin/waste fitting.

In the case of some pedestal basins, a bracket has to be included to fix the basin to the pedestal. This is done by placing the metal bracket between a washer and the back nut (Step 3). The basin is secured using the two adjustable clips, which are tightened against the pre-cast lugs in the pedestal.

Moulded seals

These are supplied for pop-up wastes and combined wastes and overflow, as used on baths, plastic washbasins and sinks. Installation details are shown in Figure 10.56.

The moulded seal fits between the waste flange and the appliance. For pedestal and plastic basins, a shaped seal and 'O' ring are used together to make the joint for the shaped metal connecting washer, or alternatively the metal pedestal-fixing bracket.

Where the appliance requires an overflow assembly, a rubber ring should be fitted between the overflow flange and the back of the appliance. When fitting this set-up:

- do not cross-thread the overflow grill and the fitting to the rear of the appliance
- do not over-tighten the fitting
- make sure you do not damage the flexible overflow tube when tightening the overflow fitting.

Pop-up wastes

Pop-up wastes are available for washbasins baths and bidets. On a bath, the plug unit is raised by twisting the circular control knob

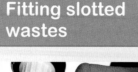

Fitting slotted wastes

Step 1: Mould a thin strip of compound or sealant around the flange of the waste

Step 2: Line up slots and apply more compound or sealant around the bottom of the basin and the thread of the waste

Step 3: A plastic washer should be pressed into the compound, and the back nut is tightened to complete the joint

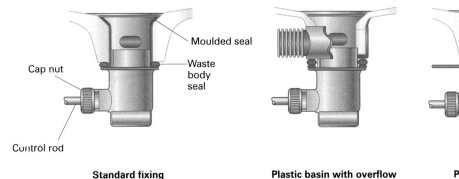

Cap nut

Moulded seal

Waste body seal

Control rod

Standard fixing

Plastic basin with overflow fitment

Pedestal basin fixing

Figure 10.55: Moulded seals

anticlockwise; this serves as both an overflow outlet and pop-up waste control. The plug unit is lowered by twisting the control knob clockwise, which allows the plug to drop into the waste by gravity.

Checklist

Fixing a washbasin waste

- Fit the waste body into the basin outlet.
- Remove cap nut and control rod.
- Position the appropriate seals, then couple the pop-up body onto the waste tail thread.
- Tighten with the control port correctly aligned with the back of the basin.
- Slacken the lock nut on the stem of the pop-up plug and fit into the waste body.
- Position the control rod in the control port so that the end of the rod is in the hole at the end of the plug stem.
- Assemble the cap nut.
- Check the operation of the plug; it should be flush with the waste flange and should lift about 10 mm when operated.

- To reduce the lift, the plug is turned clockwise, and to increase the lift it is turned anti-clockwise.
- Remove the cap nut and control rod and take out the plug.
- Tighten the lock nut against the strainer stem of the plug and then refit the plug and control rod and tighten the cap nut.
- Fit the lift rod through the fitting and locate with the eyelet on the shackle.
- Adjust so that the left knob is just clear of the fitting when the plug is fully open.
- Tighten the screws on the shackle and check for smooth operation of the assembly.

Baths

Assembling the bath

The cradle frame or brackets and legs are usually fitted before the taps and waste, as part of the dressing process. Bracket and feet assemblies will vary depending on the manufacturer, but a typical assembly detail is shown in Figure 10.56. This is sometimes called a cradle frame. The assembly should be carried out as follows:

- Place the bath upside down on a dust sheet over a clean, flat surface.
- Position each leg in the location spigots as shown in Figure 10.56. They should be pushed in until the centre section is flush with the baseboard, making sure the sides of the brackets are plumb.
- Locate the centre leg and bracket in the centre of the baseboard.

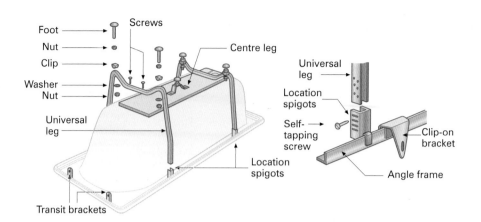

Figure 10.56: Bath fixings

- Screw all the brackets into position using the wood screws provided.
- Screw the brackets to the location spigots using self-tapping screws.
- Assemble the feet into the brackets. Stand the bath on its edge and fit the clip-on wall brackets and handles. The taps and waste are installed as described previously.

It is handy at this stage to adjust the feet while the bath is on its edge. This can be done by using a tape and getting the feet to an approximate measurement required and then 'fine tuning' the feet to the exact measurement using a straight edge and tape. Final adjustments can be carried out with the bath in place.

Installing the bath:

- Place the bath on its feet and in position.
- Check the measurement from the top edge of the bath and or bath panel to the floor. It should be the manufacturer's recommended distance.
- Check the bath is level across its length and width. Adjustment to level and height can be made via the feet.
- Once the bath is level and at the correct height, tighten the locking nut on each foot.
- Mark the position of the fixing holes for the feet and the wall fixings. If the wall is plastered, and you have been instructed to let the bath into the plaster, mark it off for the full length of the bath. Take out the bath.
- If the floor is wood, drill pilot holes to receive the wood screws for the feet. This will make fixing much easier. If the floor is solid, drill and plug the floor. Drill holes for the wall fixings.
- Chase out the plaster for the wall brackets and, if letting the bath into the plaster, chase out the plaster where you marked the line of the bath: this enables the bath to be let into the wall and provides a really good watertight seal once tiled.

- Refit the bath and screw the wall and floor brackets into position. If fixing to timber floors, make sure the length of the screws will not penetrate the underside of the floor.

- Service and waste connections are then made. It is best to prefabricate the tap connections if using soldered fittings, as this avoids using the blowtorch under the bath. The trap is connected and the waste pipe extended to the outside (if there is an external waste system).

- If the bath is fixed flush to the plaster, fill the bath to one-third full, so that the weight causes the bath to settle, and then seal the joint between bath and wall with silicone sealant, after removing the protective film from the bath's surface.

It is sometimes advisable to fit wooden battens (75 × 50 mm) under the feet of the bath. This reduces the length of the feet adjustments and so aids a more secure bath. The battens are a necessity when installing onto composite flooring like chipboard, to stop the feet puncturing the surface should it become wet or damp.

> **Remember**
>
> Bath panels should only be fitted once testing has been completed.

The method of fixing bath panels varies depending on the manufacturer, so their fixing instructions should be followed. Generally speaking, the following steps should serve as a guide:

- Check the measurement from the underside of the roll edge of the bath and the floor. Check for any adjustments needed in the event of the floor being out of level.

- If the panel has to be cut, firmly support the panel and cut it using a fine-tooth saw, then finish off with a file.

- In some cases, panel support frames or bath leg clips are supplied, and these should be fitted.

- Use a level or plumb line to mark out for a fixing batten at floor level; once you have made an allowance for any 'kicking space' on the panel, mark the position for the batten. If self-adhesive pads are to be used, an allowance for their thickness should also be made when positioning the batten (four pads should be sufficient).

- Alternatively, screws can be used. Mirror screws provide an attractive finish, and some manufacturers provide screw-head caps that match the colour of the bath panel. Pre-drill the panel to take the screws. The panel can then be positioned by inserting the top edge of the panel between the bath roll and the timber frame of the bath, and secured at the base with either the self-adhesive pads or the screws.

WC suite

Installation includes a close-coupled wash-down WC pan and siphonic action cistern.

Dressing the cistern

This will include the siphon, overflow fitting and float-operated valve. The cistern will have holes for either side inlet and outlet, or bottom inlet and outlet. Most cistern components will be plastic, and should be installed using the washers provided.

The shaped rubber sealing washer for the siphon, and flat rubber washers for the float valve and overflow union, should go inside the cistern. Carry out fitting as follows:

- Fit the siphon first, making sure that the metal bracket for the close-coupled connection is in place and that a rubber washer has been placed between the bracket and the underside of the cistern. The back nut completes the assembly. When tightening the back nut, make sure the bracket is kept in the correct position for the bolts and locating holes in the pan.
- Fit the float valve and overflow.
- Assemble the lever and linkage mechanism and check that it works correctly before installation.
- Fit the cistern to the pan.
- Before securing the cistern to the pan, make sure the soft foam-rubber close-coupling washer is fitted over the siphon nut. Then carefully place in position over the pan.
- Position the fixing bolts in the holes in the pan, and tighten the cistern to the pan using the rubber and metal washers and the bolts supplied.

Installing the WC suite

All the components have been installed in the cistern during dressing, and the cistern has been fixed to the WC pan. The installation process is very similar to the washbasin:

- Offer the cistern to the wall, check the level of the cistern and pan and mark the holes through the back of the cistern. At the same time, mark the fixing holes through the base of the pan. If a soil pipe is already in, either remove the pan connector or ease the pan outlet into the pan connector.
- Carefully take out the close-coupled suite, drill and plug the holes.
- Refit and screw back the cistern; screw down the WC pan. Again, with the cistern use soft washers, and brass or alloy screws. Some manufacturers supply screws with caps that match the colour of the suite.
- Some plumbers, once they have positioned the pan, drill through the fixing holes rather than removing the close-coupled suite to do the drilling. This is acceptable, but great care should be taken not to damage the suite with the drill, and to make sure the plug is fully located in the fixing holes.
- We have assumed that the hole for the overflow has already been prepared. If not, this should be marked out and drilled before the suite is fixed.
- Most float valves are plastic, so, when connecting the pipework with capillary fittings, do not solder the joint when it is connected to the plastic thread because it will melt and distort.

Safety tip

Close-coupled suites are heavy. Seek assistance if possible when handling the suite.

Remember

Do not over-tighten or you will split the plastic. Watch out for cross-threading, too.

- Make sure a fibre washer is in place before tightening the tap connector.
- Remember, all float operated valves must have a service valve fitted.
- Fit the overflow pipe, which is usually made from plastic.

Other appliances

The waste requirements for shower trays are straightforward and are fitted using the washers supplied with a 38 mm (1½-inch) waste fitting. On fire-clay trays, non-setting plumbing compound is applied in the same way as with the standard slotted washbasin waste installation.

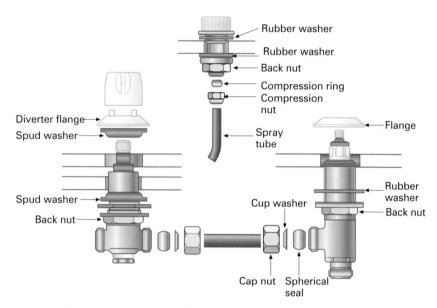

Figure 10.57: Bidet washer and dial arrangements

The waste fitting for a bidet is a 32 mm (1¼-inch) threaded unit, either slotted or solid depending on whether the bidet is fitted to an overflow; pop-up wastes are also available. Taps can be pillar or monobloc for over-the-rim applications. A typical arrangement for a douche-type (under-rim) bidet is illustrated in Figure 10.57.

The installation instructions are similar to those of the tap-and-spout assembly of the washbasin outlined previously.

Installing the washbasin

Our sample bathroom specification includes a pedestal basin. The taps and waste are in, and during dressing, the metal bracket that will hold the basin to the pedestal has also been fitted.

To install the washbasin:

- Loosely attach the clamps using the pins and washers. Do not tighten them at this stage.
- Run sealing compound along the top edge of the pedestal; this will provide a better finish to the front.

Did you know?

It is often a good idea with pedestal installations to fit the trap to the waste fitting before the basin is secured to the pedestal; this is easier than when the basin is on the pedestal.

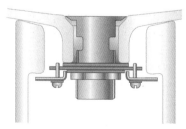

Figure 10.58: Basin fixing bracket

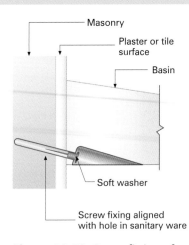

Masonry

Plaster or tile surface

Basin

Soft washer

Screw fixing aligned with hole in sanitary ware

Figure 10.59: Screw fitting of basin

- Place the basin on the pedestal. Adjust the clamps so that they rest against the pre-cast shoulders of the pedestal.
- Check the basin for level, make any final adjustments; carefully tighten the pins and then clean off any excess sealing compound.

The basin in this example has moulded screw holes at the back edges of the bowl for fixing it to the wall (Figure 10.59), as follows:

- Offer the basin up to the wall; make sure it is level and in the correct position (you might have to pack the pedestal to achieve level). Make a mark on the wall through the holes.
- Remove the basin and pedestal and place them where they will not get damaged, preferably laying them down on a dust sheet.
- The moulded holes in the basin are angled. You need to estimate this angle and drill the hole accordingly. If you drill it level, as you screw the basin to the wall the edge of the basin will crack off adjacent to the hole!
- Reposition the basin and pedestal and screw back to the wall using brass or alloy screws and soft washers. This will make the basin easier to remove for any maintenance work or replacement.
- Some pedestals are supplied with fixing holes in the base. If so, the floor should be marked at the same time as the basin. On wooden floors, pilot holes should be drilled before refitting. On solid floors, plugs are placed in the drill holes prior to repositioning the pedestal.
- Some basins are supplied without fixing holes at the back of the bowl, but have moulding holes on the underside edge of the bowl.
- Fit the bracket to the basin and tighten so that it holds firmly in the hole.
- Offer the basin to the wall; then adjust the brackets so that they are flush with the wall. Tighten the fixing bolt into the basin, and then mark the position of the bracket holes.
- The pipework installation can be prefabricated so that it fits neatly behind the basin and the pedestal. Make sure you have fitted fibre washers before tightening up tap connections for connection to pillar taps. Monobloc taps are usually supplied with compression fittings. Mark off the position of the joints between the pipework tabs and prefabricated pipe at the same time as you make the fixing holes.

Remember

Do not fully tighten at this stage because you will need to adjust the bracket.

Remember

Don't forget to use pilot holes on wooden floors.

Installing a washbasin using brackets

There is a range of brackets used for fixing washbasins – mainly in industrial or commercial situations.

When selecting a bracket for the job, you should:

- make sure the bracket is the correct pattern for the basin
- be confident that it will support the weight of the basin.

The normal fixing height of a basin is 800 mm from the floor to the front rim.

It is a good idea when marking out for the brackets to do the following:

- First, mark the centre line on the wall where the basin is going to be fixed.
- Place the basin on the floor with its rim 800 mm from a wall.
- Put the brackets in position on the basin and measure the distance to the fixing holes from the wall, and at the same time, the centres of the fixing holes between the two brackets. Halve that distance, and that will be the measurement from the centre line marked on the wall to the centre of each bracket.
- Transfer the measurements to the wall where the brackets are to be fixed. Position the brackets on the centre lines and mark the holes.
- Drill and fix the brackets, then place the basin in position and check height and level.

Alternatively, if you are working with a colleague, mark out the centre line for the basin and the rim height. Then, one person holds the basin in position, checking that it is in line and level. The other person puts the brackets in position and marks the holes.

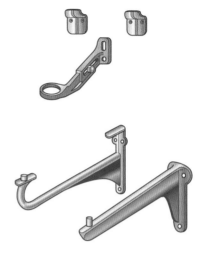

Figure 10.60: Selection of wall-mounted basin-fixing devices

Countertop basins

Usually, countertop basins are supplied with a cutting template. You must follow this carefully when cutting out the hole. Use a jig-saw blade with fine teeth and a downwards cutting action, as this will reduce any damage to the finish of the worktop.

Once you have cut the hole, seal the exposed edges with a waterproof varnish to prevent any water/moisture damage to the chipboard.

The joint between the lips of the bowl must be sealed using a recommended sealant, and the bowl is then secured using the brackets supplied with the basin.

The edge of the hole for an under-countertop basin is finished off with a strip of laminate in the same material and colour as the countertop.

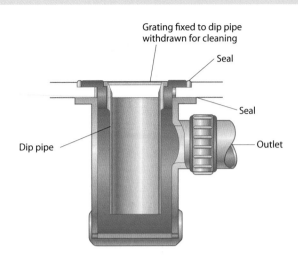

Grating fixed to dip pipe withdrawn for cleaning

Seal

Seal

Outlet

Dip pipe

Figure 10.61: Typical shower tray trap

Shower tray and enclosures

The waste is fitted to the tray during dressing. It will help with future cleaning and maintenance if a removable waste is fitted. The waste trap on a shower tray should always be accessible for maintenance purposes. Failure to leave access is a common mistake made by plumbers, which can cause real problems and potential embarrassment a few years down the line.

The most important thing to ensure when fitting the tray is that it is level in all directions. If it is not, you will face big problems when fitting the enclosure.

Fixing to timber floors:

- First, check whether the trap and waste pipework can be accessed under the floor. If not, you may have to raise the tray in a similar way as for an installation on a solid floor.
- If the pipework can be accessed, make sure the floor is clean and dry and coat the underside of the tray with a recommended sealant; alternatively the tray can be bedded on the tile cement.
- Position the tray and press firmly into place. Make sure the tray is level in all directions.
- Once the bed is dry or cured, the tray edges can be sealed to the adjoining walls.

Fixing to solid floors

On a solid floor installation, it is likely that you will have to raise the shower tray to gain access to the trap and waste pipe. This is done by building a sub-frame out of external grade plywood and timber battens. The thickness of the battens will be governed by the amount of clearance you need, but they should be placed at 250 mm intervals. Once you have fixed the sub-frame, install the tray as you would for a timber floor.

Fitting enclosures

Shower enclosures come in all shapes and sizes, and the best advice here is to follow the manufacturer's instructions carefully. However, here are a few pointers.

- Check the walls are plumb. Most manufacturers' products make allowances for the walls being slightly out of plumb. On older properties, if the wall is seriously out of plumb, some plastering work may be required before you start.
- It is almost certain that the frame of the enclosure will be fixed to a tiled surface, and this will involve drilling the tile.
- Using the manufacturer's measurements for the frame, mark out for the fixing holes. The best way is to mark the wall the full length of the

Did you know?

Acrylic trays often have panels to allow access to the underneath; they do not need to be raised.

frame, making sure it is plumb. Then offer the frame to the line and mark out for the fixing holes.

- The remainder of the process will involve fitting the panels and the door. Make sure the doors open and close correctly. The installation should be completed by fitting any sealing strips supplied with the kit, and applying a recommended sealant as per the manufacturer's instructions.

Installing the bidet

There is little that is different about the installation of the bidet, except that it is screwed to the floor using brass or alloy screws. Marking and connection to supply pipework follow a similar pattern to the other appliances.

Wall urinals

These are usually installed at a height of 610 mm from floor level to the front lip of the bowl. Quite often, when a number are installed, one or possibly two will be installed at a height of about 510 mm for children or smaller adults. Dress the urinal bowls first, fitting the inlet spreader and waste.

Urinals are fixed on the type of bracket shown in the illustration, and the measurements can be set out in a similar way to those of washbasin brackets.

Slab urinal installation

Slab urinals are found mostly in public buildings. In the main they are manufactured in fire clay, and can be supplied to any length on one or more walls, either in one piece or in smaller slabs. The floor channel is manufactured so that the internal surface is laid to a fall. This means the actual channel block can be installed level.

Figure 10.62: Urinal bowl fixing brackets

Working life

Liam has been asked to install a range of wall urinals in a gents' toilet located in a refurbishment job at a local swimming pool. There is a total of six wall urinals to be installed, fed by one cistern. He accepts delivery of the goods and checks them over on delivery. However, when he opens the packaging he notices that it is a 6-litre flushing cistern, which is too small for the requirements laid down in the Water Regulations and it has a siphon with it, which is operated by a flushing handle.

Discuss the following points:

- What type of system and flushing arrangement should have been delivered?

- Should all the urinals be hung at the same height? If not, why is this?

- What is the normal height of urinals from the floor to the rim of the appliance in either case?

- What other specialist pipework and fitments should the delivery have regarding the flushing of the urinals?

- The pool is closed overnight for eight hours every day of the week. What measures should be taken to prevent undue wastage of water?

- What do the Water Regulations specify about the flushing requirements of urinals? Will the installation provide enough volume for the required flushing arrangements?

Pipework installation, access and materials

This section overlaps with some of Unit 5, pages 208–30, such as materials and methods of jointing. This section, however, shows how the methods and materials apply to above-ground drainage systems.

Preparing to install above ground drainage system pipework

You may have to drill or cut your own pipework holes for external discharge pipework. If not, make sure the bricklayer knows in advance where they should be and when they should be cut. On some larger housing jobs, the holes may be left by the bricklayer as the brick/blockwork proceeds, and then made up afterwards.

The installation and jointing of above ground drainage system pipework

For domestic plumbing, plastic is generally the material used for above ground drainage systems. For push-fit waste systems, polypropylene to BS 5254 is used. For solvent-welded systems PVC and ABS plastics are used. These meet the requirements of BS 5255. Compression fittings are available which are suitable for both types of pipe. Waste pipe is available in 32 mm, 40 mm and 50 mm diameters, in 3m lengths, and in a variety of colours.

The most common plastic for soil and vent pipes and fittings is PVC-U to BS 4514, which uses ring-sealed push-fit joints. Pipe diameters for domestic use include 82 mm and 110 mm (this is the outside diameter of the pipe). They are available in lengths of 2.5 m, 3 m and 4 m, either as plain ended pipes or single-socket pipes.

Range of fittings

Waste systems

These come in three different types: push-fit, solvent welded and compression. There is an overlap with the design of the fittings, so Figure 10.63 only illustrates the main features.

Soil and vent systems

Figure 10.64 shows a selection of soil and vent fittings.

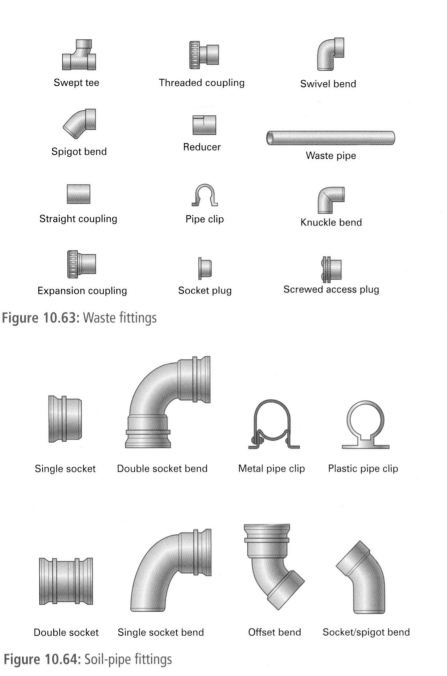

Figure 10.63: Waste fittings

Figure 10.64: Soil-pipe fittings

Access fittings

Building Regulations require you to provide access to pipework, so that it can be cleaned and any blockages removed. Access fittings are designed to allow this.

Joints to soil and vent pipes

There are a number of options here. As shown in Figure 10.65, some soil and vent fittings are supplied with a boss access already in place, ready to install the waste pipe. Others have blocked ends in the pipe which can be drilled out and fitted with a rubber boss adaptor.

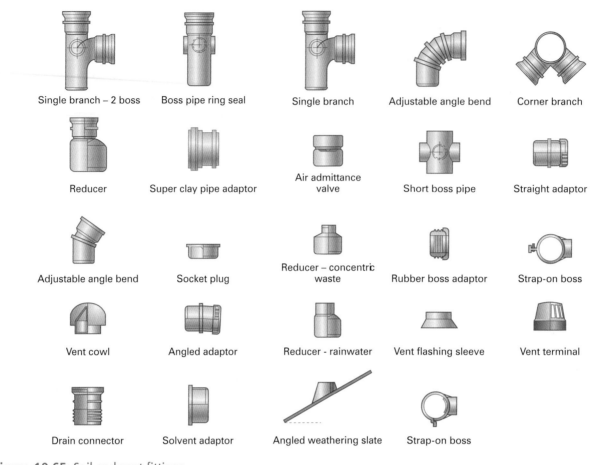

Single branch – 2 boss   Boss pipe ring seal   Single branch   Adjustable angle bend   Corner branch

Reducer   Super clay pipe adaptor   Air admittance valve   Short boss pipe   Straight adaptor

Adjustable angle bend   Socket plug   Reducer – concentric waste   Rubber boss adaptor   Strap-on boss

Vent cowl   Angled adaptor   Reducer - rainwater   Vent flashing sleeve   Vent terminal

Drain connector   Solvent adaptor   Angled weathering slate   Strap-on boss

Figure 10.65: Soil and vent fittings

Boss pipe fittings

Sometimes, these fixed connections may not suit the installation requirements. In these cases you can drill your own fixing position in the soil and vent pipe and then fit a strap boss. The best way of doing this is to use a hole saw. Make sure it is the correct diameter for the strap boss.

The strap boss is fixed using gap-filling cement.

Fixing details

Plastic waste pipes should be clipped at the intervals given in Table 10.5.

Pipe size (mm)	Horizontal (m)	Vertical (m)
32	0.8	1.5–1.7
40	0.9	1.8
50	1.0	2.1

Table 10.5: Waste-pipe sizes and clipping distances

Soil and vent pipes are normally fixed vertically every 2 m.

Working life

Kostas is installing a new bathroom but the existing primary ventilated stack system (single stack system) is to be retained. He checks outside and finds the existing plastic pipework has many bends, no clips at all and steep gradients. Due to the lack of clipping, the pipes have sagged badly and have evidence of leaks due to bad alignment starting to pull out of the push-fit fittings.

The strap bosses used to connect to the stack are also showing evidence of long-term leakage and in fact are not sealed to the stack effectively.

Discuss the following points:

- What should Kostas do now? His employer did not warn him of the issues, as the bathroom replacement should

really be a straight swap. Should he inform anyone? If so, who?

- Are there any issues with installing bends in waste pipework? If so, what are they?
- Are there any issues with having steep gradients in the waste pipework? What are these and what conditions would the issues potentially cause?
- What are the requirements for clipping horizontal and vertical waste pipes from a bath and a washbasin?
- What is the correct procedure for sealing (making them watertight) the strap bosses to the stack?

Progress check

1. When first-fix pipework is installed, what should be done to the open-ended pipes? Why is this necessary?
2. You are replacing an existing bathroom suite. Which piece of sanitary ware should you install first, and why?
3. Why do most washbasins require a slotted waste fitting?
4. When carrying out the dressing process on a bath, which components are usually fitted first?
5. What should be placed over the siphon nut before securing a close-coupled WC cistern to the pan?
6. You are fitting a siphonic WC pan. Which type of valve will you fit between the WC siphon outlet and the WC pan inlet?

7. What are the two recommended fixing heights for wall urinals (measured from floor level to the front lip of the bowl)?
8. What diameters and lengths are plastic waste pipes available in?
9. What diameters and lengths are plastic soil pipes available in?
10. How is a strap boss secured and made watertight to a soil pipe?
11. What are the recommended horizontal and vertical clipping distances for 32 mm and 40 mm plastic waste pipework?
12. What is the recommended vertical clipping distance of plastic soil and vent pipework?

5, 6 & 7. Know and be able to apply the service and maintenance, decommissioning, inspection and soundness requirements of sanitary appliances and connecting pipework systems

Sanitary appliances discharge large volumes of waste water, so it is important that systems are watertight. The test should be carried out in accordance with BS EN 12056. You also need to know how to make

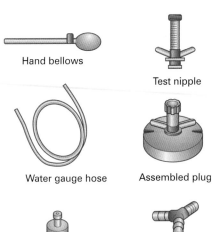

Hand bellows

Test nipple

Water gauge hose

Assembled plug

Hollow plug

Plug testing Y piece

Figure 10.66: Test kit

sure people are aware that testing is going on and know when it is safe for them to use the appliances.

Soundness testing, maintenance and decommissioning for systems are covered extensively in Unit 6 on pages 291–302. This section will show you some of the requirements that are unique to working with sanitary appliances and their pipework systems.

Soundness testing of above ground drainage systems

Figure 10.66 shows a test kit. Look at it and then read the testing checklist, referring to Figure 10.66 as you read.

Checklist

Testing

- First, seal the system using hollow drain plugs. Fix a test nipple in the one to be used for the test. The bottom test plug can be inserted through the access cover at the base of the stack. If this proves awkward, a testing bag can be inserted and inflated in position, as shown in Figure 10.68.

- Fill the traps with water, and cover the test plugs to make sure they are fully airtight.

- The rubber hose, bellows and hand pump are connected to one end. Air is pumped into the system to give a water head of 38 mm. Once this is reached, the plug cock is turned off and the test continued for three minutes.

- Where a pressure drop is found, all the joints should be tested using leak detection fluid.

- Smoke tests are also useful to locate leaks, but are not advisable because of the chance of affecting the plastic pipework and rubber seals.

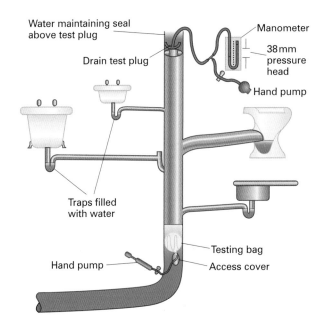

Water maintaining seal above test plug

Manometer

38 mm pressure head

Drain test plug

Hand pump

Traps filled with water

Testing bag

Hand pump

Access cover

Figure 10.67: System tests

Figure 10.68: Air test in action

Testing an installation

Jenni was asked to test and charge the installation shown below.

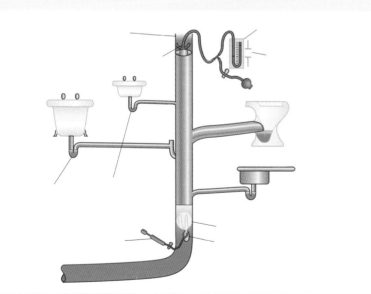

Discuss the following points:

- Label the diagram to indicate the relevant parts she must know to carry out this work and adopt a procedure.
- What procedure should she adopt to test the system?
- What procedure should she adopt to charge the system?
- What equipment will be needed?
- How could she ensure the trap seals are adequate?

Charging procedures

Once the testing equipment has been removed:

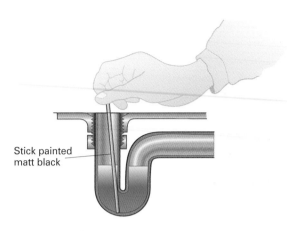

Figure 10.69: Performance testing trap

- Fill all the appliances to their overflow levels and release the water; the WC should be flushed at the same time.
- Check that the trap seals are not less than 25 mm deep. Refer to Figure 10.69 and carry it out three times.
- Measure the depth of seal as shown in Figure 10.70.

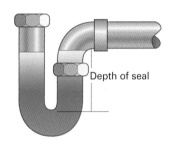

Figure 10.70: Depth of seal

Maintenance of sanitary appliances

Previous maintenance sections highlight the importance of regular maintenance and cleaning in order to make sure the system works to maximum efficiency. Above-ground drainage systems are no different, and while these systems do not have as many controls as central heating systems, for instance, they will eventually deteriorate if they are not looked after.

As a plumber, you may be called out to a blocked drain or discharge pipe, so this is covered as well.

Figure 9.20 on page 386 shows some equipment for unblocking drains but remember that more sophisticated equipment is now available.

Basic maintenance and cleaning

In domestic dwellings, this is usually restricted to the manual cleaning of pipework. Traps often accumulate hair, soap residue, toothpaste and other objects that are small enough to fall through the grid of the waste hole. Integral overflows are also prone to this, particularly on Belfast sinks. Traps should be cleaned through the access points, if fitted; if not, the trap should be broken at its joints or removed completely.

Overflows can be rodded with wire and flushed. Chemical cleaning agents are also available that can be used for both traps and overflows. When using cleaning agents, always follow the manufacturer's instructions.

Access covers to soil and vent pipes should be checked to make sure they are operating properly, and a visual inspection should be made of waste traps and fittings for signs of leakage.

Blocked pipes and drains

Blockages in discharge pipework on sinks, washbasins and baths can often be cleared using a 'force cup' or plunger as follows:

- Fill the appliance with water.
- Block the overflow with a cloth or similar, then press down repeatedly on the handle of the force cup. This creates a positive pressure on the blockage, and a partial vacuum when it is withdrawn. This is usually enough to remove the blockage. Always check the trap for signs of leakage when you have finished.

Access to blocked drains can be through a drainage inspection chamber or, on newer jobs, through a rodding eye installed for that purpose.

Flexible metal pipe and sink cleaners are also available, which work like mini drain-cleaning rods.

A blocked WC can be cleaned using a drain rod with a rubber plunger. These can also be used to clear blocked drainage pipework. Sections of the rods are joined together until the location of the blockage is reached.

Additional attachments are available. A double-worm screw is used for pushing through obstructions or for catching an object (e.g. a cloth) and pulling it back to the operator. A drop scraper is a tool with a hinged steel blade: it is flat when travelling forwards but the blade is lifted when the rods are being pulled back. This can be used for pulling rubbish back to the operator. Clearing wheels help by guiding rods around bends or over obstructions.

> **Safety tip**
>
> Broken vitreous china has edges as sharp as broken glass, so it should be handled wearing thick gloves.

Decommissioning systems

Decommissioning above-ground drainage systems will normally involve stripping out old appliances and pipework to replace with new.

When removing appliances, care should be taken not to damage them.

Removing a cast-iron bath, particularly if it has to be carried downstairs, requires careful handling. Some plumbers break the bath into four pieces for easier removal. A club hammer is the best tool for doing this, and it should be done wearing full-face protection, ear protection and gloves. Once the appliance has been removed it should be stripped of any scrap metal to be taken for recycling.

On externally mounted pipework, you will have to chop out the mortar between the pipe and the masonry. Take care in doing this: it will mean less 'making good' after you have installed the new pipework.

> **Remember**
>
> Old pipework systems could be made of lead, so you must take the usual precautions when handling this material.

Taking down cast-iron soil and vent pipes can be dangerous due to their weight, so they need careful handling. It is best to try to take down short sections by partially cutting them with an angle grinder and then tapping the pipework with a hammer, which will cause it to shear; a rope should be tied to each section in turn so that they can be lowered to the floor. Make sure no one is in the area where you are working. Fixing lugs can be broken from the joint, and the nails prised out using a wrecking bar.

Once the stack is removed, make sure the joint to the drain is covered or the joint capped; this will prevent anyone tripping on it, prevent debris from entering the drain and prevent obnoxious smells from escaping from the drain. A range of blanking plugs for decommissioning work is invaluable.

Don't forget to use rubber gloves when handling old lengths of sanitary pipework and sanitary appliances, as you could develop a health problem if you come into contact with the contents. Also be careful when dealing with blocked WCs and blocked pipework, as these can be a collection point for potentially dangerous items, such as hypodermic syringes deposited by drug users or incorrectly disposed-of feminine hygiene products. Don't just go exploring with the hands; make a visual inspection first!

Working life

Colin is asked to replace and upgrade an old two-pipe system of above-ground drainage to a detached property. The old system is a mixture of cast iron and asbestos cement; the whole system is showing signs of leakage and general deterioration.

There are rules laid down in the Building Regulations Part H1 and BS EN 12056 Part 2, which Colin must adhere to in installing the new above-ground drainage system. He decides on the Primary ventilated stack system (single-stack system).

Discuss the following points:

- What steps must Colin make to safely remove the redundant pipework, including any legislation, which he must abide by and any PPE required?

- What is the criterion that Colin must abide by? Make a comprehensive list with regard to all the points required under the legislation.

- Colin has noticed that there is one problem: the bedroom next to the bathroom has a washbasin installed in it, and the washbasin is 2.5 m away from the stack. What could Colin do to get over this problem?

- Does Colin need notify anyone of the changeover from the old system, and if so who?

- How could Colin determine which type of below-ground drainage system he is going to be connecting the new system into?

Progress check

1. What is the recommended water head (in millimetres) when soundness testing above-ground drainage discharge pipework?

2. How long should the test last when soundness testing above-ground drainage system pipework?

3. If a pressure drop is found during the test, how can you identify the leak?

4. What is meant by the term 'charging procedures'?

5. After a charging procedure has been carried out, the trap seals on all appliances should be checked. What is the minimum seal permissible?

6. How many times should the charging procedure be carried out?

7. How can blockages on sinks, washbasins and baths often be cleared?

8. What attachment tools for drain rods are typically available to aid the removal of blockages and obstructions?

9. You are removing a soil stack. What should be done to the joint to the drainage system, and why?

Check your knowledge

1. What is the maximum recommended length of a 40 mm waste pipe from a bath. which is connected to a single stack system?
   a 3 m
   b 4 m
   c 5 m
   d 6 m

2. Which British Standard covers workmanship on a building site?
   a BS 6700
   b BS 8000
   c BS 9000
   d Not covered

3. Which of the following primary ventilated stack system (single-stack system) components is used to connect a waste pipe to a vertical stack, directly opposite a connection from a sanitary fitting?
   a Parallel junction
   b Cleaning eye tee
   c Long radius bend
   d Multi-branch fitting

4. What is the minimum recommended seal for a bath trap?
   a 19 mm
   b 40 mm
   c 50 mm
   d 75 mm

5. Which type of trap should be installed when taking discharge from a range of wash hand basins?
   a Running trap
   b Walking trap
   c Bottle trap
   d S trap

6. How long should an air test hold when testing an above-ground waste discharge system?
   a 1 minute
   b 3 minutes
   c 4 minutes
   d 2 hours

7. What is the most important PPE to wear when removing a WC pan?
   a Helmet
   b Ear defenders
   c Industrial gloves
   d Dust/particle mask

8. How many times should a check for trap seal loss be carried out on a performance test?
   a 2
   b 3
   c 4
   d 5

9. What is the recommended number of WC pans that should be stacked on top of each other for storage?
   a 4
   b 5
   c 6
   d 10

Getting ready for assessment

The information contained in this unit of the book, as well as the continued practical assignments that you will carry out in your college or training centre, will help you with preparing for both your end-of-unit test and the diploma multiple-choice test. It will also support you in preparing for the practical assignments you will need to complete to demonstrate your understanding in and enabling you to carry out and apply domestic above-ground drainage system installation and maintenance techniques.

There are opportunities throughout the unit for you to test your progress in and understanding of the required underpinning knowledge; this will enhance your preparation for the forthcoming assessments, so make good use of them.

This unit will be assessed by the following assessment methods:

- externally set knowledge assessment
- externally set assignments.

With regard to the mechanical services industry, you will need to know:

- the uses of sanitary appliances and their operating principles
- the types of sanitary pipework systems and there layout requirements
- the site preparation techniques for sanitary appliances and connecting pipework systems and be able to install them
- the installation requirements of sanitary appliances and connecting pipework systems and be able to install them
- the service and maintenance requirements of sanitary appliances and connecting pipework systems and be able to carry them out

- the decommissioning requirements of sanitary appliances and connecting pipework systems and be able to apply them
- the inspection and soundness testing requirements of sanitary appliances and connecting pipework systems and be able to carry them out.

Check it out – using your knowledge from this unit, practice and test yourself on the following key terms:

Air admittance valve, air/smoke test, baths, bidets, blockages, brackets and fixing devices, British Standards, checking trap seal loss, Codes of Practice, commissioning and decommissioning, defects and blockages, drain connection, drainage materials, drainage systems (combined, separate and partially separate), expansion/contraction, flushing cisterns, gullies, hazards, industry standards, inspection and testing, joints (ring seal, solvent weld and compression), limits to lengths and slope of pipework, maintenance records, manufactures instructions, outlet and pipework sizes/diameters, pipework materials/fittings, PPE, preparatory work, primary ventilated stack, protection of customers property, purpose of sanitary appliances, remedial work, replacement/renewal defects, risk assessment, routine checks and maintenance, safe access/exit, safe handling of materials which may be contaminated, secondary ventilated stack, self sealing valve, showers/cubicles, sinks, siphons, soundness testing, statutory regulations, stub stack, temporary capping, terminals, tools/equipment, trap seal loss prevention, traps (P, S, bottle, running, re-sealing/anti-siphon), urinals, ventilated branch discharge system, visual inspection, warning notices, wash basins, waste/soil fittings and WC's.

Good luck!

Understand and apply domestic sheet lead weathering installation and maintenance techniques

Plumbers on site are sometimes required to work on weatherings to chimneys, soil-vent-pipe penetrations through roofs, and weatherings between building surfaces such as outbuildings. This is done by fabricating or working sheet lead, either by bossing, or by welding it together using oxy-acetylene equipment.

This unit is optional as this work is often carried out by roofers or using other materials.

This unit will cover the following learning outcomes:

- Know the properties of sheet lead weathering used on dwellings
- Know the types of sheet lead weathering and component layout requirements
- Know, and be able to apply, the site preparation techniques for sheet lead weathering

- Know, and be able to apply, the prefabrication requirements for sheet lead weathering components for installation
- Know, and be able to apply, the installation requirements of sheet lead weathering components
- Know, and be able to apply, the maintenance requirements of sheet lead weathering components.

1. Know the properties of sheet lead weathering used on dwellings

The covering of complete roofs and the weathering of building details in sheet lead have traditionally been the job of the plumber. Over recent years, much of this work has been undertaken by specialist roofing contractors. However, some plumbers still carry out sheet weathering in lead on domestic dwellings. This usually takes the form of:

- simple abutment flashings
- chimney flashings
- lead slates.

Types of sheet lead

Sheet lead is categorised by the way it is manufactured. There are two types:

- Cast sheet lead – sheet lead was originally produced by casting and it is still made this way as a craft operation. Production, by specialist lead-working firms, is in relatively small amounts created by running molten lead over a bed of sand. There is no British Standard for this material and its sheet sizes and thicknesses vary; it tends to be used on historic monuments.
- Rolled (milled) sheet lead – formed by passing a slab of lead back and forth on a rolling mill between two closing rollers, until it has reduced to the required thickness. The sheet is then cut to a standard width ready for distribution. Rolled sheet lead is used by plumbers for sheet weatherings. It is manufactured to British and European Standards (BS EN 12588) and will not vary in thickness by more than 5% at any given point. The standards also set down requirements for its chemical composition.

Figure 11.1: You may need to work with sheet lead

Did you know?

Rolled sheet lead was first used as an alternative roofing material to cast sheet lead at the turn of the twentieth century.

Characteristics of sheet lead

The following characteristics of sheet lead make it ideal as a weathering material:

- malleability
- resistance to fatigue and creep
- thermal movement
- durability
- resistance to corrosion
- patination
- resistance to fire
- 'recyclability'.

Malleability

Lead is the softest of all plumbing materials. It is malleable and can be easily worked, using hand tools and a process called bossing, into the most complicated of shapes, or dressed to fit the many types and shapes of roof tiles.

Fatigue and creep resistance

Fatigue is a loss of strength in the lead due to thermal movement, which eventually leads to cracking. The strength of lead relies on the grain structure of the metal. The chemical composition of lead is governed by BS EN 12588, which effectively controls the grain structure to make the lead sheet more resistant to thermal fatigue without affecting malleability.

Making sure you correctly size and fix individual pieces will reduce the risk of **creep**.

Thermal movement

Lead has a high coefficient of expansion at 0.0000297 for 1 °C. It is important, therefore, to include regular expansion joints in lead flashings to allow for expansion and contraction due to changes in temperature. Flashings may also be secured into a wall by lead wedges, which restrict their movement, so expansion joints would be needed here as well.

Durability

When specified and fitted correctly, sheet lead will provide maintenance-free weather protection for many years. It is extremely resistant to atmospheric corrosion, and will withstand severe weather conditions – hot and cold.

Corrosion resistant

Sheet lead is resistant to most forms of corrosion found in a roofing situation. However, you need to heed the following precautions:

- Mortar – unprotected lead damp proof courses and cavity trays may corrode in the presence of some mortars.
- Lichen growth – the acid run-off from lichen or moss on a roof may cause small holes to appear in lead sheeting under the drip-off point from tiles or slates. A 'sacrificial' flashing may be fitted to the lead gutter, or the growth may be treated with a chemical fungicide.

Fire resistant

Lead is incombustible, but melts at 327 °C.

Key terms

Malleability – the ability to be worked without fracture.

Creep – describes the tendency of metals to stetch slowly over the course of time.

Abutment flashing – used where a pitched roof needs weathering against a vertical wall.

Figure 11.2: Abutment flashing

Did you know?

Lead can corrode very quickly when brought into contact with aluminium in seaside locations, because of the high level of salt in the atmosphere.

'Recyclability'

Recyclability is an important property of any material. Sheet lead is fully recyclable. In the UK scrap lead is recovered from buildings due for demolition by a national network of reclamation merchants. It is then returned to manufacturers, where it is carefully refined for re-use in the rolling mills.

Because of this well-established recovery network, the sheet lead industry has a solid reputation for environmental awareness.

The effects of direct sunlight

As lead flashings are installed externally, they are subjected to a number of weather conditions. Lead flashings are vulnerable to changes caused by the heat from direct sunlight, which causes the lead to expand and then contract as it cools. We give this heating and cooling process a number and refer to it as the 'coefficient of linear expansion'. For lead this has a value of 0.0000297 for 1 °C.

Example:

If the temperature changed by 30 °C and the length of lead was 3 m:

$$3 \text{ m} \times 30 \times 0.0000297 = 0.002673 \text{ (about 3 mm)}$$

While this does not seem extreme, in some situations where lead is installed, the temperature difference can be much higher. If allowance for movement is not considered then the lead will end up distorted or fatigue will occur, resulting in the lead eventually splitting.

Expansion and contraction cannot be prevented but should be considered before fixing any lead flashings, as you will need to relate this to the location and the maximum size to keep expansion to a minimum. As an example, when fixing clips, nails and so on, you need to ensure these are not used excessively as they prevent movement caused by expansion and contraction.

Common sheet lead weathering component materials

Codes and sizes of sheet lead

BS EN 12588 lays down the production standards for sheet lead; in particular, the thickness (code) of the material and hence its approximate weight per square metre. Plumbers use the code number shown in Table 11.1 to order the thickness of sheet lead required for a particular job.

Sheet lead for flashings is supplied in rolls, in widths from 150 mm to 600 mm in steps of 30 mm: 150 mm, 180 mm and so on. It is usually supplied in 3 m or 6 m lengths. Larger width and lengths are available.

Did you know?

The Lead Sheet Association (LSA) produces detailed technical guidance on the use of rolled sheet lead on the outside of buildings.

The following types of lead are most commonly used:

- Code 3 sheet lead is normally used for producing soakers.
- Code 4 sheet lead is normally used for producing lead-welded components and in some lighter lead bossing applications.
- Code 5 sheet lead is normally used for lead bossing applications.

The above lead-working terms will be covered in greater detail later in this unit.

BS EN 12588 code no.	Thickness (mm)	Weight (kg/m²)	Colour code
3	1.32	14.97	Green
4	1.80	20.41	Blue
5	2.24	25.40	Red
6	2.65	30.05	Black
7	3.15	35.72	White
8	3.55	40.26	Orange

Table 11.1: Rolled sheet lead codes

> **Did you know?**
>
> Before metrication, lead was defined by its imperial weight per square foot, e.g. 3 lb, 4 lb. These weights are where today's code numbers originate from.

Patination oil

With time, lead develops a strongly fixed patina (or sheen), which is silver-grey and won't dissolve. However, in rainy or damp conditions, new lead sheet flashings will produce an initial, uneven white carbonate on the surface. Not only does this look unsightly but the white carbonate can be washed off by rain, causing further staining of materials (e.g. brickwork) below the flashings. The use of patination oil, applied evenly with a cloth as the job progresses, will prevent this.

Tools and equipment used for sheet lead installation

If you take this optional unit, you will be assessed in fabricating and fixing sheet lead components. The illustrations here show what these components **may** look like.

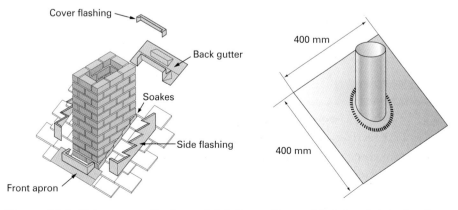

Figure 11.3: Chimney weathering set detail   **Figure 11.4:** Lead slate detail

Figure 11.5: Lead to brick chimney

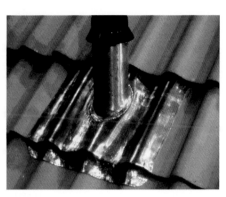

Figure 11.6: Lead slate in position

You fabricate these components either by lead **bossing** or lead **welding**.

Originally made of boxwood, the main tools for lead bossing (see Table 11.2) are now available in durable plastic.

<table>
<thead>
<tr><th></th><th>Tool</th><th>Use</th></tr>
</thead>
<tbody>
<tr><td>1</td><td>Dresser</td><td>To make the metal lie flat on a surface (remove any humps)</td></tr>
<tr><td>2</td><td>Bossing stick</td><td>To boss the lead into shape; used with a mallet</td></tr>
<tr><td>3</td><td>Bending stick</td><td>To boss the lead into shape; used with a mallet</td></tr>
<tr><td>4</td><td>Setting in stick</td><td>To reinforce or sharpen folds or angles</td></tr>
<tr><td>5</td><td>Mallet</td><td>To boss the lead in conjunction with the other bossing tools</td></tr>
<tr><td>6</td><td>Chase wedge</td><td>For setting-in a corner, fold or crease</td></tr>
<tr><td>7</td><td>Snips</td><td>To cut the lead sheet to shape (straight or curved cuts)</td></tr>
<tr><td>8</td><td>Plugging chisel</td><td>To remove mortar from existing brickwork joints; used with a club hammer</td></tr>
<tr><td>9</td><td>Club hammer</td><td>To remove mortar from existing brickwork joints; used with a plugging chisel</td></tr>
<tr><td>10</td><td>Lead knife</td><td>To cut lead sheet to the required shape (straight cuts)</td></tr>
<tr><td>11</td><td>Spirit based marker pen</td><td>To mark out the sheet lead for cutting and bending. Do not use pencils or other sharp objects</td></tr>
<tr><td>12</td><td>Ruler</td><td>To measure sheet lead dimensions (a folding ruler is commonly used as this can also measure the angle of components to be cut)</td></tr>
</tbody>
</table>

Table 11.2: Lead bossing tools

Key terms

Bossing – working sheet lead into shape using a range of mallets and dressers.

Welding – jointing sheet lead materials by applying heat to the two surfaces to be jointed, normally with the addition of lead filler material.

Progress check

1. There are two types of sheet lead. What are these and where are they used?
2. What does 'creep' mean?
3. How does sunlight affect lead flashing?
4. What is 'bossing'?
5. What is 'welding'?

2. Know the types of sheet lead weathering and component layout requirements

Types of sheet lead weathering components and uses

Sheet lead has to be formed into the various shapes we refer to as flashings. This can be done in two ways: bossing and welding.

Most sheet lead work on larger sites will be done using welding techniques. There will be occasions, however, when you might not have access to welding equipment, and you will need to know how to shape the lead by bossing.

Bossing is the use of tools to shape lead by hand. This can be done relatively easily as lead is soft and does not harden after work.

Welding is the process of joining two pieces of lead by melting the edges so that they stick together, with a filler rod of lead used to make sure the link is strong.

Different techniques are used for different types of sheet lead weathering component. This section covers the techniques used for:

- lead slate
- abutment flushings
- chimney weathering sets.

Lead slates

Figure 11.6 shows what a lead slate looks like in position. Lead slates can be formed by bossing, but are much quicker and cheaper to fabricate by lead welding.

This type of flashing is used where a pipe penetrates a roof covering. For most domestic dwellings this is likely to be the plastic soil and vent pipe. A lead welded slate is normally produced using Code 4 sheet lead.

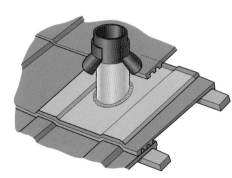

Figure 11.7: Pipe penetrating lead slate

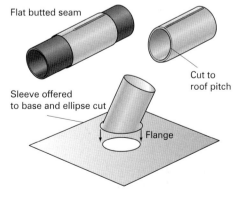

Figure 11.8: Cutting and fitting the upstand

The size of the base will vary depending on the roof covering. A typical slate for a 100 mm pipe will be 400 mm wide. The base should extend 150 mm from the front, and be no less than 100 mm under the slate. The height of the upstand should not be less than 150 mm (measured at the rear of the lead slate).

Weatherings to abutments

Apron flashings at abutments

Where a pitched roof joins onto, or abuts, a wall, this should be weathered with lead sheet. The lead sheet should have a minimum upstand of 75 mm and a turn into the brickwork joint of 25 mm. The apron should be extended down the roof to at least 150 mm over the tiles or slates (this dimension should increase to 200 mm with a low pitched roof or exposed locations). Code 4 sheet lead is normally used to form an abutment flashing.

The lead is wedged into the brickwork joint at maximum centres of 450 mm. The maximum length of a single piece of sheet lead used with an apron flashing is 1.5 m. If the length of sheet lead is greater than this distance then one or more lap joints will need to be used. Detail is shown in Figure 11.9. If a lap joint is required, then the minimum overlap distance is 100 mm.

Abutment flashings with soakers and step flashings

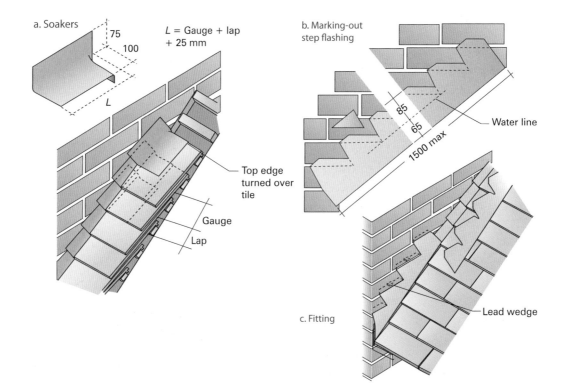

Figure 11.9: Abutment flashing with soakers and step flashings: a. soakers; b. marking-out step flashing; c. fitting

This type of flashing is used with slated or double-lap tiled pitched roofs. The flashing includes a **soaker** and separate step flashing arrangement, as follows:

- Separate soakers (produced from Code 3 sheet lead) are provided based on the tile or slate details. The formula for calculating the length of soaker is: length (L) = gauge + lap + 25 mm.
- The minimum soaker width is 175 mm with a 75 mm upstand and a 100 mm projection onto the roof.
- The top edge of the soaker should be turned over the tile to prevent slippage occurring.

The step flashing details (produced from Code 4 sheet lead as a minimum) should ensure a minimum cover over the soaker upstand of 65 mm. This is known as the water line measurement and is the minimum height of coverage to guard against capillarity. The maximum length of each piece of step flashing is 1.5 m and, where lap joints are necessary, the minimum overlap should be 100 mm. The steps are secured in position using lead wedges.

Abutment step and cover flashings to single lap tiles

With this type of weathering, no soakers are used. The step and cover flashing is formed in one piece of sheet lead. The minimum width of sheet lead to be used is normally 300 mm. The lead should project at least 150 mm onto the tiled surface. With deeply profiled roofs, or those with a pitch below 25°, it may be necessary to project 200 mm onto the tiled roof surface. The free edges of the lead will need to be clipped to prevent wind lift.

Code 4 is the minimum thickness of sheet lead to be used for this type of step-and-cover flashing. However, Code 5 may be necessary when dressing over deeply profiled tiles.

Abutment flashing with secret gutter

A secret gutter may be included in the roof design to protect against surcharging of rainwater and the risk of blockage from leaves. It may be used with a single step-and-cover flashing arrangement or a separate soaker-and-step flashing arrangement.

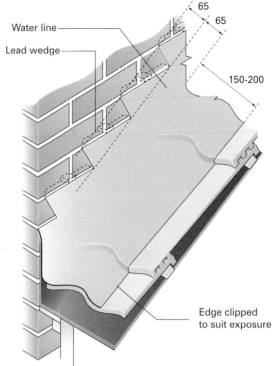

Figure 11.10: Step-and-cover flashing over single-lap tiles

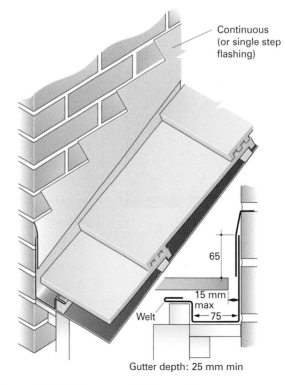

Figure 11.11: Abutment flashing with secret gutter

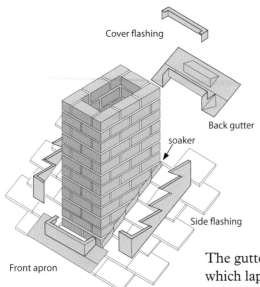

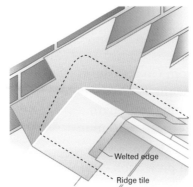

Figure 11.13: Bossed lead saddle at an abutment

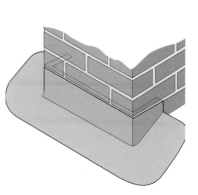

Figure 11.14: Front apron

Figure 11.12: Chimney weathering

The gutter lining pieces must not be greater than 1.5 m in length, after which lap joints will need to be used.

The use of a saddle at an abutment

The type of flashing to be used at a ridge to an abutment is known as a saddle. The saddle can be formed by lead bossing or welding. The saddle should extend at least 150 mm along the ridge and at least 150 mm down the roof on each side.

Chimney weathering sets

This section covers the details of side chimney flashings for both slate and tile.

Front apron

The lead joint in the brickwork should be at least 75 mm above the surface of the tiles or slates and an extra 25 mm should be added to turn into the mortar joint. The side of the apron needs to be turned at least 100 mm around the side cheek of the chimney. A lead welded front apron is normally manufactured using Code 4 sheet lead. Code 4 sheet lead may be used to manufacture a lead bossed apron although Code 5 is commonly used.

Figure 11.15: Side flashings

Side flashings

There are two applications for side flashings:

- Side flashings using soakers – these are used on roofs covered with slate or double-lap plain tiles.
- Side and cover flashings (combined) – these are used where it is not possible to incorporate soakers, such as over contoured tiles.

Soakers (if required) are produced in exactly the same manner as previously described for abutment flashings. Soakers are manufactured from Code 3 sheet lead whereas cover flashings are normally manufactured from Code 4 sheet lead.

Back gutter

Forming the back gutter by lead welding is a much easier process than bossing this complicated shape. Code 4 sheet lead is normally used for a lead welded back gutter.

Working life

Emma has recently qualified at Level 3 and decided to take the sheet lead option, seeing this as a potential to enhance her future career. She has recently been employed by a plumbing company who, until her appointment, have not become involved with any sheet lead work. Previously, they had no one to carry out the work.

Emma has been asked by her supervisor to provide information about the fabrication and installation of several chimney-flashing sets for a small housing project, to enable an estimate to be prepared to secure the contract.

- What methods of fabrication could Emma use to carry out the work? Select the method that would be most suitable and explain the reasons for your choice.
- Produce a list of tools, equipment and safety equipment (excluding access equipment) that would be needed. Decide on a code of sheet lead to use and the colour coding for soakers, front apron, step flashing, back gutter and cover flashing.

Progress check

1. How can you reduce 'creep' when installing lead components?
2. How is thermal movement accounted for in sheet lead installations?
3. What effect does 'lichen growth' have on sheet lead?
4. What is the melting temperature of sheet lead?
5. What is the term used to define the amount by which sheet lead expands and contracts per degree centigrade?
6. What is patination oil used for?
7. Where would a 'lead slate' be fitted. What is its purpose?
8. What is 'abutment flashing'?

3. Know, and be able to apply, the site preparation techniques for sheet lead weathering

Site preparation is extensively covered in Unit 5 on pages 199–254.

Working with sheet lead weathering will mean you will need to work at height, to gain access to chimneys and roofs. Refer to Unit 1, pages 64–70 to make sure you are working safely at all times. The dangers of working with lead are also covered on pages 15–16 and on page 27.

Fire safety when working with lead is covered on page 463.

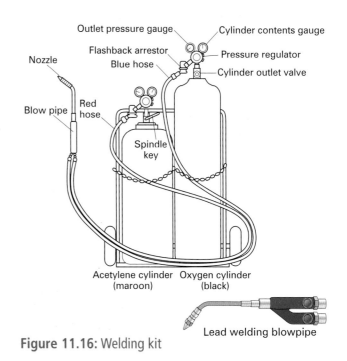

Figure 11.16: Welding kit

Welding equipment and safety

When using a welding kit, some basic safety procedures **must** be observed.

Checklist

- Gas cylinders should be stored in a ventilated area on a firm base. If possible, store oxygen and acetylene separately. Empty and full bottles should also be stored separately.
- Acetylene gas bottles should be stored upright to prevent leakage of liquid.
- Oxygen cylinders are highly pressurised. They should be stored and handled carefully to prevent falling. If the valve is sheared, the bottle will shoot forward with great force.
- Keep the oxygen cylinder away from oil or grease as these materials will ignite in contact with oxygen under pressure.
- Check the condition of the hoses and fittings. If they are punctured or damaged, replace them. Do not try to repair or piece them together.
- Do not allow acetylene to come into contact with copper. This produces an explosive compound.

- Make sure the area where you are welding is well ventilated.
- Erect signs or shields to warn and protect people from the process.
- Always have firefighting equipment to hand.
- Wear protective clothing: gloves, overalls, goggles. Clear goggles are fine for lead welding but you would be well advised to wear a clear face mask, thus gaining extra protection.
- Make sure that hose-check valves are fitted to the blow pipe and flashback arrestors to the regulators. This prevents any possible flashback on the hoses and the cylinders.
- Allow the acetylene to flow from the nozzle for a few seconds before lighting up.
- In the event of a serious flashback or fire, plunge the nozzle into water, leaving the oxygen running to avoid water entering the blowpipe.

Working life

Brett and Stephen's employer has asked them to fabricate 50 lead slates for a contract they have recently secured. Their only experience of lead welding comes from their time at college.

Their supervisor points them to a wooden site hut at the bottom of the builders' yard where an oxy-acetylene kit (including the bottles) is stored, and instructs them to carry out the fabrications in the hut so they are out of the way. In the hut they find lots of equipment, including an old engine which has leaked oil onto the wooden floor. The welding kit is very old.

The kit had been just dropped in the hut. The gauges on the regulators have broken glasses and the hoses have perished. The bottles had just been left lying on the floor, the acetylene bottle indicating signs of having been in a fire at some stage.

- Should Brett and Stephen 'make do and mend'?
- Is the environment (location) suitable for the fabrication of the lead slates?
- What equipment should Brett and Stephen be asking to have renewed and available? Produce a safety checklist highlighting causes for concern and steps to ensure equipment is better looked after in the future.

4. Know, and be able to apply, the prefabrication requirements for sheet lead weathering components for installation

This section looks at what is required to form sheet lead into the various shapes we refer to as flashings. This can be done in two ways: bossing and welding.

Most sheet lead work on larger sites will be done using welding techniques. There will be occasions, however, when you might not have access to welding equipment, and you will need to know how to shape the lead by bossing.

Preparing sheet lead for fabrication

Production of templates

Lead is a costly material and, before forming it into shapes, it may be beneficial to first produce a template from either paper or card. This will be cheaper than cutting lead first and then checking if the angle is correct. It may not always be possible to use a template but this should always be considered when it is. Also consider using a template when forming back gutters, side flashing, front aprons and corners. A good example of a template would be the upstand for the lead slate described below.

Lead slate

Marking out slate

Cut a piece of lead wide enough to give the height of the upstand, and as wide as the circumference of the pipe plus about 5 mm for tolerance. For a 100 mm pipe this is calculated as follows:

$$3.142 \times 100 \text{ mm} = 314 \text{ mm} + 5 \text{ mm for tolerance}$$
$$= 319 \text{ mm (round to 320 mm)}$$

The edges of the upstand are prepared for a butt weld and then the lead is turned around a rigid pipe and butt welded.

One end of the upstand is cut to the pitch of the roof. This can be done using a bevel and taking the actual angle from the roof. Another method is to 'develop the piece' using a drawing.

Once the upstand is cut to the required angle, the edge is dressed to form a flange. This is then placed on the base, a hole is marked and cut, and the upstand and base are prepared and welded together.

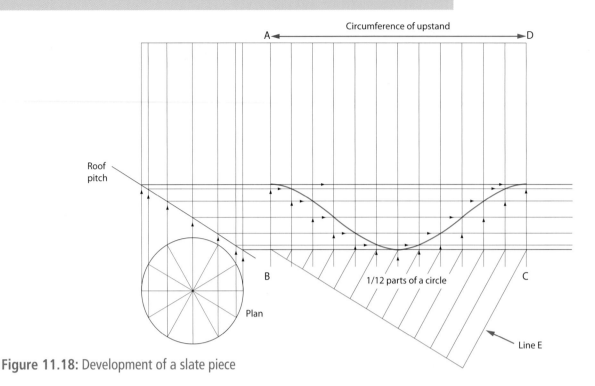

Figure 11.18: Development of a slate piece

Development of lead slate:

- Draw a rectangle ABCD, so that AB is at least equal to the height of the lead slate up-stand. Length AD is the circumference of the pipe plus 5 mm.
- From point B draw a line any length/angle and divide it into twelve equal parts.
- Project the last mark, draw a line E to point C and continue to project the other points up to line BC retaining the same angle formed by line E to point C.
- Project vertical lines from the points made on line BC and horizontal lines from the intersection at the roof pitch line, through rectangle ABCD.
- Following the arrows, mark each intersection with a cross.
- The true shape of the slate is obtained by joining the intersections by hand.

Front apron

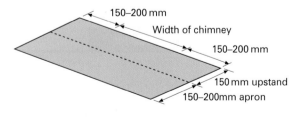

Figure 11.18: Marking out for a lead-bossed front apron

The bossed part of the apron

The Lead Sheet Association recommends that the piece of lead used for the joint apron should be no less than 300 mm wide: 150 mm for the upstand against the chimney plus 150 mm for the apron over the tiles. With a low-pitched roof this apron distance over the tiles should be at least 200 mm.

The length of the piece will be the width of the chimney plus a minimum of 150 mm for each side. If the roof is covered with deeply contoured tiles, increase this to 200 mm. This will form the basis of the dimensions for this example.

The bossed corner is set out and bossed as shown in a previous section on fabrication techniques.

The lead welded part of the apron

This will be set out based on the same dimensions as the bossed section.

Side flashings

Setting out the step flashing

The step flashing is 150 mm in width. The length will be the full length of the side of the chimney. For the overall length, you need to add a minimum of 75 mm for turning around the front of the chimney and over the apron.

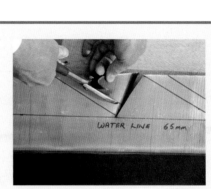

Figure 11.20: Marking out for a lead welded front apron

Preparing step flashing

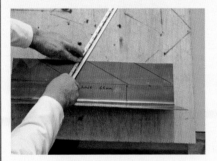
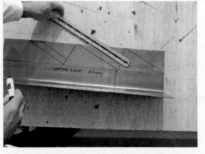

Step 1: Mark a water line of 65 mm on the length of the lead before offering it against the brickwork.

Step 2: Place the lead in position at the side of the chimney and, using a folding ruler, mark lines that correspond to the bottom of the mortar joint.

Step 3: Mark the lead out and cut as shown.

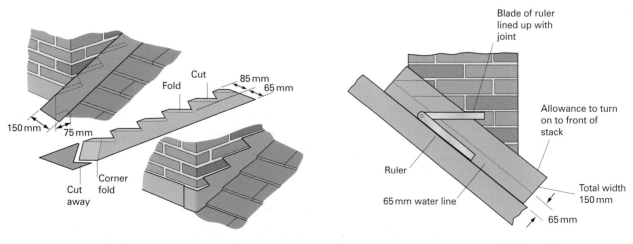

Figure 11.20: Step flashing

Figure 11.21: Setting out step flashing (1)

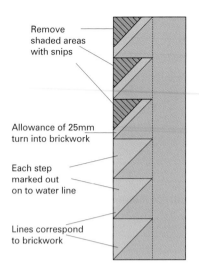

Remove shaded areas with snips

Allowance of 25mm turn into brickwork

Each step marked out on to water line

Lines correspond to brickwork

Figure 11.22: Marking lead for step flashing

When installing step flashing on slates or double-lap plain tiles, the allowance for each side of the chimney is 150 mm. If the roof is covered with single-lap tiles, a measurement of 200 mm is required.

Side and cover flashings

Flashings for both sides and cover require the same techniques as setting out a simple side flashing but now, because the lead will be extended over the roof covering to a width of no less than 150 mm, the overall width of lead will be greater.

In addition, the piece of lead that will be turned around the joint of the chimney will have to be fabricated. This can be done by bossing or lead welding. The details are illustrated in Figure 11.23.

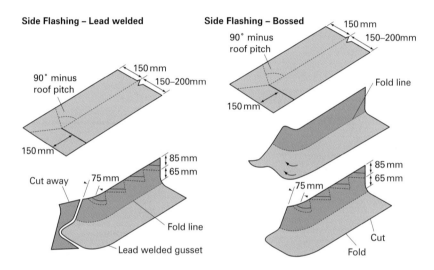

Figure 11.23: Bossing and welding around the joint of a chimney

Back gutter

Setting out back gutter

When installing a back gutter on slates or double-lap plain tiles, the allowance for each side of the chimney is 150 mm. If the roof is covered with single-lap tiles, a measurement of 200 mm is required.

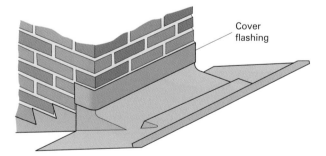

Cover flashing

Figure 11.24: Completed chimney weathering with back gutter

Once cut, fold the back gutter into shape and insert and weld the gussets.

Fit a cover flashing over the upstand of the back gutter.

Allow a minimum of 100 mm over the width of the chimney on each side for trimming around the corners, and leave the bottom of the flashing about 5–10 mm from the gutter base.

Working life

Anjum is given a task of installing some chimney flashings. She has decided to fabricate front aprons by bossing and the back gutter by welding. She visits the site and uses the access equipment correctly, recording the measurements of one of the chimney stacks and also the pitch of the roof. Anjum returns to the yard with the information she needs, to decide what to do next.

• What should Anjum do with the information she has collected? To help, take measurements from an available chimney stack in your building and produce templates for the purpose of fabrication.

Sheet lead bossing

This section explores the way in which lead bossing tools are used.

Lead at ambient (ordinary surrounding) temperatures is only 300 °C below its melting point. Compare this to copper: at ambient temperatures it is 1056 °C below its melting point. You can therefore imagine that, in many ways, lead at lower temperatures behaves in the same way as harder metals at higher temperatures.

Lead is an outstanding metal for bossing because:

- it is the softest of the common materials
- it is ductile – it will stretch quite a lot before fracturing or splitting
- it does not harden much when it has been worked (work hardening).

Bossing techniques

The best way to learn bossing is by watching an experienced lead worker, and then practising. The main aim is to achieve the required shape without making the lead too thin or too thick.

Figures 11.25 and 11.26 show bossing an internal corner and an external corner.

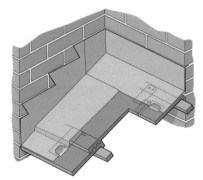

Figure 11.25: An internal corner to an abutment flashing

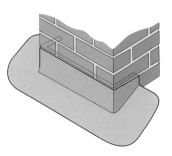

Figure 11.26: An external corner to a chimney front apron

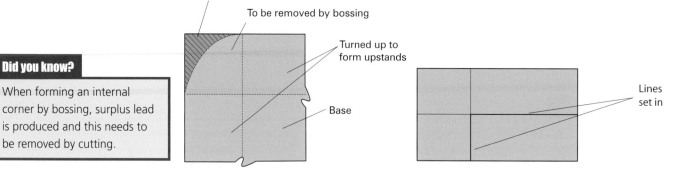

Area of lead to be removed

To be removed by bossing

Turned up to form upstands

Base

Lines set in

Figure 11.27: Marking the lead   **Figure 11.28:** Lines to 'set in'

Internal corner

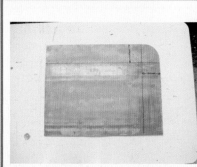

Step 1: Using a spirit-based marker pen, straight edge and tape or ruler, set out the corner. Never use sharp pencils or other objects as this could weaken or split the lead. Check the dimensions with a square.

Step 2: Turn the upstands up by 90° using a timber former. A piece of timber about 600 mm long × 100 mm wide × 50 mm thick is ideal.

Step 3: Then 'set in' the angles using the setting-in stick. This is done by placing the blade of the setting in stick on the line and tapping it with a mallet. This process fixes the position of the upstands, while the bossing is in progress.

Step 4: To get a good square base to the corner and to fix its position, use the mallet to raise a slight groove in the base of the sheet. Lead is then worked using a bossing mallet on the inside of the corner and a bossing stick.

Step 5: As the process goes on, surplus lead builds up; this can be trimmed off if it gets in the way. Keep checking that the corner is flush and square.

Step 6: Check the dimensions of the upstands; when finishing off, mark off any excess material and trim off with the snips.

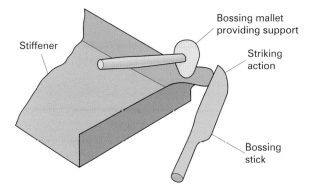

Stiffener

Bossing mallet providing support

Striking action

Bossing stick

Figure 11.29: Partially finished corner

Surplus lead folded outward

Upstand

Base

Worked up to corner position

Figure 11.30: Surplus lead

External corner

The principle of bossing an external corner is the same as for the internal corner – that is, moving surplus lead around. The marking out is slightly different, however.

External corner

Step 1: Use your marker pen, tape measure and straight edge as for the internal corner. For an external corner, such as the apron on a chimney breast, the lead has to be gained as shown in Figures 11.32–11.33.

Step 2: Fold the upstands to the angle of the roof pitch.

Step 3: Trim the return end to the angle of the chimney breast (usually 90°). It should now look something like Figure 11.33.

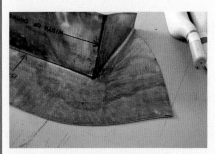

Step 4: As shown, a hump is formed, which is worked into the desired position using a dresser, bending or bossing stick.

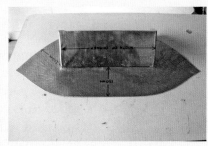

Step 5: The final shape is formed.

Step 6: Completed front apron shown in situ.

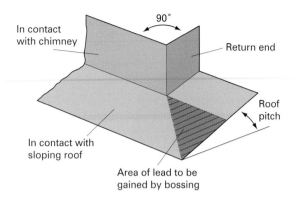

Figure 11.31: External corner – lead gain

Figure 11.32: Bossing an external corner

Remember the following tips for bossing sheet lead:

- As the job proceeds and surplus lead is removed, there's a tendency for it to thicken and crease. If creases are allowed to form they will cause cracking.
- When bossing an external corner, you have to make sure you do not stretch the lead into position. This will thin the lead, and could cause splitting.

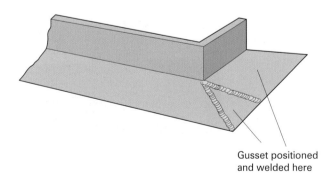

Gusset positioned and welded here

Figure 11.33: External corner to chimney front flashing

Lead welding

Lead welding is a process of joining two pieces of lead by melting the edges of the lead together (called the parent metal) while a filler rod of lead is added. It is called fusion welding. The technique can be used to form an internal or external corner, as an alternative to bossing them into shape.

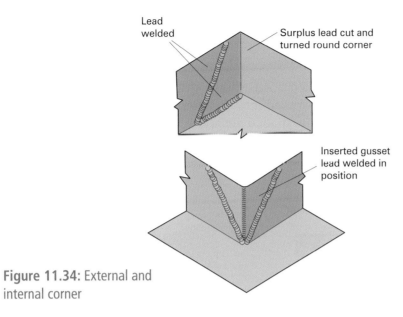

Figure 11.34: External and internal corner

Figures 11.33 and 11.34 show what the finished jobs look like.

Setting out the external and internal corner for welding

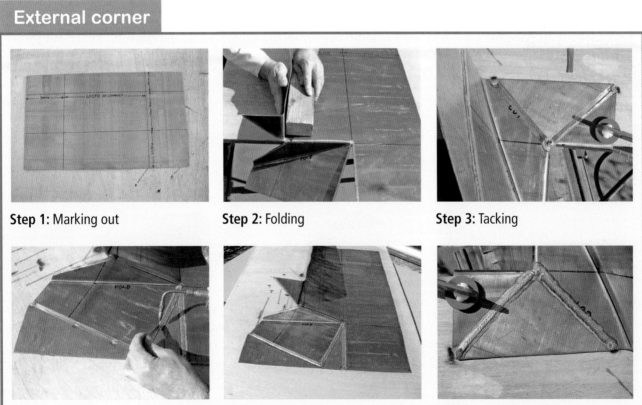

External corner

Step 1: Marking out

Step 2: Folding

Step 3: Tacking

Step 4: Welding

Step 5: Nearly complete

Step 6: Completed welded external corner

External corner

Figure 11.36 shows the setting out of an external corner. It will be used to form part of a front apron.

The apron will be marked out as shown and the joint cut between (B) and (D). The apron is worked into the required position and a gusset cut to fit as shown by the shaded area. This fits flush with the edges of the metal at (B) and (D) and is called a butt joint because the two pieces of metal are butted together.

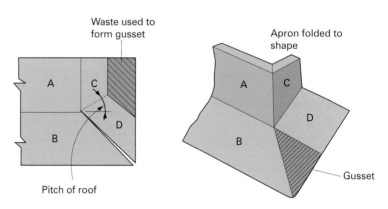

Figure 11.35 Setting out of an external corner

Internal corner

- Mark out the corner in the same way as the bossed corner but this time cut the lead as shown in Figure 11.36 (a).
- Fold the corner as in Figure 11.36 (b).
- This creates an overlap on the two upstands and it is the diagonal lap that will be welded as in Figure 11.36 (c). This type of joint is called a lap joint because the two pieces of metal overlap each other.

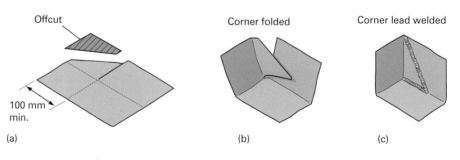

Figure 11.36: Fabricating an internal corner

The welding process

Some plumbers still refer to it this as lead burning, a term that dates from when crude welding techniques were used for jointing lead.

Like lead bossing, this is a very skilful job that will take time to master. Once you have done so, however, it is a very rewarding aspect of the job.

There are two types of lead welded joint:

- butted seam
- lapped seam.

Figure 11.37: Butted seam

Figure 11.38: Lapped seam

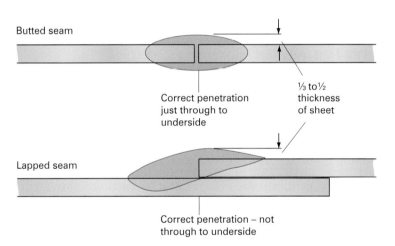

Figure 11.39: Butted seam and lapped seam

Welding can be carried out in different positions. These include the following:

- **Flat position** – the simplest, best, most natural and the most convenient position for welding is flat (down hand). This position creates excellent welded joints at a fast speed and with minimum fatigue to the welders. If the structure does not permit welding in a flat position, it may be rotated during work.

- **Inclined position** – in this case slope and rotation can vary from 10° to 45° and 0° to 90° respectively. The workpiece is in an inclined plane.
- **Vertical position** – the plane of the workpiece is vertical and the deposited weld bead is also vertical. In the vertical position the slope remains between 45° and 90° and rotation may have any value from 0° to 180°.
- **Horizontal position** – the plane of the workpiece is vertical, and the deposited weld bead is horizontal. In this position, the slope does not exceed 10°, but rotation can vary from 10° to 90°.

Lead welding tools and equipment

You will need:

- oxy-acetylene welding equipment with a lead-welding torch. These torches have interchangeable nozzles numbered 1 to 5. In general, nozzle sizes 2 and 3 are used for Lead Sheet Codes 4 and 5. For upright seams, Size 1, 2 or 3 nozzles would be used depending upon the lead's thickness
- snips
- flat dresser
- shave hook
- steel ruler.

Creating the welds

It is essential that the meeting points of the sheet lead and its faces (points where the weld adheres to the sheet) are clean and any oxide is removed. This cleaning process is carried out using a shave hook.

Mark a width of about 10 mm and, using your metal straight edge as a guide, shave the surface of the metal with the shave hook.

Cut strips to use as filler rods. This is done by cutting a thin strip of lead about 3–5 mm thick and 300 mm long. Again, it needs to be shaved clean. Alternatively, lead rods up to 6 mm can be obtained from a supplier.

Once everything is prepared, you can set up the welding equipment. Remember to check for any leakage using a leak detection fluid. Once you are confident that this is not an issue, light the blow pipe:

- Turn on and light the acetylene first.
- Then feed in the oxygen. A pressure of 0.14 bar (2 lb/sq in) for both the oxygen and acetylene is required.
- You need to achieve a neutral flame to get the best results. An incorrectly adjusted blow pipe will produce either a carburising (excess of acetylene) or an oxidising flame (excess of oxygen), both of which are unsuitable for lead welding.

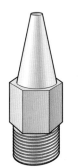

Did you know?

A shave hook is used for cleaning the edges of the lead sheet and the lead filler strips.

Figure 11.40: Lead welding nozzle

Remember

Cleaned lead will tarnish if touched or left over a period of time. This would then require further cleaning.

Remember

For a lap joint, don't forget to clean the piece underneath where the faces of the lap joint meet.

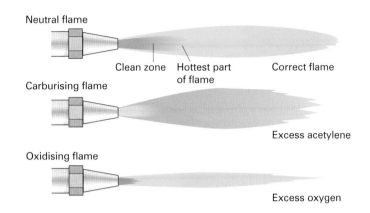

Figure 11.43: Blow pipe flames

Lead welding – good practice

Flat-butted seam welds should fully penetrate through the thickness of the lead sheet. Lapped seam welds should penetrate the surface of the lead, but not through to the underside.

The thickness of the seam should be between one-third and one-half the thickness of the sheet that is built up. The width of the weld will depend on the thickness of the lead and the seam pattern. When using Code 4 lead, the minimum width of a flat-butted seam should be 10 mm.

Beware of undercutting: reducing the thickness of the lead at the side of the weld. This causes a weakness that can result in cracking along the weld line. The main cause of undercutting is holding the flame for too long on the vertical surface.

Safety tip

Whether jointing lead using a butted or lapped seam, you should tack the pieces together. This prevents any movement and keeps the surfaces in close contact.

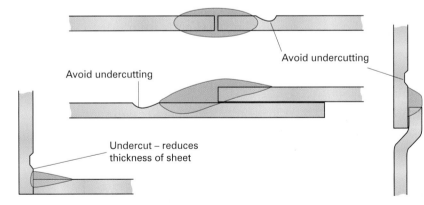

Figure 11.42: Examples of undercutting

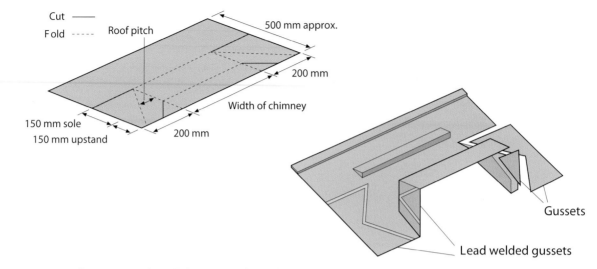

Figure 11.43: Joining flat supported lead sheet using butted seams

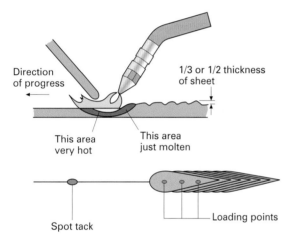

Figure 11.44: Position of blow pipe

Welding flat-butted seams

This seam is used for jointing pieces of flat supported lead sheet as shown.

The position of the blow pipe and filler rod should be the same as that used in Figure 11.45.

- The tip of the cone of the flame should be just clear of the molten lead.
- Lead is melted off the welding rod into the weld area.
- A seam that is between one-third and one-half the thickness of the sheet is built up.
- The flame is directed into the centre of the seam and is moved forward, either in a straight line or slightly from side to side. This will set the pattern of the seam.
- The weld should just penetrate through the underside of the lead.

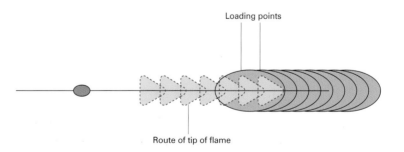

Figure 11.45: The flame is moved forwards to set the pattern of the seam

Welding a flat-lapped seam

This technique is an alternative to butted seams. It is often preferred when working onsite where there is a risk of fire during the welding process, because the flame will not make contact with the material beneath the lead. Even experienced plumbers will admit that this technique is slightly easier.

Figure 11.47 shows a typical example of a welded flat-lapped seam used on a lead slate. The preparation and process of welding a lapped seam is the same as for a butted seam; the only difference is that when welding thicker lead, two loadings are used.

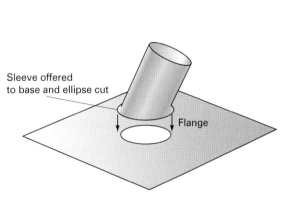

Sleeve offered to base and ellipse cut

Flange

Figure 11.46: Welded flat-lapped seam

> **Remember**
>
> Prefabricate lead components before fixing wherever possible.

Inclined seam on a vertical face

If you can, prefabricate sheet lead components before fixing them. There will be occasions, however, when you may have to weld a joint in position. One such situation is an inclined seam on a vertical face.

Although this joint can be made without using filler rods – by using the overlapping edge to make the seam – it is recommended that a filler rod should be used.

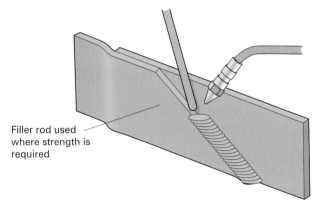

Filler rod used where strength is required

Figure 11.47: Inclined seam

> **Safety tip**
>
> Small portable kits are also available, but the same safety precautions will apply.

Fire safety when working with lead

If working on site, always try to use lapped joints. This reduces the risk of fire, which exists when making a butted seam against combustible surfaces.

On occasions, butt joints may be unavoidable. When this is the case, you should either wet the timber area beneath the weld or place a non-combustible material beneath it.

> **Remember**
>
> Acetylene cylinders are maroon and oxygen cylinders are black in colour.

Working life

Carla is a Level 2 plumbing apprentice. She has received some instruction at college with regard to the basics of lead welding but is by no means proficient. Her employer asks her to continue some welding in the workshop while Jack, another apprentice, runs an errand. Although she has limited experience, she doesn't want to make herself look foolish by saying she can't do it.

Jack has started welding, leaving the blowtorch lit. The extract fan over the bench isn't running so Carla pushes the start button and nothing happens. Ignoring this, she sweeps back her hair over her shoulders and starts to weld. Suddenly, the acetylene line connection to the torch ignites. This startles

Carla who lets the tip of the nozzle drop into the molten pool of lead, which spatters back into her face and right eye. Her supervisor makes the equipment safe, renders first aid and takes her to hospital. Carla is off sick for two weeks but luckily there isn't any lasting damage to her eye.

- There are many mistakes made here. Produce a list indicating where Carla went wrong.

- As Carla is off work due to the accident for two weeks, is there anyone the employer should contact? If so, who would this be? Who, if anyone, would be held responsible for the accident?

Progress check

1. Name the two types of lead-welded joint.
2. How many nozzle sizes are commonly for lead welding torches?
3. Which nozzle sizes would be most suitable to weld Code 4 and 5 lead sheets?
4. Name the tools (other than the oxy-acetylene kit) required as a minimum to fabricate sheet lead components by the welding method.
5. Which gas is turned on first when igniting an oxy-acetylene blowtorch?
6. Which type of flame should be used to carry out lead welding operations?
7. Why should lead sheets be 'tacked' before beginning the welded joint?
8. What colour is an oxygen cylinder?
9. What colour is an acetylene cylinder?

5. Know, and be able to apply, the installation requirements of sheet lead weathering components

Many of the installation requirements for sheet lead weathering have been covered in the earlier parts of this unit.

Flashings to abutments are fixed into position using a turned-in section of lead cover flashing that is secured in position at the brickwork, blockwork or stonework joint (seam). The joints in the brickwork will therefore need to be 'raked out' to a depth of at least 25 mm prior to the sheet lead component installation. On new properties a friendly builder will often leave the joints raked out. For existing properties raking out will normally be carried out using a club hammer and a pointing chisel, sometimes referred to as a joint chisel.

The sheet lead is secured into the brickwork joints using lead wedges which are strips of sheet lead between 20 and 25 mm in thickness that are folded several times to a dimension that is slightly thicker than the joint. One edge of the lead is flattened slightly to form the wedge which

Figure 11.48: Lead wedges

is then driven into the joint between the sheet lead turn-in and the upper part of the joint, to secure the lead in position.

Lead wedges are normally spaced 300 to 450 mm apart, depending on the building material.

With standard brickwork/blockwork joints of up to 18 mm, the joint is normally pointed with a mastic-type sealant. These sealants are either polysulphide or silicone based. If a silicone sealant is used, it must be neutral cure, as the acid cure variety can cause corrosion on the surface of the sheet lead. Any pointing material selected must be suitable for pointing sheet lead materials. Guidance can be sought from the material manufacturer.

Making fixings to wide joints

There are occasions when plumbers come across masonry walls such as stonework with joints wider than 18 mm, in which case a different fixing method is identified.

Here the sheet lead is turned up the back of the joint and fixed using corrosion resistant screws and washers (normally stainless steel) positioned at approximately 450 mm centres. As an alternative, large-headed copper or stainless steel nails may be used.

Mortar is normally used as the sealing material, but to overcome excessive cracking caused by thermal movement, a masking tape is applied, as shown in Figure 11.49.

Fixings to free edges of sheet lead abutment flashings

The free edges of raking (sloping) or horizontal abutment flashings require fixing. This is normally carried out using clips. The number of clips to be used depends on the degree of exposure of the roof surface:

- severe exposure – every tile joint
- moderate exposure – one clip at every second tile joint
- sheltered location – one clip at every third tile joint.

The maximum clip spacing distance is normally no greater than 500 mm.

Clips used at raking (sloping) abutments may be manufactured from copper or **terne-coated** stainless steel, whereas horizontal flashings should only be fixed using soft copper clips. The clips are usually about 50 mm wide and soft copper of 0.6 mm thickness is normally used.

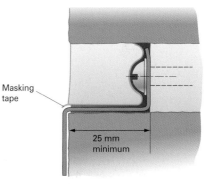

Figure 11.49: Making fixings to joints wider than 18 mm

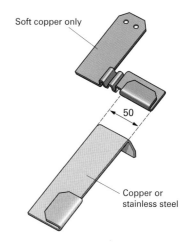

Figure 11.50: Clip

6. Know, and be able to apply, the maintenance requirements of sheet lead weathering components

When completed, a sheet lead installation needs to be tested to ensure that it is watertight. The easiest way is to apply water to the completed installation using a hosepipe or bucket. Then check the roof internally for any leaks.

Maintenance does not form a large part of the work on the components that we have looked at in this unit, but you do need to establish the basic procedures for checking systems that have been installed. Areas covered should be based on the following points.

- Establish adequate safety procedures for completing work, including safe roof access and working platforms.
- Check the internal roof space to establish whether there was any water penetration around the components.
- Check the key components for size and dimension (establish that the components were installed correctly and, in particular, could not give rise to capillary attraction).
- Check the roof coverings adjacent to components – repair as necessary.
- Check individual roof components for soundness (no splits/cracks, etc.) – repair as necessary.
- Check component fixings to ensure the soundness of the fixing – refix (where required and repoint) and check to establish whether additional fixings are necessary in the event of wind lift.
- Test components for leakage on completion of maintenance activity.

Refixing lead components with wedges

Fixing lead and using wedges have already been covered. However, if lead flashing comes loose, you may be required to refix it using the same concepts. Therefore, to do this you will need to follow this process.

1. Remove any pointing material and old lead wedges.
2. Ensure the lead returns in to the joint 25 mm and it is in sound condition. Replace if it does not meet either of these requirements.
3. Place the lead back in position ensuring any overlaps are correct, so as not to allow water penetration.
4. Refix with new lead wedges into the mortar joint to trap the lead, ensuring the wedges are spaced between 300 mm and 450 mm apart.

If there is no builder available, you may have to repoint the joint with an approved sealant.

Check your knowledge

1. How recyclable is lead?
   a It is not
   b A little
   c 50%
   d Totally

2. What is the recommended 'turn' to brickwork for a lead flashing?
   a 18 mm
   b 25 mm
   c 35 mm
   d 50 mm

3. What is the minimum distance from the water line to the roof tiles when installing lead flashings?
   a 65 mm
   b 120 mm
   c 165 mm
   d 200 mm

4. Where can details of sheet lead weathering systems be found?
   a Sheet lead guide to good practice by the SLA
   b Control of leadwork regulations
   c Construction health, safety and welfare regulations
   d Lead Sheet Association guide to good practice

5. What precautions must be taken when working with lead?
   a Wash hands, arms and face at the end of each working session
   b Remember to have a wash when you get home
   c Wash all your tools before you start work
   d Wash the lead before you start working with it

6. Which code of lead does BS EN 12588 recommend for soakers?
   a Code 3
   b Code 4
   c Code 5
   d Code 6

7. What is the recommended minimum vertical upstand against the chimney for a back gutter?
   a 100 mm
   b 125 mm
   c 150 mm
   d 350 mm

8. What is the minimum vertical upstand at the back of a lead slate?
   a 150 mm
   b 200 mm
   c 300 mm
   d 450 mm

9. Where is a secret gutter located?
   a From the front of the chimney
   b From the back of the chimney
   c At an abutment weathering
   d On a lead slate

10. When lead welding with oxy-acetylene, which one of the following flames gives the best results?
    a Aerating
    b Carburising
    c Neutral
    d Oxidising

11. Which of the following tools is **not** used for lead work?
    a Dresser
    b Lead hammer
    c Bossing stick
    d Lead knife

12. Which of the following is the correct formula to work out the number of soakers required for an abutment flashing?
    a Length of roof divided by gauge
    b Length of roof multiplied by gauge
    c Length of roof multiplied by lap
    d Length of roof divided by lap

13. Which of the following could be caused by exposure to lead with insufficient personal protection?
    a Liver disorders
    b Weil's disease
    c Cold symptoms
    d Psoriasis

14. What is the BS EN 12588 colour reference for Code 5 lead?
    a Green
    b Red
    c Brown
    d Blue

Getting ready for assessment

The information contained in this unit of the book, as well as the continued practical assignments that you will carry out in your college or training centre, will help you with preparing for both your end-of-unit test and the diploma multiple-choice test. It will also support you in preparing for the practical assignments you will need to complete to demonstrate your understanding in and enabling you to carry out and apply domestic sheet lead weathering installation and maintenance techniques.

There are opportunities throughout the unit for you to test your progress in and understanding of the required underpinning knowledge; this will enhance your preparation for the forthcoming assessments, so make good use of them.

This unit will be assessed by the following assessment methods:

- externally set knowledge assessment
- externally set assignments.

With regard to the mechanical services industry, you will need to know:

- the properties of sheet lead weathering used on buildings
- the types of sheet lead weathering and component layout requirements
- the site preparation techniques for sheet lead weathering and be able to apply tem
- the prefabrication requirements for sheet lead weathering components for installation and be able to apply them
- the installation requirements of sheet lead weathering components for installation and be able to apply them

- the maintenance requirements for sheet lead weathering components for installation and be able to apply them.

Check it out – using your knowledge from this unit practice and test yourself on the following key terms:

Abutment flashing, back gutter, bending stick, bossing, bossing stick, British Standards, butted seam, cast sheet lead, chase wedge, chimney flashing, chimney weathering set, club hammer, codes of lead, Codes of Practice, coefficient of expansion, copper nail, corrosion, cover flashing, creep, defects, dresser, durability, external/internal corner, eye/face protection, fatigue, fire safety, fixing clips/cleats, flat dresser, flat/inclined/vertical/horizontal positions, front apron, hazards, industry standards, lapped seam, lead knife, lead slate, lead wedges, lichen/moss growth, maintenance records, malleability, mallet, manufactures instructions, mortar, neutral/oxidising/carburising flames, nozzle, oxy-acetylene, patination, penetration, plugging chisel, PPE, preparatory work, pressure regulator, properties of sheet lead, protection of customers property, recyclability, remedial work, resistance to fire, risk assessment, rolled(milled) sheet lead, routine checks and maintenance, saddle abutment, safe access/exit, safe handling of materials, safety, secret gutter, setting in, shave hook, side flashing, snips, soakers, soundness testing, spirit based marker pen, steel ruler, step and cover flashing, step flashing, templates, terne-coated stainless steel, thermal movement, undercutting, visual inspection, weathering's, welding and wide joints.

Good luck!

Index

Key terms are indicated by **bold** page numbers